Concepts in Algebra

A Technological Approach

Concepts in Algebra

A Technological Approach

Contributing Authors:

James T. Fey *The University of Maryland*

M. Kathleen Heid *The Pennsylvania State University*

with

Richard A. Good *The University of Maryland*

Charlene Sheets *Western Michigan University*

Glendon W. Blume *The Pennsylvania State University*

Rose Mary Zbiek *The University of Iowa*

Janson Publications, Inc
Dedham, Massachusetts

The Computer-Intensive Algebra Project was funded
by grants from the National Science Foundation under
award numbers DPE 84-71173 and MDR 87-51500 to
The University of Maryland (Principal Investigator:
James T. Fey) and award number MDR 87-51499 to
The Pennsylvania State University (Principal Investigator: M. Kathleen Heid). Any opinions, findings, conclusions, or recommendations expressed herein are
those of the authors and do not necessarily reflect the
views of the National Science Foundation.

Library of Congress Cataloging-in-Publication Data

Concepts in algebra: a technological approach/contributing
    authors, James T. Fey, M. Kathleen Heid, with Richard
    A. Good ... [et al.].
       p. cm.
    Includes index
    ISBN 0-939765-73-X
    1. Algebra.   [1. Algebra.]
I. Fey, James Taylor.   II. Heid, Mary Kathleen.
QA152.2.C66 1995
512.9--dc20   94-26696
                            CIP
                            AC

Printed in the United States of America

9 8 7 6   4 5 6 7 8 9

Design Malcolm Grear Designers

Contents

Preface

Concepts in Algebra: A Technological Approach (CIA) provides a technology-intensive approach to beginning algebra that focuses on developing a rich understanding of fundamental algebraic ideas in realistic settings. From their first exposure to algebra, *CIA* students explore mathematics in real world situations through the development and critique of mathematical models. Within the context of these models, students develop mathematical representations of data, explore relevant questions, and formulate and defend predictions. Through their use of fundamental mathematical ideas and methods to reason about important applied problems, students develop a rich understanding of variables, functions, relations, systems, and equivalence as they investigate realistic questions. Computing tools facilitate students' explorations of algebra by providing easy access to numerical, graphical, and symbolic representations of mathematical ideas.

Content Features

Concepts in Algebra suggests a fundamentally different approach to the content, teaching, and learning of beginning algebra. It is a curriculum that emphasizes mathematical models and representations, variables and functions, symbolic reasoning rather than symbolic manipulation. The focus of the curriculum is conceptual rather than procedural knowledge.

Mathematical modeling Instead of concentrating on classical "word problems" as illustrations of by-hand symbolic manipulation techniques, *CIA* engages students in extended explorations of real world situations through graphical, numerical, and symbolic representations. Instead of the word-problem approach of asking a single question about a single equation or formula, *CIA* takes a situations approach which asks students to study several aspects of a relation between variables, to see how answers to questions change under different problem assumptions, and to see how different models provide different predictions. Through these explorations, students develop their understanding of the central concepts and processes of mathematical modeling.

- Identifying variables and relations among them
- Representing the relations among variables in numerical, graphical, and symbolic forms
- Drawing inferences about modeled relations
- Recognizing limitations in application of mathematics models to real-life situations.

Variables and functions Traditional approaches to algebra present the variable as a definite but unknown number whose fixed value is being sought. The important problems in a traditional algebra course are finding the values of variables that make an equation or set of equations true.

When algebra is used to provide symbolic models of quantitative relationships, however, variables are quantities that really do change as the situation changes. The important problems include not only finding a specific combination of values of those variables to satisfy an equation, but also determining the effect of changes in one variable on the values of related variables or finding the values of one variable that produce maximum or minimum values for some other variable.

In mathematical modeling it is usually the case that questions of interest involve relations among two or more variables. These relations are frequently describable in terms of input variables being used to predict the values of output variables; the outputs are *functions* of the inputs. Focusing beginning algebra on mathematical modeling suggests a functions approach to modeling. *Concepts in Algebra* is a functions/modeling approach to algebra that engages students in constructing and interpreting tables and graphs of function rules while developing an intuition for and understanding of several important families of functions—linear, quadratic and other polynomial, exponential, and rational. Spreadsheets (or function-table generators), function graphers, curve fitters, and symbolic manipulation programs are used throughout the course to develop these new understandings and skills.

Symbolic reasoning Instead of focusing on by-hand symbolic manipulation techniques, *Concepts in Algebra* gives students several technological approaches to working with symbolic expressions and equations, and inequalities. The *CIA* curriculum concentrates on developing a *symbol sense*, that is, graphical, numerical, and symbolic meaning for the algebraic symbols they use.

Teaching and Learning

The new content that characterizes *Concepts in Algebra* is accompanied by features of the *CIA* approach related to the context in which the content will be learned. These features also relate to the ways in which the mathematical content is developed (through multiple representations and making full use of technology) as well as to the classroom structures (collaborative group work and other foci on communication) that should facilitate this development.

Multiple representations From the beginning of the text, students encounter graphical, numerical, and symbolic representations of functions. They reason within and among the various representations as they explore realistic situations represented by function rules. It is through this flexibility in use of representation that students' understanding of the fundamental concepts of functions and particular families of functions will grow.

Technology intensive *Concepts in Algebra* is a technology-intensive curriculum. After their initial introduction to functions and variables in Chapter 1, students will need daily classroom access to computing tools. They should be encouraged to use their computing tools to explore mathematics through a variety of representations. Because students may not necessarily have the complete array of needed computing tools available to them at home, the "Exercises" have been designed so that they can be completed with access only to a scientific calculator. For the most part, "Exercises" ask the students to analyze data they have gathered in their technology-aided explorations or to do some preparatory thinking for an exploration to come.

Communication Written and oral communication are central to the implementation of the *Concepts in Algebra* curriculum. The text requires written responses instead of single-number or single-word answers, and instructors should expect students to work on communicating their ideas in writing clearly and completely. Instructors should also encourage students to work in collaboration with partners to afford them the opportunity to talk with others about mathematical concepts and problem solving. Learning to communicate about mathematics is important in and of itself, and communicating about mathematics as they explore should enhance students' conceptual understanding.

Technology Required

A variety of computing tools are fundamental to use of the *Concepts in Algebra* curriculum. Students should have access to the following.

- A function-tables program or a spreadsheet that will produce tables of ordered pairs for one or more user-defined functions over some specified domain.

- A function-graphing program that will produce graphs of one or more user defined functions of a single variable. Ideally, this program should allow for easy rescaling and have a coordinate-reading cursor.
- A curve-fitting program that accepts ordered pairs of data as input and finds parameters for one or more function rules that fit the data well.
- A symbolic manipulation program that supports production of equivalent forms for algebraic expressions, solutions of equations, *etc.*

There are numerous ways in which computer software, graphing calculators, or other computing devices can be combined to achieve an adequate tool environment, given a variety of different hardware configurations.

In addition to the computing tools listed above, the *CIA* text makes use of three special purpose programs to support specific lessons. These programs, as well as a function grapher, a function-tables program, and a curve-fitting program are available with the text and are in executable form for most IBM-compatible computers. For text users in other computer environments, spreadsheet versions of these three programs are presented in the teacher's guide.

Summary

Concepts in Algebra is a new type of beginning algebra curriculum whose content and pedagogy reflect and take advantage of the opportunities provided by computing technology available to students today. We believe that teachers and students using *Concepts in Algebra* curriculum materials will experience a mathematics course that uses computing technology in important new ways as tools for learning and problem solving. We believe that use of those tools will help students gain a rich understanding of the variable and function concepts that are at the heart of mathematics and that they can learn to use those understanding to solve significant quantitative problems.

Acknowledgements

Production of the student text materials, resources for teachers, computer software, and achievement tests for *Concepts in Algebra: A Technological Approach*, formerly known as *Computer-Intensive Algebra*, has been a collaborative effort drawing on the insights and energies of many mathematics educators. Curriculum development, teacher preparation, evaluation, and research activities have been based at the University of Maryland and the Pennsylvania State University. The authors are particularly grateful to several individuals who worked on numerous important aspects of the curriculum development over a period of two or more years: Paul Christmas, Philip D. Larson, and Mary Ann Matras.

In addition, a large number of mathematics educators have contributed in myriad ways to the project. Staff from the University of Maryland who contributed to research, development, and testing include: Monica Boers, Neil Davidson, Carol Smith, Ronald Steffani, David Dyer, Susan Ricciardi, James Menasian, Cynthia Hatheway, Orian Hight, Deborah Slade, Phil Steitz, Debra Munley, Ann Munro, and Eric Boesch. Staff from the Pennsylvania State university who contributed to research, development, and testing include: Wilhelmina Mazza, Sarah Eaton, Pete Johnson, Janice Engelder, Anne Lauver, Paul Wilson, Scott Stull, Eric Speight, and Ron Hoz.

Although earlier versions of the *CIA* curriculum have been tested by over sixty teachers from twenty different states, the initial field tests occurred at two specific sites. Field test teachers from Northwestern High School in Hyattsville, MD, were: Peter Fischer, Sarah Green, Joan Lynch, Donna Bettcher, Mike McCrae, and Carolyn Nicoli. Field test teachers from State College Area Schools in State College, PA, were: Wendel Rojik, Greg Somers, Helen Diethorn, JoAnne Maurer, and Sylvia Pezanowski.

Through the work of the project, but especially in the early planning stages, the project curriculum development, field-test, and evaluation activities were guided by advice from the following members of our advisory board: Richard Andree (University of Oklahoma), Jerry Johnson (Western Washington University), Daniel Kunkle (Midland School, Los Olivos, CA), Glenda Lappan (Michigan State University), Roy Myers (The Pennsylvania State University), Anthony Ralston (State University of New York at Buffalo), Jane Swafford (Illinois State University), and Larry Washington (The University of Maryland).

The development of *Computer-Intensive Algebra* was funded through a series of grants from the National Science Foundation program in Applications of Advanced Technology. The project has benefited enormously from the advice and encouragement provided by our program director, Dr. Andrew Molnar.

In addition to the funding provided by the NSF grants, the project received invaluable administrative support from the Department of Mathematics at the University of Maryland and from the Department of Curriculum and Instruction and the College of Education at the Pennsylvania State University. We would like to convey our special thanks to Delores Forbes at the University of Maryland and Jane Kurzinger, Claire Markham, and Linda Haffly at The Pennsylvania State University for careful attention to the many clerical, administrative, and financial matters involved in the project.

Finally, a very special thanks to Eric Karnowski for his fine editorial work and to Barbara Janson for her encouragement and interest in publishing a beginning algebra text that does not fit the mold and that breaks new ground for the teaching of algebra in a technological world.

To the Student

The textbook you are holding in your hand, *Concepts in Algebra*, will provide you with a fresh and exciting approach to algebra. Mathematics is an active science. This textbook will include you as an active participant by guiding you to explore realistic and interesting situations in which algebra is used. As you work through the explorations in the text, you will use computing technology to help you understand and make sense of these realistic situations. The textbook is designed to be read, and reading it will help you understand the situations you are exploring. We hope and expect that through your experience in this course, you will learn a great deal about the fundamental ideas of algebra—ideas you will find important in your daily life, and ideas you will draw upon again and again in the future. Good luck!

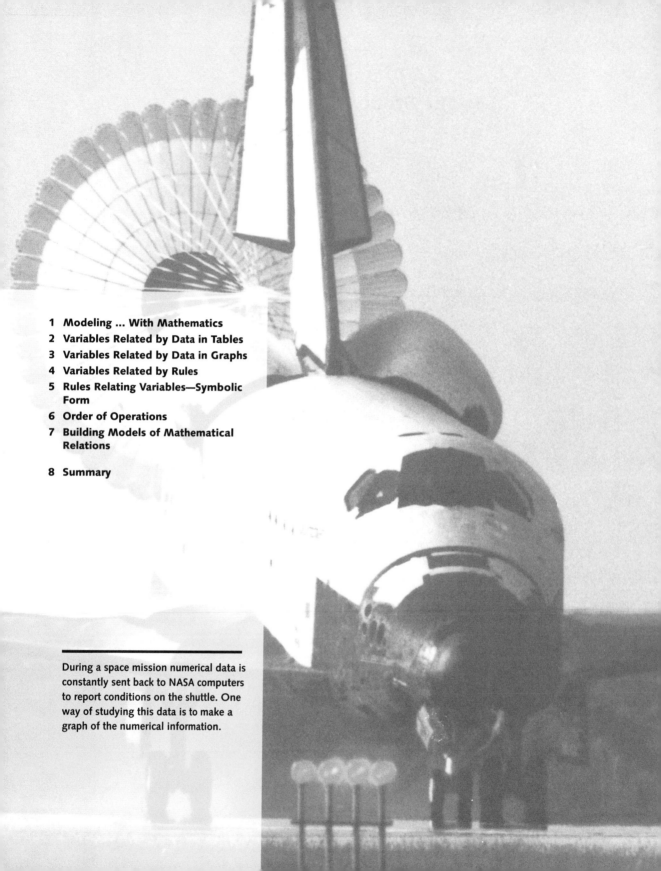

During a space mission numerical data is constantly sent back to NASA computers to report conditions on the shuttle. One way of studying this data is to make a graph of the numerical information.

1 Variables and Functions

If you were asked to predict the attendance at a school football game, you might well respond,

"That depends ...."

The attendance depends on many factors such as the weather, ticket price, the quality of opponent, the record of the home team, etc. These factors are called **variables** because they change from one game to the next. It is common to say that attendance **is a function of** those variables.

One of the most important problem-solving tools provided by algebra is a language for describing and predicting relations between numerical or quantitative variables. This chapter introduces two key algebra concepts, variable and function, and methods for using them to solve practical problems.

1 Modeling ... With Mathematics

In 1987, over 600,000 incorporated businesses began operations in the United States. That number does not count the thousands of school fund-raising projects or church bingo games or the millions of young people who went to work babysitting or mowing lawns.

Every business begins with an idea: a product or service that the owners believe will bring them fame and fortune, or at least some spare cash to enjoy life! Soon after the decision to enter business, there are other decisions to be made and problems to be solved. Many of the problems involve numbers. To deal with the several factors that shape a business's chance for success, simple arithmetic is seldom sufficient. Careful planning usually requires the language and methods of algebra to study questions of the form

"What would happen if ... ?"

The first step in your algebra course is learning to recognize situations in which algebraic reasoning is helpful and important.

 # Exploration I

The following situation is very much like the problems that face a new business. After you attempt to answer the questions that are given, you will have the opportunity to work with a mathematical model of the situation. This experience will give you a first look at the uses of algebra and technology in complex problem-solving and decision-making tasks.

SITUATION 1.1

At Northwestern High School the student Talent Show is one of the highlights of the school year. Students get a chance to perform and to see their classmates perform. The show is a fund-raising event, but its success depends on careful planning of many activities. Students and teachers must make a number of decisions to assure that things go smoothly and that a good profit is made.

Task 1. Think about the things that might happen before and during a talent show at school. Make two lists, one of some decisions that must be made and another of some jobs that must be done to produce a successful show. Use the following headings for your lists.

Decision   Jobs

Task 2. Each decision that must be made in planning the talent show will have consequences. You can choose the price for tickets, but if you charge too much, attendance will be low. If you charge too little, you will have very little income. The situation involves **variables** such as ticket price and attendance. Variables are **quantities whose values may change or vary** according to circumstances.

Look back at your list of decisions and jobs. Some involve **numerical** or **quantitative variables** because numbers can be assigned to them as values. For example, the price of a ticket is a numerical or quantitative variable; prices are numbers in dollar value form.

Identify any numerical variables involved in Talent Show planning and make a new two-column list. In column 1, list those that you can decide on, such as ticket price. In column 2, list those that are affected by your decisions, such as ticket sales. Then draw arrows from each variable in column 1 to any variable in column 2 that it influences. Use the following form for your list.

Column 1 (Values you set)	Column 2 (Values that follow)
1. ticket price	1. ticket sales
2. _____	2. _____
3. _____	3. _____
4. _____	4. _____
5. _____	5. _____

Task 3. For two important pairs of variables that you think are related, write brief statements of what you think the relationship is. For example, you might write, "If ticket price is raised, ticket sales will go down. If ticket price is reduced, ticket sales will increase."

Exploration II

For a business studying the decisions it must make, the next step after finding variables is to find relationships among those variables. If we set the ticket price at $3.00, **how many** tickets will we sell? If we spend $500.00 on advertising, by **how much** will sales increase?

The managers of large businesses commonly put variables and relationships together into a **mathematical model** of their business situation. Then they try various options to see what profit would be predicted for each. Technology is often used for these "What if ... ?" calculations. The model for the business decision-making is called a **simulation**—it is similar to the real thing.

Run the program named TALENT SHOW (your teacher may direct you to use a spreadsheet instead). This program is a simulation of the decision process that you began in Exploration I. The variables and relations built into that model might be different from what you would have chosen, but they should seem reasonable.

Run the simulation several times to become familiar with the choices it offers. Then run a series of tests, changing the values of only a few variables each time. Study the cause-and-effect relations that you suspect have been built into the program, and look for the best choice of value for each variable.

To help keep track of what changes affect which variables and how, copy and complete the table on the following page to record your trials. Two suggestions are given to help you begin.

Making Connections

In the simulation you had to make several decisions about values for numerical or quantitative variables, such as the amount to charge per soda. These variables are also called **input** or independent variables. Other numerical variables, such as the profit, were influenced by your choice of inputs. These variables are called **output** or dependent variables. As you work through this text, you will learn more about how changes in an input variable can directly and predictably influence the values of an output variable.

After working with the simulation, you should have ideas about how it works. Which decisions (inputs) influence which results (outputs)? Give a summary of your best hunches by answering the questions below.

1. Should a celebrity master of ceremonies (MC) be hired? What effect does that seem to have?

2. What ticket price seems best? What does the ticket price seem to affect?

3. What numbers of newspaper ads and posters seem best? What do those decisions seem to affect?

Table 1

Decisions	1	2	3	4	5	6
Hire a celebrity MC?	Yes	Yes				
Price of a ticket	$1.00	0.50				
How many ads?	5	5				
How many posters?	50	25				
How many sodas?	400	200				
Price of a soda	$0.30	0.25				
How many candy bars?	400	200				
Price of a candy bar	$0.25	0.30				

Results

	1	2	3	4	5	6
Attendance						

Revenue — Money taken in

	1	2	3	4	5	6
Ticket revenue						
Soda revenue						
Candy revenue						
Total revenue						

Costs — Money paid out

	1	2	3	4	5	6
Ticket printing						
Program printing						
Cost of posters						
Cost of ads						
Chaperones						
Cost to buy sodas						
Cost to buy candy						
Master of ceremonies						
Total costs						

	1	2	3	4	5	6
Profit						

4. What candy and soda prices seem best? What size order for each seems best?

5. What variables seem to affect the revenue from tickets, soda, and candy?

6. What factors seem to have been overlooked by the given model of the Talent Show planning task?

Exercises

For each of the following situations, list at least 5 numerical variables that must be considered in trying to answer the stated questions. Then pick a pair of variables that you think have a cause-and-effect relation. Describe what you think that relation is and indicate which is the **input variable** and which is the **output variable**.

1. A fast food restaurant chain is considering opening a new outlet in the area. What is the best location for the new outlet?

2. The county public service commission is trying to set rates for taxi-cabs operating in the county. What is a fair rate?

3. The Washington Redskins are setting prices for tickets, souvenirs, and concession-stand food for next season. What prices for these items will bring the highest profit?

4. The school pep club is planning to sell special t-shirts for School-Spirit Week. How much should they charge for each shirt?

5. A photographer is planning to take class pictures at Northwestern High School. How can one person take all the pictures in three days?

6. A gardener wants to grow vegetables for home use. How can a good harvest be obtained?

2   Variables Related by Data in Tables

In the Talent Show simulation, you explored relations among several variable quantities. You saw, for example, that ticket price had an effect on attendance, although you probably were not sure what the precise relationship was. One way to investigate particular relationships is to study a **table of values** containing typical pairs of values, **(input, output)**.

SITUATION 2.1

People can be found of many shapes and sizes. You may have noticed if you are taller, shorter, or about the same height as your classmates. Some people, such as doctors, have some concerns about what is average for people of various ages. Because of this, many almanacs and reference sources include tables such as the following, showing average heights of American females at specific ages.

Age (yrs)	Height (cm)	Age (yrs)	Height (cm)
0 (birth)	51	12	147
2	84	14	158
4	99	16	160
6	112	18	163
8	127	20	165
10	137		

Source: *The World Almanac and Book of Facts.* NY: Pharos Books. 1992.

As you can see, the average height changes as age increases. Age and height are numerical or quantitative variables, and we can say that height **depends on** age or height **is a function of** age.

Because height depends on age, it should seem reasonable to think of age as the input variable and height as the output variable in the relation.

According to the table, the average height of 14-year-old females is 158 centimeters. Mathematicians and scientists might use the symbolic shorthand form $H(14) = 158$ to convey this information.

We read "$H(14) = 158$" as "H of 14 equals 158." Use of notation like this indicates a relation in which one variable is a function of another. In this case, height is a function of age.

There are many interesting questions that you can answer by examining data in the table. For example:

1. How much does the average height of females increase from birth to age two? How does that increase compare to the increase between ages 16 and 20?

    These changes may be calculated by subtraction:

    $$\text{Height at } 2 - \text{Height at } 0 =$$
    $$H(2) - H(0) =$$
    $$84 - 51 = 33.$$

    $$\text{Height at } 20 - \text{Height at } 16 =$$
    $$H(20) - H(16) =$$
    $$165 - 160 = 5.$$

    Thus the average height increases by 33 centimeters in the first two years and only by 5 centimeters in the four-year period between ages 16 and 20.

2. At what age does the average height of females reach 100 cm?

    The table shows that average height is 99 cm at age 4 and 112 cm at age 6; so the average height reaches 100 cm for some age between 4 and 6 years.

3. During which 2-year period shown in the table does the average height of females increase the most? The following table shows the height differences, and the largest one has been marked.

Age	Change in Height
0 to 2	$84 - 51 = 33$
2 to 4	$99 - 84 = 15$
4 to 6	$112 - 99 = 13$
6 to 8	$127 - 112 = 15$
8 to 10	$137 - 127 = 10$
10 to 12	$147 - 137 = 10$
12 to 14	$158 - 147 = 11$
14 to 16	$160 - 158 = 2$
16 to 18	$163 - 160 = 3$
18 to 20	$165 - 163 = 2$

    The greatest increase occurs from birth to age 2.

Exploration

SITUATION 2.2

The age and height data given in the table of Situation 2.1 are averages. The actual growth pattern for any individual girl might be quite different from those averages. Consider the following data for Dawn, who found the information in health checkup records that her mother had kept.

Age (yrs)	Height (cm)	Age (yrs)	Height(cm)
0	56	12	152
2	85	14	169
4	90	16	178
6	102	18	183
8	113	20	184
10	125		

Dawn's height is a function D of her age.

1. According to the table, what was Dawn's height at age 4? That is, complete the following statement.

    $D(4) =$ _____ .

2. According to the table, at what age was Dawn's height 102 cm? That is, complete the following statement.

    $D($_____$) = 102$.

3. What information is conveyed by "$D(18) = 183$"?

4. Did Dawn grow more between the ages of 6 and 8 or between the ages of 14 and 16?

5. Was Dawn shorter or taller than the average height for females at age 4? What about at age 10? Age 12?

6. In which two-year interval did Dawn grow the fastest? By how much did she grow and how does this compare with the increase in average height for females between the same ages?

Exercises

1. **SITUATION** The table below gives the average heights of American *males* at specific ages.

Age (yrs)	Height (cm)	Age (yrs)	Height (cm)
0 (Birth)	52	10	138
2	87	12	150
4	103	14	163
6	116	16	174
8	127	18	177

Source: *The World Almanac and Book of Facts.* 1992.

The average height of males is a function M of age.
a. At what age is the average height of males 150 cm?
b. Use functional notation to write the sentence, "The average height of males at age 14 is 163 cm."
Hint: $M(\underline{}) = \underline{}$.
c. Complete the following statement.
$M(2) = \underline{}$.
d. Write in words the meaning of $M(8) = 127$.
e. In which of the two-year periods shown by the table does the average height increase the most? What is that increase?
f. How does the pattern of change in average height for males compare to that for females?

2. **SITUATION** The Talent Show advisor found the following record of profits from previous Talent Shows. Note: Negative profit means the show lost money.

Year	Profit (in dollars)	Year	Profit (in dollars)
1980	−221	1984	847
1981	−155	1985	1838
1982	−18	1986	1750
1983	233	1987	2133

Profit can be considered a function P of the year.
a. In which year did the Talent Show make the greatest profit?
b. The amount of profit is a function of the year. What does $P(1981) = -155$ tell about Talent Show profit?

c. In 1985 the Talent Show had a profit of $1838. Using functional notation, express this fact.

d. In which year did Talent Show profit increase the most from the previous year? Write the calculations needed to find each year-to-year change as follows:

$$P(1981) - P(1980) = (-155) - (-221) = 66.$$

Then use your calculator to find the results.

e. Between which consecutive years did the profit decrease?

f. Describe the overall trend in Talent Show profits shown in the table and make a prediction about profit that could be expected from the 1988 show.

3. **SITUATION** The following table gives the number of record albums sold in the United States for some recent years.

Year	Albums sold	Year	Albums sold
1973	290 million	1981	280 million
1975	333 million	1983	230 million
1977	345 million	1985	195 million
1979	330 million	1987	178 million

a. What trend, or pattern of change, do you see in the data in the table? Offer a possible explanation for this trend.

b. The number of albums sold, A, **is a function of** the year. What information does the following sentence, written in functional notation, give about album sales?

$$A(1981) = 280 \text{ million.}$$

c. Write the following statement in functional notation: 330 million albums were sold in 1979.

d. In which two-year period did the reported sales change the most? What was that change?

e. What estimate do you believe is most reasonable for album sales in 1989?

4. SITUATION A fast-food restaurant found that the number of orders of french fries that it sells depends on the price of an order of fries.

Price (dollars)	Orders of fries per day
0.50	1000
0.60	850
0.70	700
0.80	550
0.90	400
1.00	250

a. Describe the relation between price and number of orders that you see in the table and explain why it is or is not reasonable to you.

b. Write functional notation for each of the following.
Hint: Since the number of orders depends on the price, a reasonable way to write the relation might be $N(\text{price}) = \text{number of orders}$.

   i. When the price is fifty cents, 1000 orders of fries are sold.

   ii. Only 250 orders are sold when one dollar is charged.

c. Explain in words the meaning of each of these symbolic statements:

   i. $N(0.80) = 550$.

   ii. $N(0.60) = 850$.

d. If the manager wants to sell at least 500 orders each day, what price(s) should the restaurant charge?

e. If the restaurant chain will only allow the price of fries to be between seventy and eighty cents an order, how many orders of fries will be sold each day at this location?

5. SITUATION Suppose that the manager of the restaurant in problem 4 decides to try a new large size order of fries. Using the table and any trends that you observed with regular size fries above, estimate the number of orders of large fries that might be sold at the following prices.

Price (dollars)	Orders of fries per day
0.80	
0.90	
1.00	
1.10	
1.20	
1.30	

a. List some factors besides price that might affect the number of orders of large fries sold.

b. Suppose the manager wants to use a special to bring in more customers. List some factors in the business that might be changed.

c. Pick two variables from part b that might have a cause-and-effect relationship. Indicate which you think would be the input variable and which you think would be the output variable. Explain your answer.

3 Variables Related by Data in Graphs

The ways in which two variables might be related are not always shown clearly by tables of input-output data. Patterns in the data may be lost amid all the specific numbers. However, when data are displayed in a **graph**, it is often much easier to see trends and therefore to make predictions.

Age (yrs)	Height (cm)
0	51
2	84
4	99
6	112
8	127
10	137
12	147
14	158
16	160
18	163
20	165

Source: *The World Almanac.* 1992.

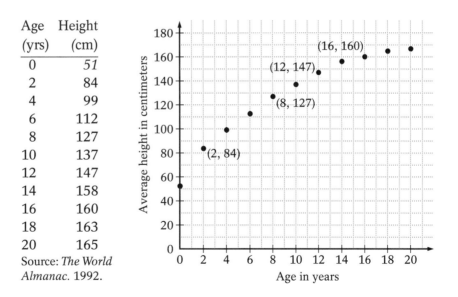

The graph displays the same data as the table, but to interpret and construct graphs with skill, it is important to understand exactly how the two forms of data correspond.

1. The table displays **ordered pairs** of data (age, height). Each pair also can be considered as **coordinates** which locate exactly one point on the graph.

   For example, the point with coordinates (16, 160) represents the fact that at age 16 the average height of females is 160 centimeters.

2. It is customary to indicate values of the input variable on the **horizontal** (left to right) **axis** and values of the output variable on the **vertical** (up and down) **axis**.

   In our example, that means age is on the horizontal axis (each tic mark represents one year), and height is on the vertical axis (each tic mark represents 10 centimeters).

3. To locate or interpret data pairs that do not hit given grid points exactly, it is necessary to **estimate** position and coordinates.

   For example, the data pair (2, 84) corresponds to a point located about halfway between (2, 80) and (2, 90). This means that the average height of females of age 2 is about midway between 80 and 90 centimeters.

4. A graph also shows **patterns of change** in related variables by the horizontal and vertical movement needed to get from point to point.

   For example, the graph of the function relating age and height shows that the average height of females increases by 20 cm from age 8 to age 12. The difference in height coordinates of (8, 127) and (12, 147) is $147 - 127 = 20$.

5. You can also estimate pairs, (age, height), not given in the table.

   For example, the following graph shows an in-between point with coordinates approximately (7, 120), suggesting that the average height of 7-year-old females is about 120 centimeters.

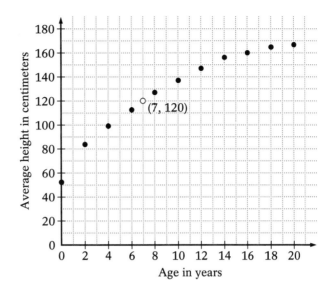

Exploration I *Constructing Graphs*

The exploration and homework exercises will sharpen your skills in constructing and reading graphs.

SITUATION 3.1

A drought in the African veldt causes the death of many animals. For example, a typical herd of wildebeest might suffer losses like those indicated in the following table.

Length of drought (months)	Number of wildebeests
0	500
1	400
2	300
3	250
4	200
5	175
6	150
7	125
8	100
9	90
10	75

1. Reproduce the grid shown here. Then, plot the ordered pairs of data given in the table above.

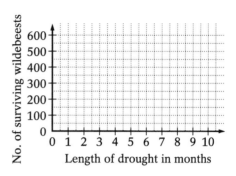

2. Give the coordinates for the points that answer the following questions:

   a. What was the population after 3 months of drought?

   b. How long had the drought been going on when the herd reached a population of 125?

   c. For what length of time did the population remain above 200?

3. What pattern in the graph shows the drop in population?

SITUATION 3.2

When NASA sends a space shuttle on a mission, scientists and engineers monitor conditions on the shuttle with many different instruments. For example, heat sensors in the nose cone steadily send temperature readings to NASA computers. These readings are watched closely from several hours before liftoff until the shuttle lands safely. The table below gives a sample of readings, (time, temperature), for a typical mission. Temperature of the nose cone *is a function of* time in the mission. Notice that time *before liftoff* is indicated by negative numbers and that temperatures also take on both positive and negative values.

Mission time (min)	Nose cone temperature (°C)	Mission time (min)	Nose cone temperature (°C)
−25	−10	15	115
−20	−10	20	130
−15	−5	25	105
−10	−5	30	80
−5	0	35	20
0	10	40	−5
5	75	45	−30
10	90	50	−40

1. Recreate the grid shown here and plot the data pairs, (time, temperature), given in the table.

Notice that points showing "negative time" (time before liftoff) are located on the left side of the vertical axis and that points showing negative temperature are below the horizontal axis. The axes divide a coordinate grid into four "quadrants." The upper right quadrant is called the first quadrant. The upper left quadrant is the second quadrant. The point shown on this grid is in the third quadrant.

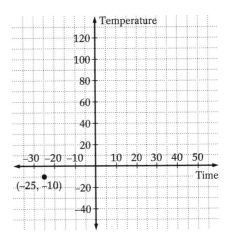

Each horizontal grid unit represents 5 minutes of time; each vertical grid unit represents 10°C temperature.

2. For parts a through d, temperature *C* (in degrees Celsius) is a function of time in minutes. For each part, do the following.
   i.   Write the value that correctly completes the statement.
   ii. Give the coordinates of the data point, (input, output), that supplies your answer.
   iii. Write the completed statement in words.
       a. $C(15) =$ _____.   b. $C(-10) =$ _____.
       c. $C($____$) = -30$.   d. $C($____$) = 90$.

3. With your classmates, discuss the factors in a shuttle flight that affect nose cone temperature. Explain the temperature pattern shown in the graph in terms of flight conditions.

Exercises

1. List the coordinates of each of the labeled points as an ordered pair. Estimate where necessary.

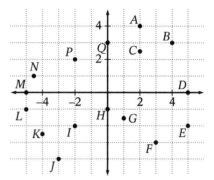

2. On graph paper, draw a pair of coordinate axes. Label the axes with appropriate units, then plot the following ordered pairs. At each point, make a heavy dot and write the coordinates.

a. (0, 0)	**b.** (–3, 8)	**c.** (–8, 3)	**d.** (–8, –3)
e. (–5, 0)	**f.** (0, –5)	**g.** (5, 0)	**h.** (6, –4)
i. (3, 3)	**j.** (7, 7)	**k.** (–5, –5)	**l.** (–7, –7)

3. List the coordinates of each labeled point as an ordered pair. Notice that each horizontal grid unit represents 10 and each vertical grid unit represents 25.

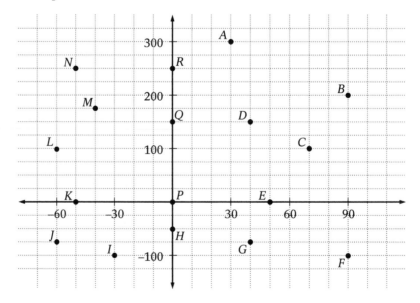

4. **SITUATION** A hot-air balloon is launched at noon, and the balloonist records the altitude in meters every 15 minutes. The recorded data are given in the following table.

Time (min)	Altitude (m)	Time (min)	Altitude (m)
0	0	90	900
15	100	105	920
30	750	120	500
45	875	135	625
60	900	150	300
75	950	165	0

a. Choose reasonable scales for the horizontal (time) and vertical (altitude) axes. Plot the data given in the table above on a pair of coordinate axes using the scales you have chosen. Copy the diagram shown here to help you begin.

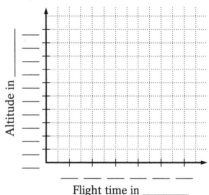

Altitude in _____

Flight time in _____

Answer each of the following questions about the flight and give the coordinates, (input, output), you used to get the answer. Estimate where needed.

b. What was the altitude 1 hour into the flight?
c. What was the altitude 0.5 hour into the flight?
d. What was the altitude 2.25 hours into the flight?
e. At what time (s) was the altitude 750 meters?
f. At what times was the altitude under 300 meters?
g. What was the change in altitude during the first hour?
h. What was the change in altitude during the second hour?
i. What was the maximum altitude?
j. At what time (s) was the balloon rising most quickly?
k. When was the balloon falling most rapidly?

5. In each of the following situations you are given data relating two variables. In each case, choose reasonable scales for axes on a graph and plot the given data. Then write a sentence describing the pattern in the graph and what it says about the relation between the two variables.

a. SITUATION The mass of an average American young person, in kilograms, is a function of the person's age in years. Here are some sample data.

Age (yrs)	0	2	4	6	8	10	12	14
Mass (kg)	3	11	15	20	26	31	38	49

Source: *The World Almanac.* 1992.

b. SITUATION A newspaper delivery person's weekly pay is a function of the number of papers delivered each day. Here are some sample data.

Papers delivered	50	100	15 0	200	300	500
Pay (dollars)	70	120	170	220	320	520

c. SITUATION Animal populations tend to rise and fall in cycles. Suppose the following data shows how squirrel populations in a central Pennsylvania city varied from 1975 to 1984.

Year	'75	'76	'77	'78	'79	'80	'81	'82	'83	'84
Population	750	700	520	680	730	650	550	625	780	700

Exploration II *Interpreting Graphs*

In Exploration I you studied relations between variables by graphing data given in tables of ordered pairs, (input, output). In many practical situations, scientific instruments produce information directly in graphic form. For example, weather service instruments produce continuous graphs of temperature, humidity, wind speed, and barometric pressure. Employees of gas, electric, and water companies monitor usage by watching continuous graphs of demand.

When information about two variables is given by a graph, it is important to be able to read specific values for input and output, but also to interpret the overall shape of the graph.

SITUATION 3.3

When the state highway patrol uses radar to catch speeding cars, an officer focuses a radar device on a car. The device displays a steady stream of readings on that car's speed as it approaches. The graph shows the record of such readings for one car.

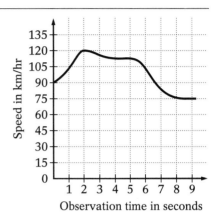

This graph shows that the car's speed is first recorded at 90 km/hr. It then increases to about 120 km/hr and holds just under that rate for about three seconds. Then it decreases to 75 km/hr, where it remains during the rest of the time that the radar device is used.

While the graph looks like a smooth curve, it is actually produced by plotting many individual points, (time, speed). For instance, the maximum speed of about 120 km/hr occurs 2 seconds after the first measurement; it is indicated by the point $(2, 120)$. The speed of 90 km/hr at the start is indicated by the point $(0, 90)$. And the speed of 75 km/hr during the last several seconds is indicated by several points with coordinates of the form $(t, 75)$.

On this graph it is easiest to read coordinates for points that represent times 1, 2, 3, ... , 9 seconds and speeds of 15, 30, ... , 135 km/hr. However, the graph can also be used to estimate times and speeds between these values. For instance, the graph shows the car's speed after 1.5 seconds is about 110 km/hr.

Using the graph, you can produce a table of data pairs, (time, speed), like this:

Time (sec)	0	1	2	3	4	5	6	7	8	9
Speed (km/hr)	90	105	120	117	115	115	100	83	75	75

Of course, the speed values are only estimates since in most cases the speed curve does not intersect one of the grid points exactly.

SITUATION 3.4

The following graph shows the temperature of a TV dinner while it is in the freezer, taken out for cooking, and served at the table. The food was removed from the freezer at time 0.

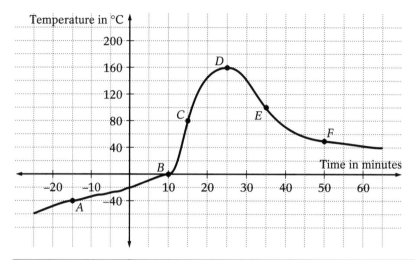

1. For each labeled point on the graph, give the coordinates and explain what probably is happening to the dinner at that time.

2. Many people buy unfrozen food at the store and put it in the freezer when they come home. On a pair of axes, sketch the pattern of a graph showing the temperature of a steak, as a function of time, before and after it is placed in the freezer at home. Let the moment it is put in the freezer be time 0.

SITUATION 3.5

In many businesses, sales of a product depends on the price charged. The diagrams below show graphs of three possible relations between price and sales of a new model car. No scales are given on the axes, but following the conventions that right and up represent positive values of price and sales, you should be able to decide which graph seems most likely to match the relation between price and sales.

1. For each diagram, explain why you think the graph does or does not fit the likely relation between price and sales.

a.   b.   c.

2. List at least 4 factors other than price that are likely to affect the sales of a new car.

3. Pick a factor from your list that is a quantitative variable. Sketch a graph similar to the ones above that shows the relation you would expect between that variable and car sales. Explain in words the relation shown by your graph. That is, tell what happens to car sales as the input variable increases, and give some possible reasons for this relation.

Exercises

1. SITUATION The following diagram shows the temperature in Frostburg, Maryland between midnight and 10:30 a.m. on a typical December day. The temperature is a function of the time of day.

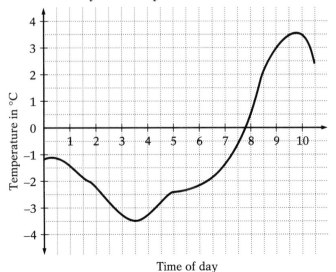

Time of day

For each of the following questions, list the coordinates of the point (s) that give the answer. Then write your answer in a sentence. Estimate where necessary.

a. What was the temperature at 4 a.m.?
b. What was the temperature at 10:30 a.m.?
c. When was the temperature −2°C ?
d. When was the temperature 3°C ?
e. What was the maximum temperature and when did it occur?
f. What was the minimum temperature and when did it occur?
g. When was the temperature rising?
h. When was the temperature falling?

2. **SITUATION** After a flu epidemic is first noticed in school, the number of students who are ill varies over the next several weeks. The diagram below shows the pattern of one such epidemic.

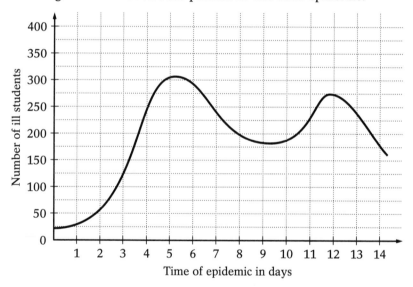

Use the graph to answer the following. Estimate where necessary.

a. How many students were ill on day 4?
b. How many students were ill on day 11?
c. On which days were at *least* 200 students ill?
d. On which day were the *most* students ill and how many were ill on that day?
e. When did the number of ill students change *most rapidly*?
f. Write a description of the *pattern* for this flu epidemic during the 14 days shown on the graph.

3. Suppose the number of ill students is a function I of the number of days since the flu was first noticed. The following statements are written using functional notation. For each, use the graph in Exercise 2 to find the value that would complete the statement, then rewrite the completed statement in sentence form.

 a. $I(6) = \underline{}$.
 b. $I(9) = \underline{}$.
 c. $I(\underline{}) = 250.$
 d. $I(\underline{}) = 300.$
 e. $I(\underline{}) > 150.$ (Recall that ">" means "is greater than".)
 f. $I(\underline{}) < 100.$

4. In each part of this problem, you are given the description of a situation involving related variables and a diagram with three graphs. Choose the graph that best fits the relation and explain your answer.

 a. When you work at a job with an hourly wage, your total income is a function of the time worked.

 b. The depth of water in a bathtub before, during, and after a bath is a function of the time since the faucet was turned on.

 c. The number of people in attendance at a water amusement park varies with the temperature of the air.

 d. The depth of water in the Chesapeake Bay changes as the tide comes in and goes out.

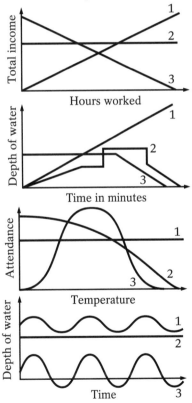

5. SITUATION Often, snow offers many opportunities for winter sports, such as sledding. For one of the most common ways to enjoy the thrills, one must climb to the top of a hill. The trip up a hill is, of course, much slower than the trip down. This diagram shows one possible graph of **speed** as a function of **time** for a sledder as he climbs up a hill and then slides down it.

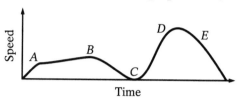

a. At which labeled point (or points) did the sledder reach the top of the hill?

b. At which labeled point(s) was the sledder headed down the hill?

c. Describe the sledder's trip further by telling what was happening at each of the other labeled points on the diagram.

6. Which graph below shows:

a. A car slowing down and then speeding up?

b. A car stopping at a stop sign and then moving on?

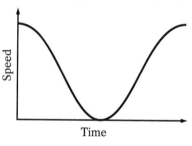
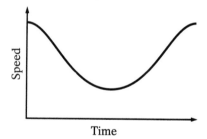

7. Sketch a graph illustrating the relation between variables in each of the following situations:

a. SITUATION When a young tree is planted, its height is a function of growing time.

b. SITUATION When a marble rolls to a stop, its speed is a function of time.

c. SITUATION When food is cooked and then set out on the table, its temperature is a function of time.

d. SITUATION During the spring and summer, the length of grass on a lawn is a function of time.

e. SITUATION At the meat counter in a supermarket, the price of a package of ground beef is a function of the weight of the package.

8. Describe two quantitative variables that you believe are related to each other. Then sketch a graph showing what you believe that relation to be.

4 Variables Related by Rules

In the preceding sections you have studied related quantitative variables in a variety of situations. You have seen ways to describe and answer questions about those relations using **tables** of data pairs, (input, output), and **graphs** of those pairs. This section introduces a third way to study related variables, using **rules** that explain how output values are calculated from input values.

SITUATION 4.1

Carla borrowed $250 from her mother to purchase a deluxe grass-catching lawn mower for her new lawn-mowing business. Her mother helped her make the table shown here to chart her summer business prospects.

Carla is planning a one-week vacation in the Poconos with her cousin. To pay for the trip, her targeted summer earnings should be no less than $500.

Number of lawns mowed	Carla's profit in dollars (Income–Expenses)
0	–250
5	–200
10	–150
15	–100
20	–50
25	0
30	50

Study the data in the table relating number of lawns mowed to profit and see if you can answer the following questions.

1. Would Carla need to list many more entries in the table to find the number of lawns needed to make a profit of $500?

2. Do you see a pattern for computing her profit for any number of lawns mowed?

3. How much do you suppose that Carla is charging per lawn?

If you can find a rule relating the **number of lawns mowed** and Carla's **profit**, you can go beyond the data given in the table. You can calculate her profit for as many as 60 or 70 lawns (or, as few as three).

You probably have discovered a relationship between the number of lawns and her profit for the summer. It appears that Carla plans to charge $10 per lawn. So, she could calculate her profit for any number of lawns:

1. by multiplying the number of lawns by 10 and subtracting 250, the cost of her mower, or

2. by adding ten times the number of lawns to the initial balance of – 250.

For instance, if she mows 60 lawns, then her profit is

$$(10 \times 60) - 250 = 350 \text{ or } -250 + (10 \times 60) = 350.$$

Instructions like these describing how to calculate Carla's profit as a function of the number of lawns she mows often are expressed as rules. For instance, the first rule written in words is: Profit equals ten times the number of lawns minus 250. How could the second rule be written?

SITUATION 4.2

Suppose that Carla decided to purchase a less expensive lawn mower—one without a grass catcher. Suppose that the mower she selects sells for $195, and that she continues to charge $10 per lawn.

1. Make a table charting her profit for 0, 5, 10, 15, 20, 25, and 30 lawns.

2. Now how many lawns must Carla mow in order to "break even"?

3. Write the calculations needed to compute Carla's total profit after having mowed 35 lawns.

4. Write a rule that explains how to calculate Carla's profit as a function of the number of lawns mowed.

SITUATION 4.3

Carla realized that with the less expensive mower she would have to take time to rake up and bag mowed grass. She decided to raise the price charged per lawn. Now her profit for mowing 0 to 30 lawns becomes:

Number of lawns	Carla's profit in dollars
0	−195
5	−135
10	−75
15	−15
20	45
25	105
30	165

1. Based upon the table above, how much was Carla charging per lawn?

2. Write a rule relating number of lawns and Carla's profit now.

You probably can see that using rules is a much more efficient way to describe the relation between two variables than long tables of data pairs, (input, output). Of course, tables and graphs have other virtues that make them useful for finding patterns and answering questions about the relationships. Furthermore, not every relation between variables can be expressed with a simple rule.

In the following sections we shall explore many similar situations in which algebraic rules provide a very useful method for describing and studying relations between quantitative variables.

 Exploration

The skill of finding rules in data that relate variables is so widely used, you will be given opportunities for practice throughout the course. Experience will suggest some useful search strategies, but there is no certain method for all problems. Be willing to guess, and test your guesses.

Activity 1. Run the program FINDING RULES TO RELATE VARI-ABLES (your teacher may direct you to use a spreadsheet instead). The program has 40 rules for tables. You can choose any of the 40 to study.

1. You will be presented with part of a table relating input and output variables—giving you clues to a rule. For example,

Input	Output	Yes/No
1	3	
2	5	
3	7	

2. Next, you will be presented with four more input values. For each, predict the matching outputs and supply your predictions. The yes or no response to your entries will tell you whether your prediction fits the rule it is using.

Input	Output	Yes/No
1	3	
2	5	
3	7	
4	9	yes
5		
10		
20		

3. Then you may supply both input and predicted output, and the response you receive again will tell you whether the pair fits the rule. When you are certain you have discovered the rule, end the program by entering "Q" instead of a numerical input. For example,

Input	Output	Yes/No
1	3	
2	5	
3	7	
4	9	yes
5	11	yes
10	21	yes
20	40	no
12	25	yes
Q		

In this last example, a rule relating input and output values can be summarized by the following statement:

To get the output, we multiply the input by two and add one to the resulting product.

We shall be exploring the use and power of finding rules like these in the sections that follow. Try a couple of rules until you get the hang of the program, then proceed with Activity 2.

Activity 2. Choose any of the 40 rules to study. When you believe you know a rule that fits the data in the table, copy the table and describe your rule. Record at least nine rules in tables like the one shown below, but try to discover as many as time permits.

Rule # _____

Input	Output	Yes/No

Exercises

In Exercises 1 through 12, study the input-output pattern in each table. Copy and complete the table, then write a rule relating the input and output values.

1.

Input	Output
1	4
2	5
3	6
4	7
5	
6	
7	
8	

2.

Input	Output
1	3
2	6
3	9
4	
5	
6	
7	
8	

3.

Input	Output
1	49
2	48
3	47
4	46
5	
6	
7	
8	

4.

Input	Output
1	48
2	46
3	44
4	42
5	
6	
7	
8	

5.

Input	Output
1	5
2	10
3	15
5	
8	
12	
17	

6.

Input	Output
1	1
2	3
3	5
5	9
8	
12	
17	

7.

Input	Output
0	1
1	11
3	31
6	61
10	
15	
21	
28	

8.

Input	Output
1	1
2	4
3	7
5	13
8	
12	
17	
23	

9.

Input	Output
20	10
18	9
16	8
14	
12	
10	
8	
6	

10.

Input	Output
1	1
2	4
3	9
4	16
5	
7	
10	
12	

11.	Input	Output
	1	0
	2	3
	3	8
	4	15
	5	24
	8	
	10	
	11	
	20	

12.	Input	Output
	1	60
	2	30
	3	20
	4	15
	5	12
	6	
	10	
	30	
	120	

13–14. Write two rules of your own and corresponding input-output tables to share with your classmates.

For each of the rules given in items 15 through 18, copy and complete the table of values. Use technology as needed.

15. Rule: Add six to the input and multiply the result by two to get the output.

Input	Output
1	
2	
4	
5	
10	
11	
20	

16. Rule: Subtract one from the input and divide the result into twelve to get the output.

Input	Output
2	
4	
5	
6	
10	
11	
20	

17. Rule: Take 5% of the input and add the result to the input to get the output.

Input	Output
25	
24	
23	
22	
21	
5	
4	

18. Rule: Find the average of the input and 24 to get the output.

Input	Output
5	
4	
3	
2	
1	
10	
11	

In Exercises 19 and 20, copy and complete the tables of values for the given rules. Note that several outputs are given and you are asked to find the corresponding inputs. Use technology as needed.

19. Rule: To get the output, add 2 to the input and multiply the result by 2.

Input	Output
	8
	10
	20
	40

20. Rule: To get the output, multiply the input by 2 and add 2 to the result.

Input	Output
	8
	10
	20
	40

21. Let's compare two possible scenarios for Carla's lawn-mowing enterprise:

SITUATION I Suppose she purchases a top-of-the-line lawn mower with grass catcher at a cost of $250, and she charges $10 per lawn. Carla is attending summer school, she is only able to mow lawns on a part-time basis. She finds that, on the average, she can mow about six lawns each week.

SITUATION II Suppose she purchases a mower without a grass catcher for $195, and she charges $12 per lawn. She finds that the additional task of raking and bagging grass reduces the number of lawns she can mow; she is only able to mow four lawns each week.

a.   For each situation, construct a table of values representing the relation between the number of lawns mowed and Carla's expected profit. For Situation I you might use inputs ranging from 0 to 60 in steps of 4 (that is, values 0, 4, 8, 12, 16, ..., 60). For Situation II you might use inputs ranging from 0 to 40.

b.   Which yields the greater profit at the end of the fourth week?

c.   Which yields the greater profit at the end of ten weeks?

d.   Write a rule relating profit to number of lawns mowed in each situation.

e.   What other variables might Carla take into account when trying to make a good prediction of her summer earnings?

5   Rules Relating Variables — Symbolic Form

When the relation between two variables can be written as a simple rule for calculating output values from input values, you can summarize the information in a long table or graph with just a few words. The language of algebra provides an even more efficient system for expressing such a rule by using symbolic shorthand.

The following illustrates this using the relation between weight on earth and weight on the moon. Due to the difference in the sizes of the earth and the moon, gravitational attraction on the moon is less than that on earth. So an object, such as an astronaut, weighs less on the moon than on earth. The following table gives the moon weights corresponding to various earth weights.

Earth weight	1	2	3	4	5	10	20	100
Moon weight	0.16	0.32	0.48	0.64	0.80	1.60	3.20	16.00

Source: *Science and Technology Desk Reference.* Carnegie Library of Pittsburgh. 1993.

In this table the moon weight of an object is treated as a function of its earth weight. The pattern in the entries suggests the rule

> moon weight equals 0.16 times earth weight.

In algebra it is common to shorten the expressions for rules by using single letters as variable names. For instance, if E is used for the variable *earth weight* and M for its corresponding *moon weight* (both measured in the same unit of weight), we can write the rule in either of two symbolic forms:

$$M = 0.16 \times E \text{ or } M(E) = 0.16 \times E.$$

To use such a symbolic rule for calculating a specific output value from a specific input value, you simply replace the variable name with a specific number and carry out the indicated operations. For example, to find the moon weight corresponding to an earth weight of 50 units:

$$M = 0.16 \times 50 \text{ units} = 8 \text{ units or}$$
$$M(50) = 0.16 \times 50 \text{ units} = 8 \text{ units}.$$

Symbolic rules like these are used throughout mathematics and its applications because they express relations between variables in very compact form. Thus it is important to develop skill in using and writing symbolic expressions for operations.

 ## Exploration I

In each of the following situations, you will be given information about a relation between two quantitative variables. Use that information to follow the given instructions. Use technology as needed.

SITUATION 5.1

During a bicycle race it is not uncommon for each wheel to average 375 revolutions per minute.

To find a rule relating the number of revolutions to elapsed time, it is helpful to make a table of values displaying several specific cases.

1. Choose some values of the input variable, *number of minutes of time*, and find each matching value of the output variable, *number of revolutions*. Record your data pairs in a table like the following.

   Input variable (min)   Output variable (revolutions)

2. Write a rule describing the relation between number of minutes and number of revolutions.

3. Use the single-letter symbols T and R for the variables *number of minutes of time* and *number of revolutions*, respectively, to write the rule relating them in the following symbolic forms.

   $$R = \underline{\hspace{3cm}} .$$
   $$R(T) = \underline{\hspace{3cm}} .$$

4. For each of the following, write the calculation required and the result.

   How many tire revolutions can be expected after:
   a. 10 minutes?
   b. 1 half-hour?
   c. 1 hour?

SITUATION 5.2

For each revolution of its wheels a bicycle covers about 1.6 meters of ground.

Find a relation between time and distance covered, assuming that the cyclist pedals fast enough for 375 revolutions per minute.

1. Calculate some specific data pairs, (time, distance), and record them in a table.

2. Write verbal and symbolic rules expressing the relation between time and distance.

3. Write the calculation required to find the distance traveled in 15 minutes. What is this distance?

4. Sketch a graph showing the pattern of the relation between time and distance traveled. The input variable should be represented on the horizontal axis, and the output variable should be on the vertical axis.

SITUATION 5.3

Kim has a part-time job selling magazine subscriptions. He is paid $20 per week plus $5 for each subscription he sells.

1. Copy and complete the table, showing his weekly pay as a function of the number of subscriptions he sells.

Subscriptions sold	Weekly pay (dollars)
1	
2	
3	
4	
5	
10	
15	

2. Using single-letter names for the variables involved, write two symbolic rules for the relation between sales and pay.

3. Write the calculation required to answer each of the following questions about Kim's pay prospects. Then find the numerical result.
   a. How much is he paid if he sells 8 subscriptions in a week?
   b. How much is he paid if he sells 0 subscriptions in a week?
   c. How many subscriptions must he sell if he wants to earn $75 in a week?

Exercises

1. SITUATION Maria is planning a summer job mowing lawns. She must purchase a mower for $225. She can earn $12 per lawn.

    Write a rule that gives the relation between the number of lawns she mows and her profit for the summer. Then, using the single-letter variable labels N for the number of lawns and P for the profit in dollars, write the rule in two symbolic forms.

2. SITUATION Kate has a part-time job selling newspaper subscriptions for which she earns $15 per week plus $4.50 for each subscription she sells.

    Write a rule that gives Kate's weekly pay as a function of the number of subscriptions she sells. Then use the single letter variable labels N for number of sales and P for pay in dollars to write the rule in two symbolic forms.

3. SITUATION Suppose that Kate's job selling newspaper subscriptions changes so that she is paid $6 per sale but no weekly base.

    Write a rule in symbolic form that gives her weekly pay as a function of sales.

4. The speed of light is approximately 300,000 kilometers per second.
    a. How far does light travel in 2, 3, 4, and 60 seconds?
    b. Use the single-letter variable labels T and D to write a rule relating time in seconds and distance in kilometers.

5. Jupiter is the largest planet in our solar system, and it has the strongest gravitational force. A rule that gives the weight of an object on Jupiter as a function J of its weight on Earth E is $J(E) = 2.64 \times E$.

    Write and carry out the calculations needed to answer each of the following questions about this relation.
    a. What is the weight on Jupiter of an object with Earth weight 50 units?
    b. What is the weight on Jupiter of an object with Earth weight 0.45 units?
    c. What is the weight on Earth of an object with Jupiter weight 25 units?

6. Remember that the rule relating earth and moon weights is $M(E) = 0.16 \times E$. Using this rule, copy and complete the following table comparing earth and moon weights of familiar objects. Make estimates of earth weights that you don't know. Label each estimate with its appropriate unit.

Object name	Earth weight	Moon weight
An average 14-year-old		
Your class		
A car		
A bowling ball		
(your choice)		

7. The following table shows corresponding weights of objects on the Earth and on the planet Mars. Study the pattern in the data to find a rule giving Mars weight as a function of Earth weight (both measured in the same unit of weight).

Earth weight	Mars weight
10	4
20	8
30	12
50	20
100	40

Source: *Science and Technology Desk Reference.* 1993.

8. Write a symbolic rule relating the number of revolutions n and the time in minutes t for a bicycle tire making 300 revolutions per minute.

9. If an airplane travels at a steady speed of 800 kilometers per hour, what rule gives its distance traveled as a function of its time in flight?

10. The number of tons of garbage collected daily in a city depends upon the number of people in the city.
    a. Construct a graph of the data provided in the following table.

Population	Garbage (tons)
25,000	15.6
50,000	31.3
75,000	46.9
100,000	62.5
125,000	78.1
150,000	93.8

b.   Write at least two questions about this situation. Make sure your questions can be answered using either the table above or your graph.

c.   Write a rule (either in your own words or symbolically) relating number of tons of garbage collected daily g and number of people p.

Exploration II (optional)

1.   Run the program GRAVITY OF THE PLANETS and try to discover the symbolic rules the program uses to determine the various planet weights from Earth weights. (A calculator may be helpful.)

Record your findings in a table such as the following.

Calculating planet weights from Earth weight	Suggested symbolic rule
Mercury weights from Earth weights	
Venus weights from Earth weights	
Saturn weights from Earth weights	
Neptune weights from Earth weights	
Pluto weights from Earth weights	

2.   Run the program FINDING RULES TO RELATE VARIABLES. Write the rule in symbolic form for each of ten or more exercises in this program, as time permits.

6   Order of Operations

When you search for patterns relating input and output variables, you probably find it easier to make specific output predictions than to write a rule for all predictions. You might even have a correct idea of how to calculate an output for any input. However, writing that idea in proper algebraic form so it can be properly used with technological tools, or by other students, requires knowledge of generally accepted rules of algebraic grammar.

One of the input-output rules you may have seen in the program FINDING RULES TO RELATE VARIABLES produces the following table of values.

Input	1	2	3	4	5	10	20
Output	8	12	16	20	24	44	84

If you looked at this table and found a pattern relating inputs and outputs, your rule might be one of these:

1. Add 1 to the input and then multiply this result by 4.

2. Multiply the input by 4 and then add 4 to this result.

If we use the letter n to stand for the input variable, it seems natural to write these rules as:

1. $n + 1 \times 4 = $ output   2. $n \times 4 + 4 = $ output

These rules suggest a sequence of operations to use in order to find output values. For example, when using a calculator, you might do the following.

1. Enter the input number.   2. Enter the input number.
    Press +.   Press ×.
    Enter 1.   Enter 4.
    Press ×.   Press +.
    Enter 4.   Enter 4.
    Press =.   Press =.

However, these two procedures give very different results!
    All calculators or computer languages using algebraic logic give the following results:

Input	Output procedure 1 $n + 1 \times 4$	Output procedure 2 $n \times 4 + 4$
1	5	8
3	7	16
5	9	24
10	14	44
20	24	84
	NOT CORRECT	CORRECT

The explanation of these very different results, from what may seem to be equivalent procedures, lies in the standard algebraic rules for performing a sequence of arithmetic operations. The agreed-upon practice in algebraic logic does not simply take the symbols as they appear, from left to right. Instead, it looks at the whole expression that is to be evaluated and performs the indicated operations in the following order:

Step 1: Perform all operations within any pair of parentheses (working from the inside out if one set of grouping symbols is inside another).

Step 2: Calculate any exponents or powers, such as x^2 or 5^3.

Step 3: Perform all multiplication and division, from left to right.

Step 4: Perform all addition and subtraction, from left to right.

Using these **order-of-operations rules**, we see that:

1. $n + 1 \times 4$ means "first multiply 1 by 4 and then add this result to n." (This is not what we had in mind!)

2. $n \times 4 + 4$ means "first multiply n by 4 and then add 4 to this result." (Just what we wanted!)

We can modify the first rule to do what we had in mind by using parentheses as follows: $(n + 1) \times 4$. This now says "add 1 to n and then multiply this result by 4."

In addition to the rules given above for order of operations, there is another agreement about algebraic symbols that is commonly used to shorten symbolic expressions. When a symbolic expression indicates one or more multiplications, we often omit the multiplication symbols $\times$ and $\cdot$ since one can be confused with the variable label x and the other with a decimal point. Thus,

> $5t$ means 5 times t,
> $7(x - 5)$ means 7 times the difference $(x - 5)$,
> $18st$ means 18 times s times t.

Using this notational shorthand for multiplication, we can write the input-output rules for the table above as "$(n + 1)4$" or as "$4n + 4$".

The multiplication shorthand procedure must be used with care. In many computer languages the asterisk (*) is used to indicate multiplication and it cannot be omitted. At the keyboard you must type 5*t instead of 5t.

The four order-of-operations rules and the shorthand for indicating multiplication help users of mathematics write long or complex computational procedures in very compact form. It takes time to become expert in full use of the rules. The exploration exercises that follow will give you practice with some of the most common forms in which the rules play a useful role.

Exploration I

In problems 1 through 6, write in words the sequence of operations indicated by each of the symbolic rules. An example is given first:

$P(N) = 5N + 8$.
The rule "$P(N) = 5N + 8$" says multiply the input value by 5 and then add 8 to get the output.

1. $R(T) = 5(T + 8)$

2. $S(N) = (0.04 \times N) + N$

3. $D(T) = 80T - 125$

4. $H(A) = 125 - 80A$

5. $G(T) = 4(T + 6)/2$

6. $D(S) = -3S^2 + 5S + 2$

In problems 7 through 13, calculate the output for each given input using the given symbolic rule. Use technology as needed.

7. If $P(N) = 5N + 8$, evaluate:
    a. $P(10)$ **b.** $P(-10)$ **c.** $P(3.2)$

8. If $R(T) = 5(T + 8)$, evaluate:
    a. $R(10)$ **b.** $R(-10)$ **c.** $R(3.2)$

9. If $S(N) = (0.04 \times N) + N$, evaluate:
    a. $S(100)$ **b.** $S(50)$ **c.** $S(-4.5)$

10. If $D(T) = 80T - 125$, evaluate:
    a. $D(5)$ **b.** $D(-5)$ **c.** $D(0.5)$

11. If $H(A) = 125 - 80A$, evaluate:
   a. $H(5)$ **b.** $H(0)$ **c.** $H(-5)$

12. If $G(T) = 4(T + 6)/2$, evaluate:
   a. $G(3)$ **b.** $G(0)$ **c.** $G(-3)$

13. If $D(S) = -3S^2 + 5S + 2$, evaluate:
   a. $D(2)$ **b.** $D(3.5)$ **c.** $D(-2)$

Exploration II

One of the most common tasks in algebra is to study a situation involving related quantitative variables, to discover the relation between the variables and write it in compact symbolic form, and then to use the symbolic expression to answer questions about the relationship. The situations presented in this exploration give you practice in this process.

SITUATION 6.1

The sales tax where José lives is 5%. This means that on taxable purchases he must pay an additional 5% of the sales price. The calculator José bought was priced at $18, but the amount he actually paid was $18 plus the tax. We can calculate the total cost as follows:

total cost = sales price + tax
$$= 18.00 + (18.00 \times 0.05)$$
$$= 18.00 + 0.90$$
$$= 18.90.$$
The total cost is $18.90.

Copy and complete José's shopping list below. Then write a rule for calculating the total cost of any taxable item as a function of the selling price p. (There are several possible correct answers.)

Item	Sales price (dollars)	Total cost (dollars)
compact disc	12.00	$12 + (0.05 \times 12) = 12.60$
winter jacket	40.00	
notebook	1.20	
skateboard	17.80	
computer	1500.00	

SITUATION 6.2

In the Talent Show simulation from section 1, if you advertise well and hire a celebrity master of ceremonies for the show, the attendance is predicted by:

$$a = 1.05(800 - 50t) \text{ or } a(t) = 1.05(800 - 50t),$$

where a indicates attendance and t indicates the price of a ticket in dollars.

After the attendance a is found, the number of sodas sold s can then be calculated by the rule:

$$s = a \times (1.2 - p),$$

where p is the soda price in dollars.

Copy and complete the following table using these two rules. Include an additional two entries using ticket prices of your own.

Ticket price	Attendance	Soda price	Soda sales
t	a	p	s
3.50		0.50	
2.00		0.35	
4.00		0.35	
4.00		1.00	
4.00		0.75	
10.00		0.50	

You probably noticed that this situation is somewhat different from the preceding examples—soda sales depend on two variables, attendance and soda price. Since attendance depends on ticket price, that variable indirectly affects soda sales also.

SITUATION 6.3

There is evidence that some health problems are related to cigarette smoking. For instance, it has been claimed that the number of deaths from heart disease in a year can be estimated from the number of cigarettes smoked in a year two decades before. One possible rule for the relation is

$$d = 39 + 0.05c,$$

where c represents the number of cigarettes consumed per capita and d stands for the number of deaths per 100,000 people.

1. Copy and complete the following table of input-output values using the rule from the study cited above.

Cigarettes per capita	Deaths per 100,000 people
1500	
2000	
2500	
3000	
3500	
4000	
4500	
5000	

2. Make a rough sketch of the relation described in the study between the number of cigarettes consumed per capita and the number of deaths per 100,000 people due to heart disease.

3. Write a sentence summarizing the relation between smoking and heart disease illustrated in the rule, table, and graph.

Making Connections

You have now used tables of values, graphs, and rules to describe and study relations between variables. For each type of "thinking tool" list some features that are helpful and some features that may limit the tool's usefulness.

Exercises

In exercises 1 through 4 you are given tables with symbolic rules. Evaluate each rule for the given values of the input variable.

1.

i	$k = 5i + 7$	$m = 7 + 5i$	$n = 5(i + 7)$
-3			
0			
3			

2.

t	$p = -3t + 2$	$q = -3(t + 2)$	$r = (-3t) - 6$
-5			
0			
5			

3.

s	$d = 15(s + 2)$	$e = 15s + 2$
-5		
-2		
0		
3		

4.

a	$f = 8 - 3a$	$g = 3a - 8$
-5		
0		
5		

5. For the following evaluation, explain why each step is correct.

$$12 + 60 / (10 + 5) \times 7 - 9$$

a. $= 12 + 60 / 15 \times 7 - 9$

b. $= 12 + 4 \times 7 - 9$

c. $= 12 + 28 - 9$

d. $= 40 - 9$

e. $= 31$

6. For the following expression, perform the indicated operations, one at a time, in proper order. Indicate the reason why each of your steps is acceptable as the next calculation.

$$(-5 \times 7) + [18 / (3 + 6)] \times (-11)$$

In exercises 7 through 12 evaluate the given expressions.

7. **a.** $24 - 6 + 2$ **c.** $24 - 6 \times 2$

　　b. $24 \times 6 - 2$ **d.** $24 / 6 \times 2$

8. **a.** $5 + 2 \times 4 + 3$ **c.** $(5 + 2) \times 4 + 3$

　　b. $5 + 2 \times (4 + 3)$ **d.** $(5 + 2) \times (4 + 3)$

9. **a.** $6 + 4 \times 5 - 2$ **c.** $6 + 4 \times (5 - 2)$

　　b. $(6 + 4) \times (5 - 2)$ **d.** $(6 + 4) \times 5 - 2$

10. **a.** $10 - 3(8 - 6)$ **c.** $9 \times 5 + 6 \times 7$

　　b. $4 + 6(5 - 2)$ **d.** $(9 \times 5) + (6 \times 7)$

11. **a.** $4 \times 9 + 7 \times 8$ **c.** $(4 \times 9) + 7 \times 8$

    b. $4 \times (9 + 7) \times 8$ **d.** $(4 \times 9) + (7 \times 8)$

12. **a.** $16 + 4(7 - 4 + 1)$ **c.** $24/2 + 4/2$

    b. $(16/2) + (4/2)$ **d.** $24/(2 + 4)/2$

13. SITUATION In Computer County the number of teachers in a high school is a function of the number of students at the school. Each school should have a ratio of one teacher for 25 students.

    a. Copy and complete the following table, showing the relation between the number of students and the number of teachers.

Students	400	500	600	700	800	900	1000
Teachers							

    b. Write an algebraic rule for the relation given in your table.

    c. Based upon the rule, how many teachers do you expect there to be in a high school with 750 students?

    d. How many students do you expect there to be in a school with 37 teachers?

    e. Write an algebraic rule giving the number of students in a school as a function S of the number of teachers, T.

14. SITUATION An Indianapolis 500 race car averages 270 kilometers per hour. We would like to know how its distance traveled is related to time elapsed in the race.

    a. Identify the related variables and assign a single-letter name to each.

    b. Copy and complete the following table.

Input variable ( )	Output variable ( )
0.5	
1.0	
1.5	
2.0	
2.5	

    c. Write a symbolic rule for the relation.

    d. How many kilometers is an Indy 500 car expected to travel in 3 hours? In 2.25 hours? In 1.6 hours?

    e. Sketch a graph of the relation between number of hours driven and distance covered.

    f. If an Indy 500 car averaging 270 kilometers per hour has traveled 495 kilometers, how long has it been going?

    g. Write a symbolic rule giving elapsed time as a function of distance traveled.

15. **SITUATION** A firm renting chain saws charges $10 for the first four hours and $3 for each additional hour. We would like to find a relation between the number of hours the saw is rented and the rental cost in dollars.

    a. Identify the related variables and assign a single-letter name to each.

    b. Copy and complete the following table.

Input variable ( )	Output variable ( )	Input variable ( )	Output variable ( )
1		6	
2		7	
3		8	
4		10	
5		15	

    c. Write a symbolic rule for the relation.

    d. What is the rental cost for 1 day? For 1 week?

    e. Write a symbolic rule relating rental cost in dollars and number of days rented.

7 Building Models of Mathematical Relations

In this chapter you have used data tables, graphs, and symbolic rules to describe and answer questions about relations among variables. However, in most of the situations which you will meet outside of school, problem solving will be a somewhat more complex task. You must identify the important variables, decide how to collect and organize information about those variables, find graphic or symbolic models of the relations among the variables, use those models to answer important questions, and be careful to consider limitations of the mathematical methods you use.

Exploration I *Smoothing Data in Search of Models*

Many situations arise in business and science where data collected from experience or an experiment suggest a relation between variables, but the exact pattern or rule governing that relation is not clear. Scientific instruments are not always completely accurate, and some relations among variables do not allow perfect prediction.

For example, a radar speed gun does not always measure the speed of an approaching car to the nearest 0.1 kilometer per hour. A store manager who studies the relation between prices and sales knows that data collected during a snow storm or a heat wave may not give a picture of sales on a typical day. Despite these problems, there are sensible ways to find useful relations among variables by studying data in tables and graphs in search of rules.

SITUATION 7.1

Sharon is studying the growth of plants in different types of light. She has been growing green beans using two types of light. Four plants are in the classroom window getting natural light, while another four plants are under a special growth light away from the window.

The growth data on the eight plants are given in the table below, with day 0 being the day the plants first showed above ground.

| | Height in centimeters | | | | | | | |
| | Natural light plants | | | | Special light plants | | | |
Day	#1	#2	#3	#4	#5	#6	#7	#8
0	4	3	4	3	3	3	4	4
1	6	5	5	6	5	6	7	6
2	7	6	8	7	8	8	9	10
3	9	9	9	10	11	12	12	13
4	12	10	11	12	15	16	15	15
5	14	12	13	13	17	19	19	18
6	16	14	15	15	20	22	22	21

Sharon looked at the table searching for a pattern in the growth. It seemed clear that the beans under special light grew faster and taller, but there seemed to be no simple pattern to the growth that would suggest a rule relating time and height. Apparently beans, like people, grow at different and somewhat irregular rates. To report findings of this experiment in a way that is clearer than the table including all data, there is one very common and helpful strategy.

To summarize the data on all the plants in a group for each day, it is common to calculate an **average** of the height data. For instance, the average height of the four natural light plants on day 0 is

$$(4 + 3 + 4 + 3) / 4 = 3.5.$$

This strategy simplifies the data table and the search for a rule relating time and height for the plants under each treatment.

1. Copy and complete the following tables of the average heights for the two growing treatments. Make two graphs, one for each set of data.

Natural light growth		Special light growth	
Day	Average height	Day	Average height
0	3.5	0	3.5
1		1	
2		2	
3		3	
4		4	
5		5	
6		6	

In this condensed form, it is easier to find a general pattern that fits the ordered pairs (time, average height). It looks like beans in both treatments were about 3.5 cm tall at the start. However, the beans under natural light grew about 2 cm per day while those under special light grew about 3 cm per day. The following rules fit those patterns:

$$\text{Natural light: } N(T) = 3.5 + 2T$$
$$\text{Special light: } S(T) = 3.5 + 3T$$

These rules are models of the relation between time and height. They can be used to make predictions of growth for other bean plants under similar conditions.

Using an appropriate table or graph or rule, answer each of the following questions:

2. What height do you predict for a bean growing under natural light for 4.5 days? What do you predict for those under special lighting?

3. At what time should a bean plant reach a height of 12 cm under natural light? What about growth under special lighting?

This example of finding patterns in experimental plant growth data has given you a first look at the main ideas and procedures in finding models of mathematical relations. The homework and explorations that follow give further illustrations and practice with this important process.

Exercises

1. **SITUATION** Denise works for a car magazine. Part of her job is to test drive new cars at the track. Before the test drive, the car to be tested has attached to it a device known as the fifth wheel, which runs along the pavement at the same speed as the car's own wheels. Inside the fifth wheel is a computer that records all of the acceleration and braking done during the test drive. After the test runs, Denise takes the computer's measurements back to her office. Her next tasks are to prepare an acceleration graph from the computer data and to write an article describing the car's performance.

   Here is a table giving acceleration data for three different runs of a new car.

   Acceleration performance — 1995 Tiger

	Speed in kilometers per hour		
Time in seconds	Test run 1	Test run 2	Test run 3
0	0	0	0
5	44	42	47
10	66	66	64
15	82	79	84
20	91	89	93
25	98	98	102

a. For each time value listed in the table above, calculate the average speed for the three test runs. Record the results in a new table giving ordered pairs (time, average speed).

b. Plot the data pairs, (time, average speed), from part a on a graph with appropriate scales of your choice.

c. Describe in words the pattern in the data as modeled by the table and the graph.

d. Use either the table or the graph to answer the following questions:

i. How long does it take the 1995 Tiger to reach a speed of 50 kilometers per hour?

ii. If a 1995 Tiger is traveling at 70 kilometers per hour on an acceleration test, how much time has passed since the test started?

2. SITUATION   The data for another car that Denise tested are given in the following table.

Acceleration performance — 1995 Lion

Time in seconds	Speed in kilometers per hour		
	Test run 1	Test run 2	Test run 3
0	0	0	0
5	35	33	40
10	54	54	54
15	75	73	79
20	83	80	83
25	94	89	90
30	102	96	94

a. Find the average speed at each time and use the results to make a table of data pairs, (time, average speed).

b. Plot the data, (time, average speed), using scales that show the data well.

c. Use the table or graph model you have just constructed to answer the following questions about acceleration of the 1995 Lion.

i. How long does it take this car to accelerate to a speed of 50 kilometers per hour?

ii. In what 5-second time interval does the speed increase the most?

3. Think about the relation between official speed tests and normal driving. List several factors that might make the acceleration of a 1995 Tiger or Lion differ from that predicted by the models you made above.

4. Think back to Sharon's bean growing experiment which involved the two treatments that varied exposure to light. List several other factors that affect growth rate for a plant.

Exploration II *A Small-Group Experiment*

Many major American sports involve throwing, kicking, hitting or catching some sort of ball. To be good at those sports it is important to know how the ball rebounds from a bat, a foot, a racquet, a floor, a wall, a backboard, or even a human head (in soccer). Scientists studying sports have paid particular attention to properties of various balls that affect their bounce.

One standard way to measure the bounce of a ball is to drop it from several different heights and record the respective heights to which it rebounds. To do the experiment that follows, you will need a ball of some sort, a meter stick to measure drop and rebound heights, and some partners to help with the various steps in the experiment.

On a blank sheet of paper, record the type of ball you are using. Drop your ball from heights of 60, 90, 120, 150, and 180 cm, and watch carefully to see how high it bounces. Perform three trials and calculate the average bounce for each height.

1. Record your results in a table such as the following.

Dropping height (cm)	Height of bounce (cm)			
	Trial 1	Trial 2	Trial 3	Average
60				
90				
120				
150				
180				

2. How would you describe the relationship between the two variables, dropping height and average rebound height?

3. As the dropping height increases from 60 to 90 cm, what happens to the height of the bounce? Does it increase or decrease? By how much?

4. As the dropping height increases from 90 to 120 cm, from 120 to 150 cm, and from 150 to 180 cm, what happens to the height of the bounce?

5. Do you expect that this pattern would extend for dropping heights that are greater than 180 cm or less than 60 cm? Explain your reasoning and, if time permits, test your hypothesis experimentally.

6. Based upon the data you have collected, how high do you think that your ball will bounce when dropped from a height of 75 cm? a height of 105 cm? If time permits, test your guesses experimentally.

7. Looking at the numerical data, you might see a pattern relating dropping height and rebound height. Try to express that pattern algebraically in the form $R(d) = \_\_\_ d + \_\_\_$.

8. Add a column titled "Predicted Height" to your table and complete with values calculated using your algebraic rule for $R(d)$. Compare the predicted values with those obtained experimentally. How might you account for any differences?

In Exploration II you collected data and tried to summarize the pattern in those data by a symbolic rule. Your rule served as a model of the relationship between the two variables of interest: the height d from which your ball was dropped and its rebound height R. Compare your rule with the rules of other groups testing the same type of ball and of groups testing different types of balls.

Exercises

SITUATION Several years ago a high school student collected data in a simple ball-dropping experiment. For each of four different kinds of balls, he was exploring the relation between temperature and the elasticity of the ball in bouncing. He dropped each ball from a height of 183 centimeters and recorded the height of its bounce. Each ball was tested at three different temperatures: cool (after being in a freezer for one hour), normal (that is, at room temperature), warm (after being heated to 40°C for 15 minutes). The data are given in the following table.

Height of bounce in centimeters

Type of ball	Temperature		
	Cool	Normal	Warm
Baseball	46	51	56
Solid rubber ball	58	97	117
Golf ball	81	117	130
"Super" ball	150	152	165

1. What is the overall relation between temperature and bounce of these balls?

2. Which ball seems most affected by temperature change and which least affected?

3. In a sport like racquetball, where the ball is hit very often during play, the ball warms up from the contact with racquets and walls. How would that influence your play from early to late in a game?

When comparing the bounce of different types of balls, merely stating the height to which a ball rebounds is meaningless without telling the height from which the ball has been dropped. The elasticity of a ball can be conveniently indicated by:

$$\text{Elasticity} = \frac{\text{Rebound height}}{\text{Dropping height}}.$$

One advantage in describing elasticity in this way is that it provides a single number that can be used to compare different types of balls.

4. Compute the elasticity of each ball for which experimental data are given at the beginning of this homework. For each ball there are three elasticity values—one for each temperature: cool (after being in a freezer for one hour), normal (at room temperature), warm (after being heated to 40°C for 15 minutes). Copy the following table, then complete it with your computations.

Elasticity of ball when dropped from a height if 183 centimeters

| | | Temperature | |
Type of ball	Cool	Normal	Warm
Baseball	46/183 =	51/183 =	56/183 =
Solid rubber ball	58/183 =	97/183 =	117/183 =
Golf ball	81/___ =	117/___ =	130/___ =
"Super" ball	___/___ =	___/___ =	___/___ =

8 Summary

This chapter has introduced several of the most important ideas in algebra and its application to problem solving.

1. Solving problems and making decisions often require study of relations among **variable** quantities.

2. Relationships between variables can be modelled by **tables** of value pairs, (input, output), by **graphs** of those data pairs, by written descriptions of **patterns** in the data pairs, or by symbolic **rules**.

3. When a relationship between variables is established by table, graph, or symbolic rule, the model can be used to answer many questions about the relation.

For example, suppose that cable television costs $20 per month plus $3.50 per movie from a "pay-per-view" channel. The monthly bill for this service **is a function of** movies ordered. This relation can be modeled by the following table, graph, and rule:

Number of Movies	Cost
0	20.00
1	23.50
2	27.00
3	30.50
4	34.00
5	37.50
6	41.00
7	44.50
8	48.00
9	51.50
10	55.00

Rule: $C(m) = 20 + 3.50m$

The answer to *"What is the monthly charge if the pay-tv channel is used for 7 movies?"* is highlighted in the table and on the graph. In symbolic form it is given by:

$$C(7) = 20 + 3.50(7)$$
$$= 20 + 24.50$$
$$= 44.50.$$

4. This last calculation shows one of the rules for **order of operations—** multiplication before addition—governing the evaluation of algebraic expressions.

5. Useful relations between variables can be found in many situations where data are collected and analyzed—for example, from experiments in science or business. In measurement situations such as these, the data recorded often varies from one example to another. This variation is usually called "noise". To find patterns in noisy data, it is often useful to smooth the data by averaging many values that represent outcomes of the same experiment.

Review Exercises

1. SITUATION   Mr. Adams wants to run a soccer camp this summer for youngsters aged 9 through 12. He wants to attract as many participants as he can. The following is a list of variables that Mr. Adams might want to take into account.
    1. The amount spent on advertising
    2. The ability of the youngsters to play as a team
    3. The condition of the playing field
    4. The number of children who will attend
    5. The number of instructors per child
    6. The variety of food offered
    7. The cost of the camp
    a. Identify those that are numerical variables.
    b. Find two variables that are related so that one could be called an input variable and the other an output variable.

2. The average retail price of unleaded gasoline is a function of the year in which it is sold. Let T represent the year and $R(T)$ the corresponding price. Data for the years 1984 to 1991 follow.

Year	Average retail price (cents)	Year	Average retail price (cents)
T	*R(T)*	*T*	*R(T)*
1984	121.2	1988	94.6
1985	120.2	1989	102.1
1986	92.7	1990	116.4
1987	94.8	1991	114.0

Source: *The World Almanac and Book of Facts.* Mahwah, NJ: Funk & Wagnalls Corp. 1993.

a. Identify the input and output variables.

b. Find $R(1989)$.

c. Find the correct value for the blank in the following statement:

$$R(\underline{}) = 120.2.$$

d. Explain the meaning of $R(1986) = 92.7$.

e. Did the average retail price decrease more rapidly between 1987 and 1988 or between 1990 and 1991?

f. When was the retail price of unleaded gasoline less than 100 cents?

3. Give the coordinates of each labeled point on the graph below.

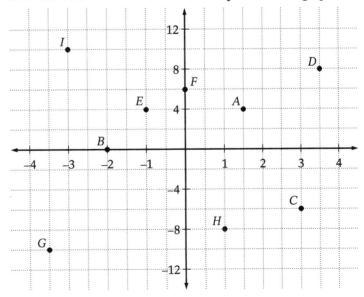

4. Plot the following ordered pairs on a coordinate system.

a. (3, 10)   c. (0, –9)   e. (–3, –8)

b. (2.75, –6)   d. (0.6, 2.5)   f. (–0.5, 7)

5. The following graph shows the quantity of gasoline in the tank of a car as a function of the time the car is on the road. The car starts out with a full tank.

   a. Identify the input and output variables.

   b. How do you interpret the jump in the graph from point A to point B?

   c. On one tank of gas the car was on urban roads; on the other tank, the car was traveling at constant speed on an interstate highway. The car consumed its gasoline more quickly in the urban area. How is this reflected in the graph?

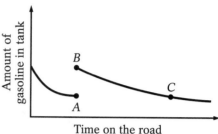

6. Sketch a possible graph of speed as a function of time for a car traveling down a street with three stop signs.

7. Take the data from Review Exercise 2 and sketch a trend curve. Use the graph to answer the following.

   a. Predict the average retail price of unleaded gasoline in 1996.

   b. In what year was the average price at a minimum? Where on the curve did you look for your answer?

   c. Between which pair of consecutive years did the price increase the most? How does the shape of the curve help you to answer this?

8. a. $R(m) = 4m / (3 + m)$. What is $R\,(9)$?

   b. $C(t) = (24 / t) - 8$. What is $C\,(3)$?

   c. There are two input variables in the expression $2x - 5y$. Write in words what this expression means. If $x = 2$ and $y = 3$, then what value does $2x - 5y$ have?

9. For each of the following tables find a rule that expresses the output as a function of the input. Use functional notation.

   a.
Input n	-2	-1	0	1	2	5	6	7
Output $f(n)$	-7	-6	-5	-4	-3	0	1	2

   b.
Input m	-2	-1	0	1	2	5	10
Output $g(m)$	-8	-4	0	4	8	20	40

10. The perimeter of a square is a function of the length of one side. Let *s* stand for the length of a side and *P(s)* for the corresponding perimeter.
   a. Write in functional notation a rule that relates the perimeter to the length of one side.
   b. Find *P* (3). What information does this give you?

11. In each problem below tell whether the parentheses are necessary. In other words, would the result be the same without the parentheses?
   a. $7 - (2 \times 2)$
   b. $(5 + 3) \times 6$
   c. $1 + 3 \times (5 \times 2)$
   d. $(2 + 6) \times 8 + 5$
   e. $4 \times (2 - 6)$
   f. $60 \ / \ (3 \ / \ 4)$

12. SITUATION   Kerri has a part-time summer job at La Salle's Boutique. She earns $55 per week plus $3 for each customer she registers for a charge account. Write a rule that gives her salary for a week as a function of the number of customers she registers. Using the single-letter variable labels *C* for the number of customers registered during a week and *P* for Kerri's pay for the week in dollars, write the rule in two symbolic forms.

13. Saturn can be seen from Earth with the unaided eye, but its rings cannot. It is one of the most beautiful objects in the sky, and it is the second largest planet in our solar system. A rule that gives the weight of an object on Saturn as a function *S* of its weight on Earth *E* is $S(E) = 1.16E$. Answer each of the following questions about this relation.
   a. What is the weight on Saturn of an object with Earth weight 40 units?
   b. What is the weight on Saturn of an object with Earth weight 20.5 units?
   c. What is the weight on Earth of an object with Saturn weight 70.76 units?

14. Only Neptune and Pluto are farther from the sun than Uranus, the seventh planet in our solar system. The following table shows corresponding weights of objects on Earth and Uranus. Study the following data to find a rule giving Uranus weight as a function of Earth weight.

Earth weight	Uranus weight
10	9.1
20	18.2
30	27.3
100	91.0
200	182.0

a. What is the Earth weight of an object with Uranus weight 136.5?

b. Do you think an object will weigh more on Uranus than on Saturn? Explain.

15. SITUATION   At a fast food restaurant, Paul is in charge of scheduling workers and deciding when to cook food. To help make these decisions, Paul collected the following data on the number of people in line each hour on the hour for three days.

Time	Day 1	Day 2	Day 3
11:00	12	12	10
12:00	16	15	82
1:00	20	10	23
2:00	8	2	9
3:00	7	2	8
4:00	8	9	8
5:00	24	26	20
6:00	32	35	31
7:00	25	24	20
8:00	12	10	11

a. Average the values for the different times and plot them on a graph.

b. Determine a curve to model these data.

c. Write a paragraph interpreting the table and the graph in this situation.

d. What other information might you want about the data and the data collection before you make ordering and scheduling decisions based on the table and graph?

e. Suppose you were told that at 1:00 on Day 2, the repair crew from the phone company had put a barrier across the driveway to the fast food place and at 12:00 on Day 3, a bus load of band students came in for lunch. Would you change your interpretation of the graph? How?

How far can you see from the top of the Empire State Building? Do you think you could see twice as far from the top of the building as you could from halfway up?

2 Functions and Computing Technology

In Chapter 1 you studied a variety of situations in which numerical (or quantitative) variables are related by patterns in tables, graphs, or rules. In many cases the value of one variable depends in some predictable way on the value of another. To describe this kind of relation of dependence we use phrases like "attendance **depends on** ticket price" or "height **is a function of** age." For each value of the input variable there is one and only one related value of the second variable, the output variable.

In mathematics it has become common to use the single word **function** to describe such a relation among variables. Thus, you will see expressions like "there is a function relating ticket price and attendance" or "there is a function relating age and height." When the relation between input and output variables is described by a rule, you will see expressions like "the function relating attendance to ticket price can be expressed by the rule $A(p) = 800 - 50p$." Often, the function is then referred to by the output variable's label. In this example, the function is called A.

In this chapter you will learn how to use technological tools to produce tables and graphs for functions, to calculate values of input and output variables, and to solve equations and inequalities that correspond to important questions about those functions.

1 Functions with Rules — By Hand

An algebraic function rule gives a compact summary of many conditions relating variables. It gives clear directions for calculating the output for any given input.

SITUATION 1.1

The Budget Box Office theater charges $4 per person for all-day admission every day. The theater manager estimates that each person spends an average of $1.50 more on candy, popcorn, or sodas. Daily operating costs for the theater include $77 for newspaper ads, $328 for theater employees' pay, and $255 for heat, lights, taxes, food, and rent of the building and movie.

The income for the Budget Box Office depends on the number of people who attend the theater each day. For each person, the theater takes in 4 dollars in admission fees and a projected 1.50 dollars at the concession stand, or 5.50 dollars per person. If daily attendance is a people, the theater takes in $5.50a$ dollars. The operating costs are fixed: no matter how many people attend, daily operating costs are $660. All of this information can be summarized simply. The theater profit, in dollars per day, is a function of the number of admissions and can be described with the rule

$$P(a) = 5.50a - 660.$$

To calculate the profit if attendance is 100, for example, we follow the order of operations to evaluate:

$$P(100) = 5.50(100) - 660$$
$$= 550 - 660$$
$$= -110.$$

So, if only 100 tickets are sold in a day, the theater ends up $110 "in the hole." That is, it loses $110.

With a rule at hand, you are well-prepared to answer many questions about the theater profit situation. One of the most important questions in any business is what the break-even point will be. That is, what is the smallest number of paying customers with which the business will not lose money?

With a function for which a rule to calculate the output for any given input is known, questions like this can be translated largely into symbols and equations or inequalities. In this case,

Find the attendance that makes profit $0

translates into

Find a where $P(a) = 0$ or $5.50a - 660 = 0$.

To answer the question, we solve the equation, $P(a) = 0$; that is, we find all values of a that make the equation a true statement. Each such value of a is called a root of the equation.

To locate the value(s), it is helpful to do some intelligent guessing. For instance, we found earlier that attendance of 100 people led to a "profit" of –110 dollars: not enough money! So you might try guessing a somewhat higher attendance for the break-even point and then refine your search, as follows:

Guess value of a	Test $5.50a - 660$	Decision
125	$5.50(125) - 660 = 27.50$	too high
110	$5.50(110) - 660 = -55.00$	too low
120	$5.50(120) - 660 = 0$	that's it!

The break-even point for the theater is 120 tickets per day. The equation $5.50a - 660 = 0$ or $P(a) = 0$ has *one* root: $a = 120$.

This same kind of guess-and-test calculation can be used to solve other equations of interest. For example, the problem "Find the attendance that results in a \$825 profit" can be translated into

Find a where $P(a) = 825$ or $5.50a - 660 = 825$.

You might start your search with a guess of 300 for attendance.

Guess value of a	Test $5.50a - 660$	Decision
300	$5.50(300) - 660 = 990$	too high
250	$5.50(250) - 660 = 715$	too low
275	$5.50(275) - 660 = 852.50$	too high
270	$5.50(270) - 660 = 825$	that's it!

The equation $5.50a - 660 = 825$ or $P(a) = 825$ has one root: $a = 270$.

Some questions about a function translate into statements involving inequalities rather than equations. For instance, suppose you wanted to know what attendance gives at least $1000 in daily profit. This question can be answered by solving the inequality

$$P(a) \geq 1000 \text{ or } 5.50a - 660 \geq 1000.$$

You can work toward solution of this problem by guess-and-test search also:

Guess value of a	Test $5.50a - 660$	Decision
300	$5.50(300) - 660 = 990$	too low
325	$5.50(325) - 660 = 1127.50$	ok
305	$5.50(305) - 660 = 1017.50$	ok
302	$5.50(302) - 660 = 1001$	ok
301	$5.50(301) - 660 = 995.50$	too low

So it looks like the solution is all numbers a that are greater than or equal to 302; that is, $a \geq 302$.

Knowing a function rule also makes it easy to calculate predicted increase or decrease in profit for any change in admissions. For example, suppose we would like to answer the following question:

How does the profit change if admissions drop from 125 to 100 per day?

"How does the profit change" means we want to know the difference in the output values for profit. "...admissions drop from 125 to 100" means we began with an input of 125 and ended with an input of 100. The change in profit is therefore

$$\begin{aligned}
\text{new profit} - \text{old profit} &= P(100) - P(125) \\
&= [5.50(100) - 660] - [5.50(125) - 660] \\
&= (-110) - (27.50) \\
&= -137.50.
\end{aligned}$$

The negative sign means the profit went down; that is, the theater profits are $137.50 less than what they were. (Do you see why we didn't sub-tract: old profit – new profit?) Similarly, if your bank account changes from $140 to $110, the change is given by $110 - 140 = -30$, showing that the account *decreases* by $30.

Exploration

Now practice using function rules to answer the questions posed in the following situation.

SITUATION 1.2

Suppose that the Budget Box Office theater decides to change its standard price to $5 per person, but all other factors in its profit picture remain unchanged.

1. Profit is still a function of attendance but with a different rule. Write the rule in symbolic form.

2. Using this function rule, write an equation that can be used to answer the question: What is the break-even point for Budget Box Office now?

    Solve the equation using a guess-and-test strategy. Record your results in a table with the following column headings.

Guess value of a	Test	Decision

3. How does the change in admission fee from $4 to $5 affect the break-even point? (Is the new break-even point lower, higher, or the same?)

    How would you expect a change in admission fee to $6 to affect the break-even point?

    How would changing the admission fee to $3 affect the break-even point? Explain your reasoning.

4. Using the function rule based on the admission price of $5 per person, write and solve an equation to answer the question: What attendance is needed for Budget Box Office to make a profit of $500? Answer the question using a complete sentence.

5. Write and solve an inequality to answer the question: What daily attendance gives profit of at least $500? Answer the question using a complete sentence.

6. What change in Budget Box Office profit results if average daily attendance increases from 100 to 125?

7. What change in Budget Box Office profit happens if daily attendance increases from 150 to 175?

8. Compare your answers to questions 6 and 7. Describe any patterns you observe.

Exercises

1. SITUATION   Suppose that the manager of Budget Box Office decides to stop advertising, cutting costs by $77 per day. The profit function then can be expressed as $P(a) = 5.50a - 583$, where a represents the average daily attendance.
    a.   Calculate profit if attendance is 150.
    b.   Calculate profit if attendance is only 50.
    c.   Write, and solve by guess-and-test, an equation that can be used to answer the question: What is the break-even point?
    d.   Write and solve an equation that can be used to answer the question: What attendance is needed to get profit of $275?
    e.   Write and solve an inequality that can be used to answer the question: What attendance gives at least $300 profit?
    f.   What is the change in profit if attendance increases from 100 per day to 110 per day?

2. SITUATION   At the Strikes to Spare bowling lanes the management offers a special deal to attract large groups on Sunday. There is a $25 charge for the group and then $1.50 per game bowled by each group member. That means that for the group the total cost of an outing at Strikes to Spare is a function of the number of games that members of the group bowl. If n represents the total number of games bowled, the rule for total cost can be written as:

$$C(n) = 25 + 1.50n.$$

    a.   Calculate and explain the meaning of $C(40)$.
    b.   Calculate $C(70)$ and explain its meaning.
    c.   Write an equation that can be used to answer the question: How many games have group members bowled when the total cost is $82? Solve your equation using a guess-and-test strategy.

    d. Write and solve an inequality that can be used to answer the question: How many games can group members bowl if the total cost is to be no more than $100?

    e. How does the cost change if the number of games bowled increases from 40 to 50?

    f. How does the cost change if the number of games bowled increases from 70 to 80?

3. SITUATION Suppose that the management of Strikes to Spare bowling lanes changes the group rate deal to $20 plus $1.60 per game bowled.

    a. Write a function rule for total cost.

    b. Find the cost if a group bowls 50 games.

    c. Find the cost if a group bowls 100 games.

    d. Write an equation that can be used to answer the question: How many games are bowled if the total cost is $83? Solve your equation using a guess-and-test strategy.

    e. Write and solve an inequality that can be used to answer the question: What number of games can be bowled if the cost is not more than $110?

    f. How does the cost increase if the number of games bowled increases from 50 to 55?

    g. How does the cost increase if the number of games bowled increases from 20 to 25?

    h. Compare your answers to parts f and g.

4. SITUATION The admission fee for groups at Queen's Dominion amusement park is given by $Q(n) = 50 + 8n$, where n is the number of individuals in the group and $Q(n)$ gives the cost for the group in dollars.

For each of the following symbolic representations, write a question which can be answered by performing the given calculation or solving the equation or inequality.

    a. $Q(40) = $ _____.

    b. $50 + 8n = 210$.

    c. $50 + 8n < 300$.

    d. $Q(20) - Q(15) = $ _____.

2 Functions with Rules — Using Calculator Functions

In some function rules the calculations needed to find data pairs (input, output), are simple enough to do in your head or perhaps with some pencil-and-paper work. However, many others involve large numbers, fractions, or complex rules for which technological tools provide a definite advantage.

SITUATION 2.1

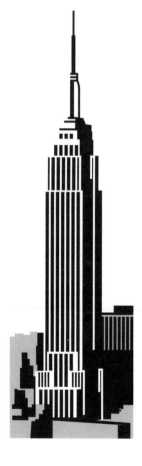

Often a popular tourist attraction in a large city is a trip to the tops of its tall buildings. In New York this means the Empire State Building or the World Trade Center; in Chicago it is the Sears Tower or the John Hancock Building.

One goal in visiting these tall buildings is to get a view of the city as far as the eye can see. The greatest, or maximum, distance you can see across flat land is a function of your height above the ground. If h is the height in meters of your viewing place, the distance in kilometers that you can see can be given by the rule

$$S(h) = 3.532 \sqrt{h}$$

To calculate viewing distance for any given height, you must find the square root of the height and then multiply that number by 3.532. In a few simple cases you can do that in your head or by hand. For example,

$$S(100) = 3.532 \sqrt{100}$$
$$= 3.532(10) \quad \text{(Remember: } \sqrt{100} = 10 \text{ because } 10^2 = 100.\text{)}$$
$$= 35.32.$$

From a height of 100 meters you should be able to see a distance of approximately 35 kilometers. Use a map of your state to find a city that is about this far from your own city.

$$S(400) = 3.532 \sqrt{400}$$
$$= 3.532(20) \quad \text{(because } 20^2 = 400.\text{)}$$
$$= 70.64.$$

From a height of 400 meters you should be able to see a distance of approximately 70 kilometers. Use a map of your state to find a city that is about this far from your own city.

For many other heights, the calculation of viewing distance is not so easy. However, nearly every calculator has the capability of finding square roots. This helps a lot.

In algebraic order of operations the square root operation takes priority over multiplication, division, addition, and subtraction. Technology with built-in algebraic logic applies this fact when carrying out the keyed instructions. However, not all technological aids follow algebraic logic. To be safe, it is a good idea to perform the calculations just as the rule for the order of operations specifies.

Exploration

Use technology and the given rule for viewing distance as a function of height to answer the following questions. Remember: $S(h) = 3.532 \sqrt{h}$.

1. The CN Tower in Toronto, Canada is 555 meters tall. It is near the shore of Lake Ontario, about 50 kilometers across the lake from Niagara Falls. On a clear day, can one see as far as the Falls from the top of the Tower? Explain your reasoning.

   What factors other than building height influence viewing distance?

2. Here is a list of some of the world's tallest buildings and free-standing towers. Calculate the viewing distance from the top of each to the nearest whole kilometer.

	Structure	Location	Height (m)	Viewing distance (km)
a.	CN Tower	Toronto, Canada	555	
b.	Sears Tower	Chicago, USA	443	
c.	World Trade Center, N	New York, USA	419	
d.	World Trade Center, S	New York, USA	419	
e.	Empire State	New York, USA	381	
f.	Amoco	Chicago, USA	346	
g.	John Hancock	Chicago, USA	344	
h.	Centrepoint Tower	Sydney, Australia	325	
i.	Texas Commerce Tower	Houston, USA	305	
j.	Allied Bank	Houston, USA	300	
k.	Eiffel Tower	Paris, France	300	

3. To improve your understanding of the relation between height and viewing distance, first copy and complete the table below. Express each output value to the nearest whole number, and then plot the data points on a grid such as the one shown.

h (m)	S(h) (km)
0	
50	
100	
150	
200	
250	
300	
350	
400	
450	
500	

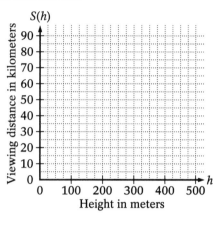

4. For each of the following questions about viewing distance, write a calculation, equation, or inequality that can be used to answer the question. Then use information from the table in Exercise 3 and guess-and-test searches to solve the problem.
    a. The Washington Monument in the Nation's Capital is 170 meters tall. How far can one see from its top?
    b. How high must a tower be in order to see at least 60 kilometers?
    c. How high must a tower be in order to see at least 25 kilometers?
    d. Advertising for Queen's Dominion amusement park claims that you can see 40 kilometers from the top of its observation tower. How high is the tower?

5. Answer the following questions to explore how viewing distance changes as height of the viewing place increases.
    a. If height increases from 100 to 200 meters, what is the change in viewing distance?
    b. If height increases from 200 to 300 meters, what is the change in viewing distance?
    c. If height increases from 300 to 400 meters, what is the change in viewing distance?
    d. Describe the trend in changes found in parts a through c.

6. Under what conditions do you think the viewing distance function might be an accurate model of real situations? Under what conditions might its viewing distance estimates be inaccurate?

Exercises

1. Use technology to produce a table of values for the function with rule $f(x) = \sqrt{x}$ for $x = 0$, 1, 2, ... , 20. Round your answers to 3 decimal places. A sample of what your final table should look like is given here.

x	$f(x)$
0	
1	
2	1.414
3	1.732
$\vdots$	
20	

2. Plot the $(x, \sqrt{x})$ data pairs from Exercise 1 on a coordinate grid with scale of 1 on each axis.

3. Use the given viewing-distance function rule to answer each of the following questions to the nearest whole number. Remember $S(h) = 3.532 \sqrt{h}$.
   a. How far can you see from the top of a 25-meter tower?
   b. If the ground is level, how high do you need to be to see 20 kilometers?
   c. If the ground is level, how high do you need to be to see 5 kilometers?

4. The altitudes of some high mountains can be expressed in kilometers more easily than in meters. For example, the highest mountain peak in the world is on Mount Everest, approximately 8.8 kilometers high. With some careful thought we can still use the viewing-distance function rule, even if the height we are considering is not in meters. All we need to do is convert the given height to meters and then use the function rule.

For the Mount Everest height:
1. Convert 8.8 kilometers to meters. There are 1000 meters in each kilometer, so 8.8 kilometers is the same as 8.8×1000, or 8800, meters.
2. Use the function rule. $S(8800) = 3.532 \sqrt{8800}$
$$\approx 3.532(93.808)$$
$$\approx 331.$$

So, from the top of Mount Everest, if our line of sight were not blocked by other mountains, we could see a distance of approximately 331 kilometers—a little less than the distance from Washington, D.C., to New York City.

a. In Africa, Mount Kilimanjaro stands on a flat plain, reaching a height of approximately 4.9 kilometers above that plain. How far could you see from the peak of this mountain?

b. Instead of first converting our measurement to meters, we can modify the formula to accept input values expressed in kilometers. If the height of the place where you are is k kilometers, write a function rule for $d(k)$, the distance in kilometers that you can see.

c. Use this new formula and your technological tool to find the viewing distance from a height of 3.2 km. Write the sequence of keystrokes and commands you needed.

Making Connections

In the first two sections of this chapter you have solved equations using a guess-and-test strategy—with and without computing technology.

List problem features that lead you to use computing technology rather than mental or paper-and-pencil computations for solving equations and inequalities.

When is mental computation, perhaps with paper-and-pencil aid, more sensible than using technology?

List things about which you need to be especially careful when using technology to help with arithmetic computation.

3 Functions with Rules — Using Programmable Tools

For functions with known rules, many interesting questions involve calculating outputs for given input values, or solving equations or inequalities. With simple rules and only a few questions, mental or paper-and-pencil arithmetic is fast and accurate enough. For more complex rules or many equation or inequality questions, technology such as calculators are very helpful with a guess-and-test method.

You have probably discovered, however, that even using a calculator often involves many keystrokes to find the output for each given input. Repeating keystrokes for several input values becomes tedious. It would save a lot of work if we could combine all the keystrokes for arithmetic operations in a function rule into one instruction. Then the calculation of an output would merely require entering the input and applying that one instruction. The combination of all the keystrokes is often called a program, even though it might be much less complicated than other programs you may be familiar with. Many technological aids can be programmed to give just this kind of help.

The next example illustrates the basic steps in getting the assistance of a programmable calculator or a computer to find many outputs for a given function in an efficient manner.

SITUATION 3.1

In the Talent Show simulation, when a celebrity is hired as MC and the show is well advertised, predicted attendance at the show is a function of the ticket price in dollars. This function can be expressed with the rule

$$A(p) = 840 - 52.50p.$$

From this rule you can answer questions such as the following:

What attendance is likely if the ticket price is set at $10?

What ticket price gives attendance of 525 people?

If you have a programmable tool to help you work with functions, the first step often is to enter the function rule. Depending on the particular tool, you may need to use a variety of special symbols for operations, such as the following:

$a \times b$ or $a * b$ or ab	(for multiplication),
$a \div b$ or a / b	(for division),
$SQR(a)$, $a\wedge(1/2)$, or $a\wedge 0.5$	(for square root), and
$a\wedge n$	(for exponents)

After the rule is entered, it is easy to evaluate the function for any input. For example, to find the attendance if a ticket costs $10, you can enter the input 10 or perhaps a function expression like $A(10)$ and press the appropriate key to execute the function rule. The machine *very quickly* multiplies 52.50 by 10 and subtracts that product from 840 to find the output, 315. Only the number 315 is displayed for you to see.

The second question above asks for a root of the equation $A(p) = 525$ or $840 - 52.50p = 525$. A guess-and-test search for the solution might be done like this:

1. Enter the rule (unless it has already been entered): $A(p) = 840 - 52.50p$.

2. Begin a guess-and-test search for values of p where $A(p) = 525$.

You enter	Response	Decision
10 or $A(10)$	315	too low
5 or $A(5)$	577.5	too high
5.5 or $A(5.5)$	551.25	too high
6 or $A(6)$	525	correct

Without changing the function rule, you can solve other equations or inequalities involving this situation by using the same method.

There are many different computer programs and programmable calculators available today, and new products appear every year. Each of these tools can produce many data pairs quickly and help with guess-and-test solutions to equations or inequalities that involve a function. One word of caution: each tool has its own methods for entering and using a function rule. Many programs provide on-screen directions for use of their methods, but you may need to consult a user's manual.

Learn the procedure for entering and using function rules in the programmable tools available to you. Then use those tools to work problems in the following exploration.

Exploration

In this exploration, use a programmable tool to answer the questions. In each case:

–record the instructions that you give to the tool;
–give the equation or inequality you are solving;
–list the guess-and-test steps that lead you to the solution; and
–write your answer to the question in a complete sentence.

SITUATION 3.2

In the Talent Show simulation, if no celebrity MC is hired, but advertising is done well, the rule predicting attendance as a function of ticket price in dollars can be written as

$$A(p) = 840 - 105p.$$

1. What ticket price leads to a predicted attendance of 525 people?
    a. Write the steps you use to enter the function rule.
    b. Write a calculation, equation, or inequality that can be used to answer the question.
    c. List the guess-and-test steps you go through, if needed. Record your steps in a table using the following column headings.
        Guess   Test   Decision

    d. Write your answer in a complete sentence.

2. What ticket price gives a predicted attendance of 567 people?
    a. Write a calculation, equation, or inequality that can be used to answer the question.
    b. List the guess-and-test steps you go through, if needed. Record your steps in a table using the following column headings.
        Guess   Test   Decision

    c. Write your answer in a complete sentence.

3. What ticket price leads to a prediction that no one will attend?
    a. Write a calculation, equation, or inequality that can be used to answer the question.

b. List the guess-and-test steps you go through, if needed. Record your steps in a table using the following column headings.

Guess   Test   Decision

c. Write your answer in a complete sentence.

SITUATION 3.3

The school track coach plans to rent a video-tape camera and player to study the team's skills. The minimum rental time at Vinnie's Video is one week. The cost (in dollars) of renting the equipment is a function of the number of days it is rented and is defined by the rule

$$c(d) = 80 + 8(d - 7).$$

Use the function rule to answer the following questions. Record any keystrokes and commands you need for your technological tool as well as your answers.

1. How much does it cost the coach to rent the equipment for two weeks?
   a. Write a calculation, equation, or inequality that can be used to answer the question.
   b. List the guess-and-test steps you go through, if needed. Record your steps in a table using the following column headings.
   Guess   Test   Decision

   c. Write your answer in a complete sentence.

2. If the coach's rental cost was $456, for how long was the equipment rented?
   a. Write a calculation, equation, or inequality that can be used to answer the question.
   b. List the guess-and-test steps you go through, if needed. Record your steps in a table using the following column headings.
   Guess   Test   Decision

   c. Write your answer in a complete sentence.

3. If the track team can spend only $175 on video studies, what is the greatest number of days for which the coach can rent the equipment?

   Note: Not all of the money available must be spent, but the coach certainly cannot spend more!

   a. Write a calculation, equation, or inequality that can be used to answer the question.
   b. List the guess-and-test steps you go through, if needed. Record your steps in a table using the following column headings.

   Guess   Test   Decision

   c. Write your answer in a complete sentence.

Exercises

Once a programmable tool has been given a rule for a function, it can be used very efficiently for guess-and-test calculations. The key skill needed is writing function definitions carefully. This is the main theme of the exercises that follow.

1. For each of the following situations, write the function in the form required by the programmable tool you have been using.

   a. SITUATION The cost in dollars of a group bowling party at Strikes to Spare lanes is a function of the number of games bowled, with rule given as $C(n) = 20 + 1.60n$.
   b. SITUATION Talent Show revenue is a function of ticket price, with rule given as $R(p) = -100p^2 + 800p$.
   c. SITUATION Talent Show profit is a function of ticket price, with rule given as $P(t) = -100t^2 + 800t - 376$.
   d. SITUATION Distance traveled by a race car at 150 kilometers per hour is a function of time in hours, with rule given as $D(t) = 150t$.

2. For each of the following functions, find the indicated values.

   a. $F(x) = 3x - 4$
      i. $F(10)$ ii. $F(-4)$ iii. $F(2.5)$ iv. $F(-1.6)$

b. $G(x) = -16/x$

    i. $G(2)$ **ii.** $G(0.5)$ **iii.** $G(-0.5)$ **iv.** $G(-8)$

c. $H(T) = 10 - 4T$

    i. $H(2)$ **ii.** $H(-2.5)$ **iii.** $H(-2)$ **iv.** $H(2.5)$

3. **SITUATION**   Satellites use the energy of sunlight for electric power. Solar cells such as the one shown here convert the sunlight directly into electrical energy. In direct sunlight, each square centimeter of solar cell produces about 0.01 watt of electric power. The function S with rule $S(w) = \sqrt{38.5w}$ relates w, the number of watts produced by a hexagonal cell, and the length in centimeters of each side of the cell. Note that the rule here is: first, multiply 38.5 by the input value w; then take the square root of the product.

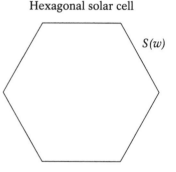

Hexagonal solar cell

$S(w)$

Answer each of the following questions about this function, using a programmable tool where necessary and recording the steps in your reasoning.

a. If the solar cell is to produce 20 watts, determine the length of a side of the hexagonal cell.

b. If the number of watts is doubled, does the cell's side length double?

c. If the dimensions of the satellite dictate that the length of a side must be no more than 42 cm, what is the greatest power output that can be expected from the hexagonal cell?

Making Connections

You have now used a number of methods for the arithmetic needed to solve equations and inequalities.

1. List some reasons for choosing to use a programmable tool rather than paper and pencil or simple calculator functions.

2. When is mental computation, with paper-and-pencil assistance, or simple calculator functions preferable to a programmable tool?

3. List some things that you must be especially careful about when using a programmable tool.

4   Programs and Function Tables

When two variables are related by a simple rule and the questions of interest are modeled by one or two equations, it makes sense to use mental or calculator arithmetic and a guess-and-test strategy to find the answers. When you are faced with a more complex function rule or questions that require looking at the overall pattern in the relation between variables, it may be helpful to produce a table of sample data pairs, (input, output), and to use that table as a guide to answering the questions.

You should not be surprised to learn that standard programs make this task very easy. In fact, with only a little help you would be able to write workable table programs yourself.

SITUATION 4.1

For a school play, revenue from ticket sales depends on many factors: the quality of the play and the performances, advertising, the price of tickets, and other input variables. A typical relation between the ticket price and the play's ticket revenue is given by the function rule

$$R(p) = -100p^2 + 720p,$$

where the price is p dollars and the revenue is $R(p)$ dollars.

For example, if ticket price is set at $3, the predicted dollar revenue is

$$R(3) = -100(3)^2 + 720(3)$$
$$= -100(9) + 720(3)$$
$$= -900 + 2160$$
$$= 1260.$$

One of the most interesting problems in this situation is finding the ticket price that gives **maximum** ticket revenue. To get an answer we really need to compare the revenues predicted for many different prices. The following table of values gives some clues to the answer.

Table 1

p	$R(p)$	p	$R(p)$
0.00	0	6.00	720
1.00	620	7.00	140
2.00	1040	8.00	−640
3.00	1260	9.00	−1620
4.00	1280	10.00	−2800
5.00	1100		

First, the table shows that as ticket price increases from $0 (free) to about $4, the predicted revenue increases. The $4 price seems likely to produce the largest ticket revenue, because as the price increases beyond $4 the predicted revenue decreases. To locate the maximum revenue point more precisely, we could use a table that focuses on prices between $3 and $5 and test increases in smaller steps of $0.20 instead of $1.00.

Table 2

p	$R(p)$	p	$R(p)$
3.00	1260	4.20	1260
3.20	1280	4.40	1232
3.40	1292	4.60	1196
3.60	1296	4.80	1152
3.80	1292	5.00	1100
4.00	1280		

The pattern in this table suggests maximum revenue at a ticket price of $3.60, because revenue rises and falls on either side of that price.

The original table also gives hints at solutions for many other equations and inequalities involving the price/revenue relationship. For instance, the question "What ticket prices give total ticket revenue of $1000?" fits the equation

$$R(p) = 1000 \text{ or } -100p^2 + 720p = 1000.$$

The entries in that table suggest that one such price is between $1 and $2. Another appears to be between $5 and $6.

More detailed tables pinpoint the prices that give ticket revenue of $1000. Table 3 gives revenue for ticket prices from $1 to $2 in steps of $0.10.

Table 3

p	$R(p)$	p	$R(p)$
1.00	620	1.60	896
1.10	671	1.70	935
1.20	720	1.80	972
1.30	767	1.90	1007
1.40	812	2.00	1040
1.50	855		

Apparently one such ticket price is between $1.80 and $1.90. A more exact value can be found through use of another table. A similar search, through tables that "zoom in" on desired data pairs, (input, output), can also locate the price between $5 and $6 that gives revenue of $1000.

As these examples show, tables of function values are very helpful in giving an overall picture of a relation between variables. By focusing on tables with progressively smaller steps, you have another method for locating points of interest by systematic search. What makes this "zooming-in" strategy very attractive in problem solving is the variety of computer programs available to help with construction of the tables.

To use a typical function table program you usually enter only a few pieces of information:

1. The function rule;
2. The beginning and ending values for the input variable; and
3. The size of each step by which the input variable changes.

Usually, having entered a function rule once, you can ask for many different tables. Simply change the starting or ending input values or the step size in order to produce a new table that enables you to focus wherever you choose.

To produce the three tables given above, enter the function rule for $R(p)$. Then the tables can be defined as follows:

Table 1
Smallest p = 0.00
Largest p = 10.00
Step size = 1.00

Table 2
Smallest p = 3.00
Largest p = 5.00
Step size = 0.20

Table 3
Smallest p = 1.00
Largest p = 2.00
Step size = 0.10

When using tables, the main problem for you is to select the smallest and largest input values and perhaps the step size. Choose a step size small enough to show as much detail as you need; however, with too small a step size, the table may be too large, and you can be overwhelmed by too much detail.

Select the starting and ending inputs for your table so that they surround the values where important features of the function appear. Your best guide in selection of a search area is to think about inputs that are reasonable in the situation. For instance, ticket prices greater than $10 are probably too high for a high school play!

 ## Exploration I

With any new tool, some practice is needed to become comfortable in its use. This exploration will give you experience using programs that generate tables, but keep your eyes open for interesting patterns in those tables.

1. On paper, copy the following table for the function with rule $f(x) = 0.5x - 1$. Generate the completed table using technology, copy its entries on your paper, then answer the questions below.

x	-3	-2	-1	0	1	2	3	4	5	6	7
$f(x) = 0.5x - 1$											

  a. Give the instructions for the rule, the smallest and largest values of the input, and the step size you entered to produce the table.
  b. Copy and complete the following sentence to describe any patterns you see in the table.
      As the value of x increases....
  c. In your table, circle the input that solves the equation $0.5x - 1 = 2$. Then check that solution by hand and describe your check.

2. Copy and complete the following tables for the function with rule $g(t) = t^2$. Then answer the questions that follow.

Table 1		Table 2	
t	$g(t) = t^2$	t	$g(t) = t^2$
−5		0.0	
−4		0.2	
−3		0.4	
−2		0.6	
−1		0.8	
0		1.0	
1		1.2	
2		1.4	
3		1.6	
4		1.8	
5		2.0	

a. Record the following information as you entered it in your technological tool to produce Table 1.
   i.   The function rule
   ii. The smallest value for t
   iii. The largest value for t
   iv. The step size of the t-values
b. Record the following information as you entered it in your technological tool to produce Table 2.
   i.   The function rule
   ii. The smallest value for t
   iii. The largest value for t
   iv. The step size of the t-values
c. Copy and complete the following sentences to describe any interesting patterns you see in the tables.
   In Table 1, as the variable t increases from −5 to 5....
   In Table 2, as the variable t increases from 0 to 2 in steps of 0.2....
d. Write two equations or inequalities that can be solved using entries in these tables. Give the solution for each.

3. Copy and complete the following tables for $h(s) = \frac{1}{s}$. Then use the tables to answer the questions that follow.

Table 1		Table 2	
s	$h(s) = \frac{1}{s}$	s	$h(s) = \frac{1}{s}$
0		0.0	
1		0.1	
2		0.2	
3		0.3	
4		0.4	
5		0.5	
6		0.6	
7		0.7	
8		0.8	
9		0.9	
10		1.0	

a. Write a sentence describing any patterns you see in the first table.

b. How is the pattern in the second table like, and how is it different from, the pattern in the first table?

c. If time permits, enter some function rules of your own design. Then produce various tables of values and look for patterns of interest. Record your findings.

Exploration II

Now use a table-generating program to answer questions in the following situations. For each question you may wish to look at more than one table in order to find an answer that seems accurate enough.

SITUATION 4.2

The **profit** for a play is the difference between revenue (total amount of money received) and expenses (total amount of money paid out). Profit generally depends on the ticket sales for the play. Suppose profit in dollars is related to the number of tickets sold t by the rule

$$P(t) = -0.02t^2 + 9t - 200.$$

1. First, describe the overall pattern in the relation between ticket sales and profit. Then record the information (largest input, smallest input, and step size) you enter in the program in order to produce the table that best shows you the pattern.

2. For which numbers of ticket sales does the play make a positive profit and for which does it lose money?

3. What number of ticket sales gives the greatest profit?

4. What numbers of ticket sales give at least $500 profit?

5. What is $P(0)$, and how do you explain the fact that it is a negative number?

SITUATION 4.3

The daily profit function for the Budget Box Office theater can be expressed as $P(a) = 5a - 660$, where a is the attendance.

1. Create a table that offers a good overall view of this profit function. Record the values you use for the smallest a, the largest a, and the step size.

2. Describe the overall pattern and any especially important values in the table you produced.

3. For which attendance numbers does the theater make a (positive) profit? For which does it lose money? For which does it break even?

4. What attendance seems likely to give maximum profit?

5. How is the profit function for the Budget Box Office different from that of the play? Does one of them, as a model for its situation, seem to be more reasonable than the other? If so, which one? Why?

Exploration III

Three equations are given in this exploration, each with a table of some data pairs, (input, output). In each case give data (smallest input, largest input, and step size) for new tables that help you locate the root(s) of the given equation more accurately. Search for values of x that give output within 0.1 of the required value.

1. Equation: $6x - 100 = 64$

   Starter table: $f(x) = 6x - 100$

x	$f(x)$
0	-100
5	-70
10	-40
15	-10
20	20
25	50
30	80

2. Equation: $x^2 + 5 = 174$

   Starter table: $g(x) = x^2 + 5$

x	$g(x)$
-10	105
-5	30
0	5
5	30
10	105
15	230
20	405
25	630

3. Equation: $3.5 - 6.8x = -12.5$

   Starter table: $h(x) = 3.5 - 6.8x$

x	$h(x)$
0	3.5
1	-3.3
2	-10.1
3	-16.9
4	-23.7
5	-30.5
6	-37.3
7	-44.1
8	-50.9

Making Connections

In this section you have used programs to produce tables of function values. List some reasons for choosing to use a tables program rather than calculator or paper-and-pencil computation to find answers to questions about a function.

When is mental computation, perhaps with paper-and-pencil aid, more sensible than a tables program?

List several things about which you must be especially careful when using a tables program.

Exercises

SITUATION

For the Talent Show simulation you studied in Chapter 1, when a celebrity master of ceremonies is hired, the function relating ticket price and ticket revenue for the show has rule $R(p) = -50p^2 + 800p$. The following table shows a sample of data pairs.

p	0	2	4	6	8	10	12	14	16
$R(p)$	0	1400	2400	3000	3200	3000	2400	1400	0

1. Give the entries that are needed for a program to produce the table above.

2. Describe the overall pattern in the above table as it relates ticket price and revenue.

3. a. Write an equation or inequality that can be used to answer the question, "What ticket price (s) give revenue of $2000?"
   b. What information does the above table tell about the solution of the equation or inequality you wrote in part a?
   c. Give the entries for a program that can be used to find a more accurate estimate for the root(s) of your equation.
   d. Make a mental estimate of a root for your equation and test that estimate.

4. a. Write an equation or inequality that can be used to answer the question, "For what ticket price(s) is the revenue at least $1000?"
   b. What information does the above table tell about the solution of that problem?
   c. Give the entries for a program that can help find a more accurate estimate for the solution.

5. If you have used several different table programs to produce the tables in the Explorations, compare them by telling what is easiest about using each and what is most difficult about using each.

5 Technology and Function Graphs

While studying the revenue obtained in a situation like a school play, you learned to use a technological tool to produce tables—tables that helped you study the pairs, (ticket price, revenue), and answer questions about the revenue function. It is often easier to answer questions about functions by examining a graph showing the pattern of values for data pairs, (input, output). Computers are very helpful in constructing function graphs.

SITUATION 5.1

Earlier in this chapter, you studied a function that possibly could be used to describe the relation between the price, p dollars, of a ticket for a school play and the revenue, $R(p)$ dollars, received from selling the tickets. The rule you used was

$$R(p) = -100p^2 + 720p.$$

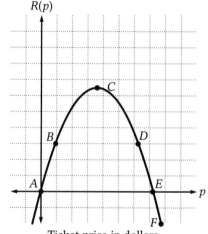

$R(p) = -100p^2 + 720p$
The horizontal (input) scale unit is 1 and the vertical (output) scale unit is 200.

Here is a graph of this revenue function. Several points have been labeled with letters keyed to the discussion that follows. Before reading ahead, think about questions that are answered by the pattern of this graph and by the specially labeled points.

The graph of a function is especially helpful if you are looking for trends in the relation between inputs and outputs or for approximate pairs of input and output values. The following examples refer to the graph of the school play revenue function.

1. From point A to point C, the **curve is rising**. This means that an increase in any relatively low ticket price gives an increase in ticket

revenue. For example, point B, with approximate coordinates $(1, 600)$, shows that ticket price of $1 gives ticket revenue of about $600. The point with approximate coordinates $(2, 1000)$ shows that a ticket price of $2 gives ticket revenue of about $1000. This increase of $1 in ticket price gives an increase of about $400 in revenue.

2. At point C, a significant change occurs. The **trend** is no longer upward and the curve begins to fall. This means that an increase in any relatively high ticket price leads to a *decrease* in ticket revenue. From point D to point E, the ticket price increases by about $1 but the ticket revenue drops by about $600.

3. Along with the change in trend that occurs at C, the diagram clearly shows that C is at the **top** of the graph. This is the point where the revenue has its **greatest** or **maximum** value.

Recognizing the exact coordinates of a point may not be a simple task. However, you should be able to obtain good **approximations**. In the case of the point C, the second coordinate appears to be at, or slightly below, 1300, while the first coordinate is just about halfway from 3 to 4, at about 3.5. As estimates, you might report that the school play can have about $1300 in ticket sales if the admission price is about $3.50. Moreover, by referring to the trends discussed above, you can report that a considerably smaller revenue should be expected if the price is significantly different from $3.50.

4. The graph meets the p-axis at the points A and E. At each of these points the revenue is zero—no money is taken in.

The input or ticket price at point E is its first coordinate, which you might guess is about a quarter of the way from 7 to 8. You might record your estimate as 7.25, or perhaps as 7.20 or 7.30. Any of these could be an acceptable approximation. The first case you might think of as rounding to the nearest quarter dollar; the second case as rounding to the nearest dime.

As you can see, a function graph displays the same data, (input, output), as a table of values. The graph gives information in a form that makes finding some conclusions, like overall trends and maximum or minimum points, easier than it would be with a table. Other conclusions, like exact coordinates of specific points, may be impossible to find from the graph. A function rule sometimes is needed to be certain of finding points precisely.

5.1   Graphing Tools

To create a useful graph with technological tools, you may need to make decisions like those you encounter when creating a table of values. As you did with the tables, you must give instructions to your technological tool based on those decisions. After the machine receives instructions, its speed and accuracy provide a powerful aid in getting the picture.

As an example, study the steps involved in graphing the revenue function $R(p)$. Guiding the graphing tool requires two main steps.

1. Enter the function rule, which in this case is $R(p) = -100p^2 + 720p$.
2. Decide where to focus your look at the graph and enter your decision.

Of these two tasks, entering the function rule is clearly the easier. If you take care in typing and remember the special symbols that may be needed, such as ^ for exponents and * for multiplication, entering the rule can go smoothly.

It is generally not so easy to decide where to focus your look at the graph and how to get the graphing tool to give you the window, or viewing area, you want. In this case, since both ticket price and revenue are greater than or equal to 0, the first quadrant seems the best place to look. However, the first quadrant is a huge expanse; only a small part of it can fit on any device. You should give the graphing tool specific instructions about which part of the graph to display. It helps to think about input and output values that make sense in this situation.

In this case, the input variable is the price of an admission ticket. Ticket prices below $0 make no sense. You might also guess that any price beyond $10 is unreasonable and not worth considering.

The output variable in this case is the revenue. To be successful a school play must take in considerable money: hundreds of dollars, perhaps as much as a few thousand. There are at least two sensible ways to estimate the range of outputs. First, you might know the capacity of the auditorium. Perhaps it is only 400, and 400 times the largest reasonable price of $10 puts a limit of $4000 on the revenue. On the other hand, you might use a tables program to calculate outputs for the rule. That method suggests that revenue can not be more than $1300.

In summary, it is quite reasonable to estimate that the inputs vary from 0 to about 10 and the outputs vary from 0 to about 1300.

Having estimated the sensible values for the variables, you need to give instructions to the graphing utility so that the significant part of the graph can be displayed on the screen. Basically this involves the selection of a scale on each axis. The most common graphing tools do this in one of two ways.

Scales set by the utility Some utilities require you to enter the smallest and largest values on each axis, then they will make tic marks on each axis and automatically calculate the value represented by each mark. In the case of the revenue function R(p), you would enter 0 and 10 as limits on the p-axis (the input or horizontal axis) and enter 0 and 1300 as limits on the R(p)-axis (the output or vertical axis). The utility might use 15 or 20 tic marks on each axis, giving scales calculated as follows:

1. $(10 - 0)/15 = 0.666...$; so p-unit $= 0.67$ might be used.
2. $(1300 - 0)/15 = 86.66...$; so $R(p)$-unit $= 87$ might be used.

Working with scales like these makes interpretation of a graph complicated. Some graphing utilities that automatically set scales round the scale values to more convenient numbers. In this case, the p-unit might be chosen as 1 and the $R(p)$-unit as 100 (or perhaps 90). In general, it makes sense to round to higher values rather than lower ones, so that all relevant points appear on the resulting display.

Some graphing utilities require only that you enter limits on the input or horizontal axis. The utility then calculates the corresponding outputs and determines suitable vertical-axis limits and scale units for both axes.

Scales set by the user For other graphing utilities, the choice of scales is up to the user. The natural decision is to let the tic marks on each axis be one unit apart. However, this choice does not always produce a useful display. In the case of our revenue function, values from 0 to about 1300 are needed on the vertical axis, and one unit per tic mark on the vertical axis would result in 1300 marks — difficult (or even impossible) to read. Usually, 10 to 20 marks are best for readability. If the vertical axis has about 15 tic marks, each mark might be worth 1300/15, which is about 87. This is not a simple unit for the mental arithmetic required for interpreting a graph, so 90 or 100 would probably be better.

In either type of graphing tool, when the necessary data have been entered, the utility makes many calculations and plots many points in order to display the graph according to your instructions. If the display you get shows more or less of the graph than you want to see, you can adjust the limits or scales and try again to see a magnified picture of some special part of the graph or a more distant view of a larger part of the graph.

Exploration I

This exploration will give you practice in making the decisions required by the use of graphing utilities. In the first situation you are given a rule and a table of values for the function. You are asked to produce and interpret a graph that fits the situation.

For some of the questions there is not just one right answer; you must make a choice, and your correct answer may be different from your classmates'. A table program can help you choose answers to items 3 and 4.

SITUATION 5.2

Daily profit at Budget Box Office theater is a function of the number of admissions sold. The rule can be expressed as $P(a) = 5.50a - 660$. A table of sample data pairs, (admissions, profit), follows.

Admissions	Profit (in dollars)	Admissions	Profit (in dollars)
0	−660	250	715
50	−385	300	990
100	−110	350	1265
150	165	400	1540
200	440		

1. Write the function rule in the form you should enter in your graphing utility.

2. Name the input variable represented on the horizontal axis and the output variable represented on the vertical axis.

3. Where will you focus your look at the graph? Hint: Are negative values of the input variable sensible? Are negative values of the output variable meaningful?

4. Give the smallest and largest values of the input and output variables that you want to show on your graph and the scale you want to use for each axis.

5. Produce a graph with a graphing utility and make a rough sketch of the result on paper.

6. Label one point on the graph that gives information that did not appear in the given table. Explain what that point tells about the daily profit at the Budget Box Office theater.

In Situation 5.2 you were given both the function rule and a table of function values to help you choose a graphing window and scales for the axes. However, in a typical situation the only information you have at the start is the function rule. In the next four situations, you will have to do some thinking about suitable sets of input and output values before displaying graphs of the functions on a graphing tool. Use Situation 5.3 as an example for your work on Situations 5.4, 5.5, and 5.6.

SITUATION 5.3

When a celebrity master of ceremonies is hired in the Talent Show simulation, the rule used to describe the profit from ticket sales as a function of ticket price can be expressed as

$$P(t) = -50t^2 + 800t - 376.$$

1. Write the function rule as you would enter it in your graphing utility.

2. Complete the following, determining a reasonable set of input values.

   Since t represents the price of a ticket, its values are positive. A safe upper estimate for the ticket price is probably about 15 dollars. Consequently a reasonable set of input values is....

3. Determine the smallest and largest values of the output variable $P(t)$ to be displayed in the window.

   Hints: One way to pick reasonable estimates is to examine a table of function values. Try using this method.

   Another way to find suitable output variable limits is to think about the situation. The function represents profit. If the admission price is low, then there is little money taken in from ticket sales, perhaps not enough to pay all the bills; so the profit might be negative. This observation is checked by glancing at the function rule and noting that input 0 gives a negative output, namely, −376.

With regard to the largest output, you have seen earlier that revenue for a talent show might be a few thousand dollars. Since expenses need to be paid, profit is certainly smaller than revenue; so you might estimate the profit to be not more than about 3000 dollars.

Notice: As in these hints, knowledge about the situation helps you work with the mathematical rule when solving problems.

4. Determine the scale for each axis. Remember to round the units if necessary.

   Hint: The scale for the horizontal axis is

   (largest t – smallest t) / number of tic marks

   and the scale for the vertical axis is

   [largest P(t) – smallest P(t)] / number of tic marks.

5. Produce a graph using your graphing utility and make a rough sketch of the result.

6. Label one interesting point on the graph. Explain what this point tells about the profit from the Talent Show.

SITUATION 5.4

Viewing distance in kilometers from the top of a tower or tall building is a function S of the building's height h in meters. The rule can be expressed as

$$S(h) = 3.532 \sqrt{h}.$$

1. Write the function rule as you would enter it into your graphing utility.

2. Which variable will appear on the horizontal axis and which on the vertical axis?

3. Choose where to focus your look. Are you interested in negative values for height? for distance?

4. Think about the variables in this situation and estimate a largest reasonable value for h. Then make a similar estimate for the largest reasonable viewing distance value, or use a table program to get ideas about a reasonable range. For each variable choose the largest value that you want to be able to display on your graph. Finally

choose a scale for each axis. Remember to round the units if it is necessary.

5. Produce a graph on your graphing utility and make a rough sketch of the result.

6. Describe any trend from the graph. Explain the trend in terms of the viewing–distance function.

SITUATION 5.5

The organizers of a talent show predict that the number of people who attend A is a function of the admission price in dollars p. This function can be expressed by the rule

$$A(p) = 840 - 55p.$$

1. Write the function rule in the form that you would enter it in your graphing utility.

2. Decide and record what numbers are reasonable values for the input variable, and then choose and record a scale for the horizontal axis.

3. Determine the values of the output variable that you want displayed in the window, and then select a scale for the vertical axis. Record your selection

4. Produce a graph on your graphing utility and make a rough sketch of the result.

5. Label two interesting points on the graph. Explain what each point tells about the relation between ticket price and attendance.

SITUATION 5.6

When a basketball is shot from the free throw line, its path toward the basket depends on the height, angle, and velocity of its release. If a free throw is released from a height of 7 feet at an angle of 65 degrees from the horizontal, the maximum height of the ball, in feet, depends only on the release speed v in feet per second. This rule can be expressed as

$$H(v) = 7 + 0.014v^2.$$

1. Write the function rule in the form that you would enter it in your graphing utility.

2. What is a reasonable set of input values? Hint: You probably have a better idea of reasonable heights that a free throw can reach than of release speed. Try several values of the release velocity, v, and check to see which of the corresponding output heights are reasonable.

3. Determine the smallest and largest values of the output variable that you want to be able to display on the screen. Explain how you made your choices.

4. Determine the scales for the vertical and the horizontal axes. Explain how you arrived at these scale values.

5. Produce a graph on your graphing utility and make a rough sketch of the result. Note: Be careful when interpreting this graph. It is *not* a picture of the path of the ball, although many students make that mistake! Why isn't the graph a picture of the path?

6. Label two interesting points on the graph. Explain what these points tell about the relation between release speed and maximum height of a free throw in basketball.

5.2   The Effects of Scales on Graphs

Graphing tools give displays that reveal quickly the overall patterns and special points of interest in a relation between two variables. However, if those programs are not used carefully, the appearance of their pictures can be very misleading.

SITUATION 5.7

In the plans for a school talent show one of the most important relations involves the two variables *ticket price* and *profit*. Experience from previous shows suggests that when a celebrity MC is hired for the show, profit $P(t)$ is a function of ticket price t. A typical rule for this relation can be expressed as

$$P(t) = -50t^2 + 800t - 1950.$$

Two students who are using a graphical method to study the profits predicted by this rule might come to very different conclusions.

For example, Jan produces a sketch of the graph for the profit function. Jan argues that a very good profit can likely be earned if any ticket price between about $4 and $12 is selected. Terry produces a different sketch of the graph. Terry concludes that the show has very little prospect of making much profit, regardless of the ticket price chosen.

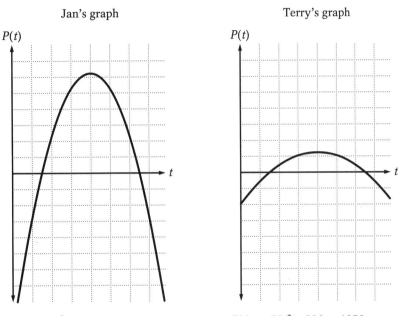

Jan's graph

$P(t)$

t

$P(t) = -50t^2 + 800t - 1950$
The horizontal (input) scale unit is 2 and the vertical (output) scale unit is 200.

Terry's graph

$P(t)$

t

$P(t) = -50t^2 + 800t - 1950$
The horizontal (input) scale unit is 2 and the vertical (output) scale unit is 1000.

If you look carefully, you notice that the only difference between the two graphs is the choice of scales for the $P(t)$-axis. Copy each graph and label points on them as indicated below.

1. On each sketch locate the point that shows the pair, (price, profit), giving the greatest profit. Label each point with the letter *M* and give its coordinates.

2. On each sketch locate the points that show ticket prices for which the talent show breaks even. Label each point with the letter *B* and give its coordinates.

If you correctly interpret the scale information, the coordinates for each labeled point on one sketch should be the same as for the corresponding point on the other. In fact, the two displays show exactly the same information for price and profit. For every point on Jan's sketch there is a corresponding point with the same coordinates, (price, profit), on Terry's sketch. However, the choice of different scales makes the sketches seem to give very different visual messages.

3. How does the change in the *P(t)*-unit scale affect the appearance of the relation between ticket price and profit?

4. Copy and complete the following sentence that describes the overall relation between ticket price and talent show profit.

   "As the ticket price increases from $0 to $16, predicted profit...."

This example is the first of many important steps in training yourself to read a graph wisely. It is important to realize that any change in scales can change the *appearance* of a pattern relating two variables. However, the pairs of values, (input, output), are *only* determined by the function rule.

 ## Exploration II

The following exercises give you practice reading graphs. Compare effects of the various scale changes. You may learn how to predict some ways that changes in scaling will affect the appearance of a sketch.

SITUATION 5.8

In the first section of this chapter you studied the profit prospects of the Budget Box Office theater. The rule $P(a) = 5.5a - 660$ gives the profit $P(a)$ in dollars during a day when a people attend the theater. The following diagrams are sketches of the graph of this function for input values $0 \leq a \leq 400$.

Notice that the only difference between the two sketches is the choice of the P(a)-axis unit scale.

1. On each sketch locate the point showing where the profit is closest to $200. Give its coordinates.

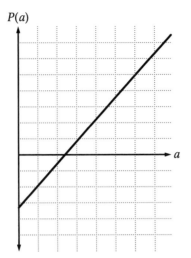

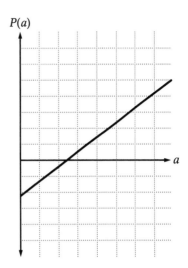

$P(a) = 5.50a - 660$
The horizontal (input) scale unit is 50 and the vertical (output) scale unit is 200.

$P(a) = 5.50a - 660$
The horizontal (input) scale unit is 50 and the vertical (output) scale unit is 300.

2. On each sketch locate the point that shows the attendance for which the theater breaks even. Give its coordinates.

3. On each sketch locate the point that shows the profit if no one comes to the theater. Give its coordinates.

4. Copy and complete the following sentence describing the overall relation between attendance and profit for a day.

   "As the attendance increases from 0 to 400, the profit...."

5. How does a change in the scale for $P(a)$ affect the appearance of the relation between attendance and profit?

Making Connections

You have been studying how the appearance of a graph is affected when the scale on the vertical axis is changed. Copy and complete the following sentence summarizing your observations about the effect of the change of scale. Explain why you think this effect occurs.

If the value of each tic-mark segment on the vertical axis is increased, the sketch of a function graph appears to _____ because _____ .

Exploration III

The preceding examples have illustrated how the appearance of a graph can depend upon the choice of scale on the vertical (output) axis. The appearance can also depend upon what scale is chosen for the horizontal (input) axis. In the following exploration, study the effect of changing the horizontal scale.

SITUATION 5.9

Revenue $R(p)$ for a school play depends upon the admission price P with a typical rule $R(p) = -100p^2 + 720p$. Both of the following diagrams are graphs of this function with the $R(p)$-unit chosen as 200, but with different p-units.

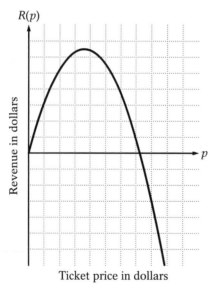

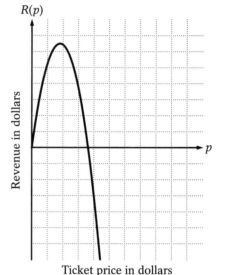

$R(p) = -100p^2 + 720p$
The horizontal (input) scale unit is 1 and the vertical (output) scale unit is 200.

$R(p) = -100p^2 + 720p$
The horizontal (input) scale unit is 2 and the vertical (output) scale unit is 200.

Copy these graphs on separate paper, then complete each of the following items.

1. On each sketch locate the point that shows the pair, (price, revenue), with maximum revenue. Label the point with the letter M and give its coordinates.

2. On each sketch locate the points corresponding to zero revenue. Label each of these points with the letter *B* and give its coordinates.

3. Copy and complete the following sentence that describes the overall relation between ticket price and revenue.

   As price increases from $0 to $10, predicted revenue....

4. How does change in the scale for input values affect the appearance of the relation between price and revenue?

This exploration shows again, in a different way, the basic message about scales on graphs: while a graph shows overall patterns in a relation between variables, you must read the scales carefully to get accurate information from the graph. When you want approximate answers, graphs can be very useful.

Making Connections

Technological aids can be valuable in studying relations between variables. You have learned to use both graphical and numerical tools.

1. List some ways in which a graph can be especially helpful for answering questions about a function.

2. List some ways in which tables of values are especially helpful for answering questions about a function.

3. What are some things about which you should be careful when using graphing tools?

4. In what ways is the information from looking at a graph like the information from looking at a table, and in what ways are there differences between the two approaches?

6 Solving Equations with Technology

When variables are related by a function, answering many of the interesting questions requires solving equations or inequalities. For instance, if $P(a) = 5.50a - 660$ is the profit function for the Budget Box Office theater, finding the break-even point (the attendance for which profit is $0, or $P(a)=0$) involves solving the equation $5.50a - 660 = 0$.

In preceding sections you have learned how to use computing technology to solve equations by a systematic search through tables or graphs of data pairs. Such technology often includes keys or commands, such as a SOLVE command, that give those solutions instantly. This section provides you with the opportunity to learn how to use these solving capabilities.

SITUATION 6.1

In most countries, the measurement of temperature uses the **Celsius** scale. In the United States temperature is more commonly reported in the **Fahrenheit** system. Temperatures in the two systems are related by the rule

$$C(F) = \frac{5}{9}(F - 32) \text{ or } C = \frac{5}{9}(F - 32),$$

which expresses the temperature C in degrees Celsius as a function of the temperature F in degrees Fahrenheit.

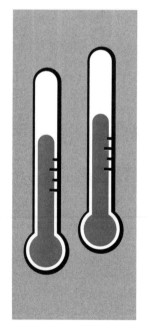

F	C(F)	F	C(F)
−100	−73.3	60	15.6
−80	−62.2	80	26.7
−60	−51.1	100	37.8
−40	−40.0	120	48.9
−20	−28.9	140	60.0
0	−17.8	160	71.1
20	−6.7	180	82.2
40	4.4	200	93.3

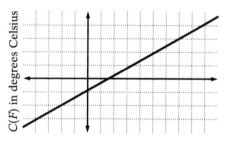

The horizontal (input) scale unit is 20 and the vertical (output) scale unit is 20.

The overall pattern in this relation is described well by the table of sample pairs (F,C) and by the graph of those values. The table and graph give help in finding *approximate* outputs for many inputs. For example, $C(10)$ is likely to fall between -17.8 and -6.7. The rule can give exact output for any input.

Reversing the process, finding the input that produces a specified output, is not quite as easy. For example, if you are visiting a foreign country where the Celsius scale is used for temperatures, you might want to know what a forecast of 25°C means on the Fahrenheit scale. This problem translates into the equation

$$C(F) = 25 \text{ or } \frac{5}{9}(F - 32) = 25.$$

The table of sample pairs (F,C) and the graph suggest that the answer lies somewhere between 60°F and 80°F. You might use a guess-and-test strategy or you might search in a table with smaller steps or a graph with smaller unit scales to get better approximations to the solution. However, even with computer help, finding very accurate answers by systematic search can require a great deal of work. You still must choose the smallest and largest inputs and outputs and the step size or graph scales, then repeat the process to improve accuracy. In finding exact roots for equations, technological aids which can solve equations directly through a single command such as SOLVE are especially helpful.

The instructions required to use these solving capabilities vary. A typical session might proceed in one of the following ways:

I. To solve a single equation:
   a. Enter the equation, *e.g.*, $\frac{5}{9}(F - 32) = 25$.
   b. Use the SOLVE key or other appropriate command to solve the equation.
   c. The result is displayed, *e.g.*, $F = 77$.

II. To solve several related equations:
   a. Enter the function rule, *e.g.*, $C(F) = \frac{5}{9}(F - 32)$.
   b. Write the equation, *e.g.*, $C(F) = 25$.
   c. Use the SOLVE key or other appropriate command to solve the equation.
   d. The result is displayed, *e.g.*, $F = 77$.

It might look to you as if the first form is simpler, but remember that once a function has been defined, it can be used for many purposes. Further, with function notation it is easy to change the equation without retyping the rule. In each case, you are asking the computer to solve the given equation for F, that is, to find the value of F that gives a true statement.

Software has been written to include methods for solving this and other kinds of algebraic equations. However, you still must be able to find the equations that match the questions in the given situation.

 ## Exploration

In this exploration, use a technological aid with equation-solving capabilities, such as a SOLVE command, to answer the following.

1. Enter the temperature conversion function. Record what you enter and the response you receive.

2. To check that you have entered the function rule accurately, calculate the following Celsius equivalents of familiar Fahrenheit temperatures. Record both what you enter and the response you receive.
   a. Temperature in degrees at which water freezes: $F = 32$
   b. Temperature in degrees at which water boils: $F = 212$
   c. Normal body temperature in degrees: $F = 98.6$

3. Use a technology tool to answer each of the following questions. Record your entry and the response as well as the answer to the question.
   a. A good temperature for baking cookies is 200°C. What Fahrenheit temperature is equivalent to a Celsius temperature of 200°?
   b. If the temperature at the stadium is 23°C, should you wear a coat to watch a football game?
   c. The chart below shows the number of days for which milk may be kept fresh at a variety of Fahrenheit temperatures.

Temperature (degrees Fahrenheit)	Time (days)
60 to 80	0.5
50 to 60	1
40 to 50	5
32 to 40	10
30 to 32	24

If your refrigerator is set at 5°C, will milk kept in it for a week stay fresh? Explain how you reached your conclusion.

d. One spring day the temperature is 96°F in Texas, while the temperature in Ohio is 10 Fahrenheit degrees cooler. What is the Ohio temperature in degrees Celsius?

e. What temperature has the same Fahrenheit and Celsius degree measure? Hint: Try $C(F) = F$.

f. What temperature has Celsius degree measure exactly 100 degrees lower than the Fahrenheit measure? Hint: Try $C(F) = F - 100$.

g. What is the rule for calculating Fahrenheit temperature from a given Celsius temperature? Hint: See if your aid will allow $C = \frac{5}{9}(F - 32)$. Solving may produce an equation of the form "$F =$ some expression using C." We call this "Solving for F in terms of C."

h. The local weather service reports the temperature for today as 56°C rather than 56°F. Is this Celsius reading likely to occur where you live?

Exercises

Write the commands needed to use a technological tool to answer the questions in Exercises 1 through 4. Indicate which tool these commands are for.

1. SITUATION The cost of a group bowling party at Strikes to Spare lanes is a function C of the number of games bowled N with the rule $C(N) = 20 + 1.60N$.
   a. What is the cost for bowling 75 games?
   b. What number of games bowled cost $68?

2. SITUATION The number of liters of water left in a particular bathtub is a function A of the number of seconds of time T since the plug was pulled. The function can be expressed by the rule $A(T) = 125 - 0.6T$.
   a. What number of seconds does it take to have 100 liters of water remaining in the tub?
   b. What number of seconds does it take to empty the tub?
   c. Would the rule change if we considered a larger bathtub? Explain.

3. **SITUATION** The number of calories in a vanilla ice cream cone depends on the number of grams G of ice cream used. This function C can be expressed using the rule $C(G) = 1.6G + 110$.
   a. What number of calories does a cone with 200 grams of vanilla ice cream have?
   b. What number of calories are in a cone without ice cream?
   c. Is it possible for an ice cream cone to have no calories?
   d. Do you expect to find a vanilla ice cream cone with 1710 calories? Explain.
   e. Is it reasonable for a vanilla ice cream cone to have 100 calories? Explain.
   f. A cone with 200 grams of chocolate ice cream may contain a different number of calories than the value found for vanilla in part a. List several reasons for this difference.

4. **SITUATION** The number of minutes for an average reader to read a novel can be expressed as a function of the number of pages in the book. One possible rule for such a function is $T(P) = 2.7P$, where the time in minutes required to complete a novel is a function T of the number of pages P in the book.
   a. Using this rule, what number of minutes does it take to read a 180-page novel?
   b. Using this rule, what additional number of minutes does it take to read a 210-page novel rather than a 180-page book?
   c. Using the same rule, can a 500-page novel be read in one day?
   d. Debbie says she can read a novel in "no time at all." If you take her literally, what does this phrase imply about the length of the novel?
   e. Myhun claims she read a 179-page book in 6 hours. Eddie says he read a 179-page book in 10 hours. Can they both be correct? List several reasons for your answer.
   f. Ken read for one hour every day for an entire year. Could he have read exactly one book during that time? Explain how you reached your conclusion.

For each of Exercises 5 through 8 write the necessary instructions for your technological tool to solve the given equation for x.

5. $11 = 3x - 7$

6. $510 = 12x^2 - 78$

7. $3 = \dfrac{12}{(x-5)}$

8. $234x + 564 = 3138$

Making Connections

List some reasons why equation-solving technology, such as a SOLVE command, is useful in answering questions about functions.

List some limitations of these capabilities in answering questions about functions.

7 Summary

The central purpose of this algebra course is to develop your skill and understanding in the ways that functions can be used to describe and study relations between quantitative variables. In the first chapter you learned how to use tables, graphs, and symbolic rules to represent functions. In this chapter you have begun learning how and when to use a variety of technological, paper-and-pencil, and mental tools for analyzing functions.

1. When a problem involving variables and functions can be solved by a few relatively simple calculations with simple numbers, it is probably best to do that arithmetic in your head, with some paper-and-pencil work to record results.

2. When a problem requires complicated calculations with "messy" numbers or difficult operations such as square roots, you may wish to use the features of a four-function or scientific calculator.

3. When a problem requires repeated calculations using the same function rule, it is usually helpful to use a programmable tool.

4. Computing technology which can produce tables or graphs of functions are especially helpful in discovering overall trends in a relation between variables.

5. Since many of the important questions about quantitative relations require solution of equations or inequalities, it is important to know how to use technological aids to produce estimates of roots by guess-and-test strategies or by the inspection of tables and graphs, or to find exact answers by the use of technology with equation-solving capabilities, such as a SOLVE command.

Your explorations in this chapter have probably also suggested a variety of other ideas about when and how the various computing tools can be used best to study relations between quantitative variables. As you apply these tools in further study of algebra, you will develop your ability to judge which tools are best for different types of problems—which give the quickest way to get answers of desired accuracy for different kinds of questions and problems.

Review Exercises

One of the important skills in problem solving is learning to choose a useful method and calculating tool for the problem at hand. For many problems, you can choose from among several suitable alternatives.

    —mental arithmetic   —table of values

    —calculator arithmetic   —graph

    —computer/Programmable   —equation-solving technology
       calculator   (*e.g.* SOLVE command)

Instructions: In each of the following review exercises:

a. Write, in algebraic form, a calculation, equation, or inequality that can be used to answer the question.

b. Solve the problem.

c. Describe the method and tool you use.

d. Solve the problem using a second choice of method and tool.

e. Compare the advantages and disadvantages of the methods you use.

SITUATION

One of the first NASA astronauts to walk on the moon took a golf club and ball onto the lunar lander. While on the moon's surface, he hit a long shot. The height of a golf ball hit on the moon is a function of its time in flight. One typical rule would be

$$H(t) = 20t - 0.8t^2,$$

where t is in seconds and $H(t)$ is in meters.

1. According to the rule above, what was the height of the ball after 10 seconds?

2. Using the same rule, find the height of the ball after 9.5 seconds.

3. Find the time, to the nearest second, at which the ball reached a height of 80 meters.

4. Find the time when the ball returned to the moon's surface.

5. At what time(s) was the height of the ball 100 meters?

6. What was the maximum height of the ball and at what time was this height attained?

7. At what time(s) was the height of the ball at least 50 meters?

8. At what time(s) was the height of the ball 45 meters?

9. What is the pattern in the height of the ball as a function of time in flight?

10. When is the ball within 5 meters of the moon's surface?

11. At what average speed (in meters per second) does the ball rise or fall during the first second of flight? During the fifth second of flight?

SITUATION

The management at the Blue Ridge Inn is planning to offer a special Winter Weekend at its resort hotel in the mountains. There will be special meals, entertainment, and outdoor recreation activities for the whole family, with all activities included for a fixed price per person. The problem is what price to charge!

Market surveys suggest that the number of customers is a function C of the price charged P and can be expressed by the rule

$$C(p) = 450 - 2.5p.$$

After itemizing expected costs, the management estimates that the profit F also depends on the price charged and can be expressed by the rule

$$F(p) = -2.5p^2 + 600p - 27\,000.$$

Use these function rules to answer the following questions.

12. What number of customers is predicted if the price is set at $100?

13. What profit is predicted if the price is set at $100?

14. If the Inn wants 300 customers, what price should be charged?

15. What price(s) will give at least 255 customers?

16. At what price(s) does the Inn break even, that is, have profit of $0?

17. For what price(s) does the Inn have a profit of $5000?

18. At what price(s) is the profit for the Inn at least $1000?

19. According to the rule, at what price will no customers come?

20. What price gives the Inn maximum profit? What is that profit?

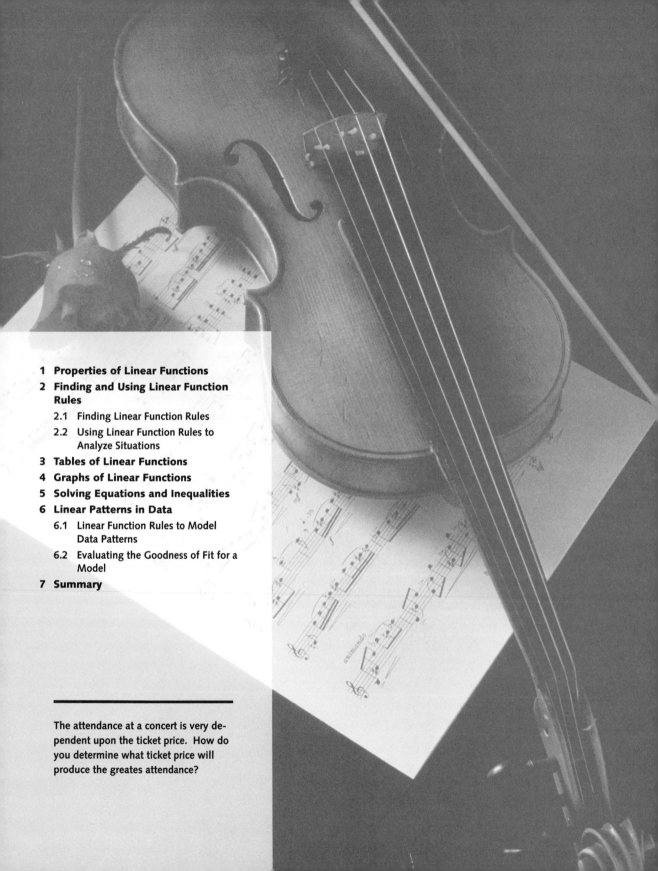

The attendance at a concert is very dependent upon the ticket price. How do you determine what ticket price will produce the greates attendance?

3 Linear Functions

The mathematical functions that relate variables in the fields of science, engineering, business, and government have many different kinds of rules and graphs. Some special types of functions occur so often that it is useful to study them carefully. This chapter focuses on the most widely used family of functions, the **linear functions**. By the end of the chapter you should be able to recognize and analyze problem situations that are described by linear functions.

1 Properties of Linear Functions

What are the special properties of functions that are called linear? What do babysitting, fast food restaurants, and jet airplanes have in common? As you study the situations in this section, look for patterns in the relations among variables being studied.

SITUATION 1.1

Until high school students reach age 16, babysitting is one of the best ways to earn spending money. If you get into that business, your pay will probably be a function of time spent sitting. A variety of mathematical questions arise.

Suppose you plan to charge $3.50 per hour. You may want to know:

1. What is the charge for sitting 2 hours, 3 hours, 4.5 hours, *etc.*?

2. What is the length of sitting time needed to earn $38.50 for concert tickets, $12.95 for a new tape, *etc.*?

You have seen in earlier work that function tables, graphs, and rules are very helpful tools in answering such questions. The following table shows the charges for 1 to 6 hours of babysitting time at $3.50 per hour.

Time (hours)	1	2	3	4	5	6
Charge (dollars)	3.50	7.00	10.50	14.00	17.50	21.00

If you extend this table you find that the charge for 7 hours is $24.50, the charge for 8 hours is $28, and so on. To earn $38.50 for the concert tickets you must sit for 11 hours. Notice that only hourly charges are given here. Would you be likely to charge for an additional half-hour? Do you think it's likely a parent would pay $7.82 for babysitting, or would the charge be rounded up to $8.00?

The table entries can be obtained in several different ways. First, you can notice that each added hour of work increases the charge by $3.50. To get the next entry, add $3.50 to the current charge.

$$3 \text{ hours} + 1 \text{ hour} = 4 \text{ hours.}$$
$$\$10.50 + \$3.50 = \$14.00.$$

This pattern is often described with the statement "charge increases at the rate of $3.50 per hour."

Second, the rate "$3.50 per hour" means that the charge $C(h)$ for h hours can be described by the rule $C(h) = 3.50h$. For instance,

$$C(8) = 3.50 \,(8)$$
$$= 28.00.$$

Another way to approach the questions of this situation is to graph the relation between time and charge. If you begin by plotting several pairs, (time, charge), you will see a pattern that suggests the graph for all such pairs. The graph on the right shows many more pairs than you may ever need, since you are likely to be paid for a rounded number of hours like 3 or 4.25 rather than times like 3.782 hours.

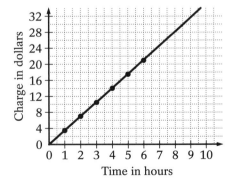

Using the graph of the relation between time and charge, you can make good estimates of the answers to questions like these:

1. What is the correct charge for 3.5 hours of babysitting?

    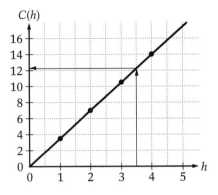

    Follow the arrows up from 3.5 until you meet the graph, and then over to meet the vertical axis. The vertical arrow on the following graph is pointing to about 12.25. This indicates that the charge should be $12.25.

2. What babysitting times give earnings of at least $35?

    The charge graph is at or above the $35 level for all values of h equal to or greater than about 10. So sitting for 10 hours or more earns at least $35.

This function relating babysitting time and charge has a simple algebraic rule. You might find that rule easy to work with in your head, with paper and pencil, or with the features of a four-function or scientific calculator. You could also use more complicated technology to produce more detailed tables and graphs.

 Exploration

Use the babysitting example as a guide in the next situations. Look carefully to see what the examples have in common.

SITUATION 1.2

To rent movies from video clubs, it is usually necessary to be a club member and to pay a daily rental charge. Suppose a club offers free membership for one year and a daily charge of $2.50 per cassette. Your cost for one year of membership is a function of the number of cassette-rental days.

Copy and complete the following table. On a grid like the one shown, graph the data pairs, and use both the table and the graph to help you answer questions about rental costs for the year.

Days	Cost (dollars)
0	0.00
1	2.50
2	5.00
3	
4	
5	
10	
15	
20	
25	

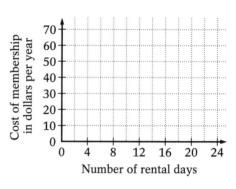

1. At what rate does the cost increase? That is, for every additional day for which a cassette is rented, by how much does the cost increase?

2. The annual cost is a function C of the number of cassette-rental days d. Write a rule which can be used to calculate the annual cost $C(d)$.

3. If you connect the plotted points on the graph of C, do all of the new graph points give meaningful information about cassette rentals?

4. What does your answer to question 3 indicate about a reasonable set of input values for the cost function?

SITUATION 1.3

Suppose the daily rental charge for another video club is $1.75, but the yearly membership is $15. Is this a better deal than the offer from the club described in Situation 1.2? Answer the questions below before you decide.

Copy and complete the following table. Then graph this new cost function *on the same coordinate diagram* used in Situation 1.2.

Days	0	1	2	3	4	5	10	15	20	25
Cost	15.00	16.75	18.50							

1. At what rate does this cost function increase? That is, what increase in cost results from an increase of 1 in the number of rental days?

2. What increase in cost results from an increase of 5 in the number of rental days?

3. The annual cost in dollars for this video club is a function V of the number of rental days. Write a rule relating the number of rental days and the annual cost.

4. Does this video club cost less than the first? Explain.

5. What factors other than cost might influence your choice of which club to join?

6. Compare the functions in Situations 1.1, 1.2, and 1.3. What are the most striking similarities and differences:
   a. in the tables?
   b. in the graphs?
   c. in the rules?

As you continue through the chapter, try to verify the patterns you are beginning to see in the tables, graphs, and rules of linear functions.

Exercises

The following situations involve linear functions with similar but not identical properties. Try to identify common patterns in the tables, in the graphs, and in the rules for linear functions.

1. **SITUATION** Consider the following plans for babysitting fees:
   Plan 1. $4.25 per hour;
   Plan 2. $7.50 minimum charge plus $2.50 per hour.
   a. Copy and complete the following table showing fees for sample times worked.

Time (hours)	Fee (dollars)	
	Plan 1	Plan 2
1		
2		
3		
4		
5		
6		
7		

    b. At what rate do the charges increase for each plan?

    c. Graph the two functions on a single coordinate diagram.

    d. Write a function rule relating time and fee under each plan.

    e. Write at least three questions that would be of interest in comparing the two plans for babysitting charges. (For example: "For what times does Plan 1 yield less income for a babysitter than Plan 2?")

2. **SITUATION** On December 17, 1903, the Wright brothers made the first airplane flights. The longest was 852 feet, lasting 59 seconds. Today commercial jets fly non-stop from Los Angeles to Sydney, Australia—over 10,000 kilometers in 15 hours.

Planning fuel requirements for flights is an important problem for airlines. A Boeing 727 jet that has been refueled immediately before take-off hold about 28,000 liters of fuel and uses about 5000 liters per hour of flight. Although other factors often have an affect, we can consider the amount of fuel remaining on this plane as a function of time in flight.

A rule giving fuel remaining as a function of time can be found by studying a few typical data pairs, (input, output). For example,

Time t (hours)	Fuel remaining $f(t)$ (liters)
0	28 000
1	23 000 = 28 000 - 5000 × 1
2	18 000 = 28 000 - 5000 × 2

The pattern suggests the rule $f(t) = 28\ 000 - 5000t$. Use this rule and any necessary technology to answer the following.

    a. Copy and complete the following table giving fuel remaining in the jet as a function of time elapsed in the flight.

Time t (hours)	Fuel remaining $f(t)$ (liters)	Time t (hours)	Fuel remaining $f(t)$ (liters)
0.0	28 000	3.5	
0.5	25 500	4.0	
1.0		4.5	
1.5		5.0	
2.0		5.5	
2.5		6.0	
3.0			

b. On a sheet of graph paper, draw a pair of axes and label them as shown here. Plot the data pairs, (time, fuel remaining), from your table on your grid. (You could check your result later with a graphing tool.)

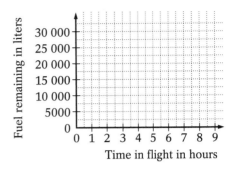

Use the rule, table, or graph to answer the following questions about fuel use in the jet. Where appropriate, write an equation or inequality that can be used to answer the given question.

c. How much fuel remains after 4.5 hours of flight time?

d. How long does it take for half the fuel to be used?

e. At what rate does the plane's remaining fuel decrease? That is, what is the decrease in fuel for each additional hour of flight?

f. What flight times leave at least 5000 liters of fuel in the plane (for safety)?

g. If the plane travels 800 kilometers per hour, what is the longest trip it can make, allowing a 5000-liter fuel safety margin?

3. SITUATION Most companies try to predict the quantities that they can sell as functions of the prices they charge. To increase sales they might choose to lower prices. To increase profit they might choose to increase prices, but then they risk losing sales.

A function predicting sales of a product from prices is called a **demand function**. It shows the relation between the price and the amount of the product a company can expect to sell at that price.

In the TALENT SHOW simulation in Chapter 1, attendance depended on the ticket price. With no celebrity master of ceremonies hired, the demand equation met the following conditions.

    i.   If tickets are free, attendance would be 800.

    ii. Each $1 increase in ticket price causes a decrease of 100 in attendance.

a. Copy and complete the following table relating ticket price and attendance for prices from 0 to 10 dollars in steps of $1.00.

Ticket price p (dollars)	Attendance $a(p)$
0.00	800
1.00	$800 - 100 \times 1.00 = 700$
2.00	$800 - 100 \times 2.00 = 600$
3.00	
4.00	
5.00	
6.00	$800 - 100 \times 6.00 = 200$
7.00	
8.00	
9.00	
10.00	

b. Write an algebraic rule relating price p and attendance $a(p)$.

c. Graph the data, (price, attendance), from part a.

d. Write (in words) three different questions that can be answered using the demand function.

4. Each of the situations in Exercises 1–3 has been described by a function whose graph lies on a straight line. Such a function is called a **linear function**. Some of the functions you have seen in previous chapters are not linear. You will learn more about linear functions as you complete this chapter, but first, you need to be able to identify a linear function. In this exercise, try to figure out how to tell whether a function is linear by looking at its graph, its table, or its rule.

The table and graph shown here are for a function that is not linear.

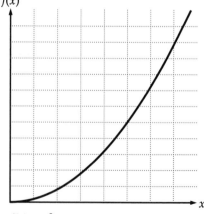

x	$f(x) = x^2$
0	0
1	1
2	4
3	9
4	16
5	25
6	36
7	49
8	64

$f(x) = x^2$

The horizontal (input) scale unit is 1 and the vertical (output) scale unit is 5.

Consider the function with rule $f(x) = x^2$ (see the table and graph) and the function with the rule $C(h) = 1.50h$ (from Situation 1.1). What similarities and differences do you notice:

a. in the tables of data pairs, (input, output)?
b. in the graphs?
c. in the rules relating input and output values?

Making Connections

The situations of this section have shown functions relating several different kinds of variables. The function tables, graphs and rules for most examples shared some important properties that give them the name **linear functions**.

1. In a table of values, (input, output), for a linear function, the output variable increases (or decreases) at a constant rate as the input variable increases.

2. The points of a linear function graph lie on a straight line.

3. The rule relating inputs and outputs for a linear function can be written in several different symbolic forms, but each is equivalent to the standard form:

$$f(x) = mx + b,$$

where m and b are numbers set by facts of the situation.

For example, the function relating ticket price and attendance at the talent show can be represented by a rule,

$$a(p) = 800 - 100p \text{ or } a(p) = -100p + 800$$

($m = -100$ and $b = 800$), and by a table or graph of typical pairs, (price, attendance).

Ticket price p	Attendance $a(p)$
0.00	800
1.00	700
2.00	600
3.00	500
4.00	400
5.00	300
6.00	200
7.00	100
8.00	0

$$a(p) = -100p + 800$$

By the end of this chapter you should be able to recognize linear relations among variables, and you will have been given many opportunities to use linear functions to analyze real-world problems.

2 Finding and Using Linear Function Rules

Using mathematics to solve real-world problems often involves finding mathematical functions that model the behavior of related variables. This task involves making choices, because there are many different kinds of behaviors. The first part of this section demonstrates some strategies for finding linear rules. The second part gives you practice in analyzing and solving problems when the model is a linear function.

2.1 Finding Linear Function Rules

When a situation involves a linear function, it is very useful to find a rule that relates the variables. The rule can be used to generate a table of values and a graph of the function. The rule, table, and graph can then be used to answer questions about the linear function.

SITUATION 2.1

When the Northwestern High School girls' volleyball team was planning a three-day trip for games in Pennsylvania, the coach checked on costs to rent a fifteen-passenger minibus. The best deal was $50 per day plus $0.30 per kilometer traveled. Under this arrangement, the cost for the three-day trip is a function of the distance to be driven.

To find a rule for this function, it is helpful to look for a pattern in some typical pairs, (input, output).

Distance d (kilometers)	Cost $C(d)$ (dollars)
0	150.00
1	$150.30 = 150 + 0.30 \times 1$
2	$150.60 = 150 + 0.30 \times 2$
3	$150.90 = 150 + 0.30 \times 3$
.	.
.	.
.	.
100	$180.00 = 150 + 0.30 \times 100$
200	$210.00 = 150 + 0.30 \times 200$
d	$C(d) = 150 + 0.30 \times d$

To create the table above, we first needed to generate and organize data, observe a pattern, then find a rule that describes the pattern. A second strategy to write a rule for the relation makes use of the observation that here the cost is in two parts—a **fixed cost**, $150.00 for the 3 days, and a **variable cost**, $0.30 per kilometer.

Using either of these strategies, it is easy to write rules for different cost deals. For example:

1. If the fixed cost is reduced to $120 and the variable cost for distance is increased to $0.50 per kilometer, the cost function rule changes. Let's call this second rule C_2:

$$C_2(d) = 120 + 0.50 \times d.$$

2. If there is no fixed cost, but variable cost for distance is $0.75 per kilometer, the cost function rule becomes:

$$C_3(d) = 0.75 \times d.$$

Exploration I

In the following situations, find algebraic rules for the functions.

SITUATION 2.2

Suppose you begin working as a babysitter charging $3.75 an hour.

1. Write a rule relating the number of hours worked, t, to your pay in dollars, $P_1(t)$.

2. Suppose you raise your babysitting rates to $4.25 per hour. Write a new rule for the function relating pay to time worked ($P_2(t) = ?$).

3. When you babysit on a "football Saturday" your rates change to $4.75 per hour, and the parents pay you an extra $7.50 to fix a meal for the children. Write a rule for the function relating your pay on a football Saturday to the time worked ($P_3(t) = ?$).

4. During homecoming weekend, your customers offer to double your usual hourly rate for football Saturdays and still give you the extra $7.50 to fix lunch for the children. Write a new rule relating pay to time worked ($P_4(t) = ?$).

SITUATION 2.3

Membership in health and fitness clubs is now very popular among young working people. Typical club membership costs include an annual fee, fees for each use of the facility (or time spent in the club), and for special services.

1. Suppose that the Zodiac Health Club charges $195 annual membership fee plus $2.50 per day that you use the club exercise rooms. Total cost for a year is a function Z of the number of days d that you use the club. Write a rule for $Z(d)$.

2. The costs to join the Olympia Health Club are $150 for an initiation fee and $58 for regulation leotards and tights. After that, each member pays $3.50 per visit. What is the function rule $O(v)$ that relates total cost to number of visits v?

3. The Healthy-and-Fit Club charges $250 for its one-time initiation fee. Members are then entitled to use all of the facilities but must pay $2 for every class they attend. The total cost is a function H of the number of classes attended c. Write a rule for $H(c)$.

4. The Premier Health Club has a special arrangement for long-time customers. The charge is a flat fee of $395 for the first two years with the next two years free, and $20 per year after the first four years. Suppose a person has been a member for more than four years. Write a rule for the function P that relates total cost to the total number of years y for which the person has been a member.
   a. Does your function rule work for $y = 2$? Explain.
   b. What input values are not suitable for your function?

SITUATION 2.4

Commercial airplanes cruise at very high altitudes. Their descent to landing is gradual. Suppose you are in a plane that starts its glide to landing at 10,000 meters above ground level, dropping at a rate of 450 meters per minute. The plane's altitude in meters is a function A of time t in minutes since the descent for landing began.

1. Write a rule for the function A.

2. In contrast to a plane approaching a landing, consider the situation where a plane takes off from an airport runway. Suppose a commercial airplane climbs at a steady rate of 500 meters per minute after take-off. The plane's altitude in meters is a function H of the time t in minutes since take-off. Write a rule for the function H.

SITUATION 2.5

For small planes, a more typical cruising altitude is 1500 meters above ground level. On approaching an airport, such a plane usually drops to a distance of about 300 meters above ground level. It then flies in a rectangular pattern at this "pattern altitude" until it is ready to begin its approach to the runway.

1. Suppose you are in a small plane at an altitude of 1500 meters above ground level. The plane starts a glide toward a pattern altitude of 300 meters at a rate of 120 meters per minute. The plane's altitude $A(t)$ in meters is a function of time t in minutes since the descent began.
   a. Write a rule for the function A.
   b. Describe the strategy you used to obtain your rule.
   c. For what input values is your rule reasonable?

2. Small single-engine planes can cruise safely at altitudes as high as 3000 meters above ground level. Suppose one plane drops from 3000 meters to a pattern altitude of 300 meters at a rate of 150 meters per minute. The plane's altitude in meters is a function B of time t in minutes since the descent began. Write a rule for B.

3. Once a single-engine plane has reached its pattern altitude, it usually maintains that altitude until it is within a few kilometers (horizontal distance) from the intended point of contact with the ground. At that time it often begins its descent to landing at a rate of about 160 meters per minute. Suppose that a single-engine plane has a pattern altitude of 240 meters and descends at a rate of 160 meters per minute. The plane's altitude in meters is a function C of time t in minutes since its descent to landing began. Write a rule for C.

While the preceding examples cover quantitative variables in very different situations, we hope that you are beginning to see the following common features that lead to linear function rules.

Feature 1. The steady rate of change in outputs, as inputs increase.
Examples: Charge per kilometer travelled in a rental car
Hourly rate of pay for babysitting
Fee for each event at a health club
Rate of climb or descent of an airplane

Feature 2. The fixed value that occurs as output when 0 is the input.
Examples: Fixed cost of a car rental, regardless of distance driven
Introductory membership fee for a health club
Altitude of an airplane at the start of its descent or climb

The next section asks you to use these features of linear function rules in writing and solving equations that can be used to answer questions in the situations.

2.2   Using Linear Function Rules to Analyze Situations

With a function rule available, you can answer various questions about a situation. Technological tools can be used to calculate function outputs for various inputs and to find solutions for equations and inequalities (using a command such as SOLVE).

Reconsider, for example, Situation 2.1 in which a school team rents a minibus for a three-day trip. The cost in dollars of renting a bus is a function C of the distance d in kilometers traveled and can be expressed using the rule

$$C(d) = 150 + 0.30d.$$

You could start by defining the function in your programmable tool then use the function rule to answer a variety of questions about the bus rental costs. There are a number of questions that can be asked. For example:

1. What is the cost to travel 435 kilometers?

   This can be answered by the calculation of $C(435) = 150 + 0.30(435)$. Using paper-and-pencil arithmetic or a calculator, you should find that

$$
\begin{aligned}
C(435) &= 150 + 0.30(435) \\
&= 150 + 130.50 \\
&= 280.50.
\end{aligned}
$$

   If you have defined $C(d)$ in a programmable tool, you can also find the answer using the input 435.

2. How far can the team travel on a budget of $360?

   You can find the answer to this question by solving the equation

$$C(d) = 360 \text{ or } 150 + 0.30d = 360$$

   using the equation-solving capabilities of a technological tool, such as a SOLVE command, or a technology-aided guess-and-test search. A session of a technology-aided guess-and-test search might generate the following table.

Guess d	Test $150 + 0.30d$	Decision
600	330	too low
800	390	too high
700	360	right

3. How far can the team travel and still keep the cost under $360?

   This requires solution of the inequality

$$C(d) < 360 \text{ or } 150 + 0.30d < 360.$$

You might choose to begin by solving the equation $150 + 0.30d = 360$ by a guess-and-test search or by using equation-solving technology, such as a SOLVE command. Then, since a 700 kilometer trip costs $360, any trip with mileage less than 700 has a cost under $360.

You might then informally check your reasoning about the inequality by calculating $C(d)$ for values of d near 700. That is, if you find that 700 satisfies the equation $150 + 0.30d = 360$, you can evaluate the nearby values $C(699)$ and $C(701)$.

 ## Exploration II

The following questions address the situations from Exploration I. Using the function rules you found, write calculations, equations, or inequalities that can be used to answer the given questions. Then use a technological tool to find answers to the questions. Record your answers in a complete sentence.

1. Recall from Situation 2.2 that pay for babysitting is a function of time spent sitting, with different rules for different rates of pay.
   a. Suppose you are paid $3.75 per hour for babysitting. Write a function rule for your total pay $P_1(t)$.
   What might a babysitter expect to be paid for 3.25 hours of work?
   b. Suppose you are paid $4.25 per hour. Write a function rule for your total pay $P_2(t)$.
   If you babysit for 5.75 hours, how much more do you make at this rate than at the rate of $3.75 per hour?
   c. On a normal "football Saturday," your babysitting rates are $4.75 per hour with a $7.50 bonus for fixing a meal. Write a function rule for your total pay in this case $(P_3(t) = ?)$.
   How long must you babysit on a football Saturday if you want to earn at least $35?
   d. Your babysitting rates on homecoming Saturdays are doubled from the usual $4.75 per hour and they also include a $7.50 bonus. Write a function rule for your total pay on a homecoming Saturday $(P_4(t) = ?)$.
   In part c, you were asked how long you would have to babysit to earn $35. If you were to babysit for the same length of time on a homecoming weekend, how much would you earn?

     e. If Sergine's father wants to pay her babysitter no more than $175 for a month, how much time can he use if the babysitter charges $4.25 per hour?

2. Recall from Situation 2.3 that the annual cost of membership in a health club depends on membership fee and other fixed costs as well as the number of times that you use the club. Refer to your answers to Exploration 1 for cost function rules of the clubs. Remember to record your methods for finding the answers to questions. Describe the function rule as well as the calculation, equation, or inequality you use.

     a. What is your total annual cost at the Zodiac Health Club if you use the facilities on 45 days?

     b. A member of the Olympia Health Club paid $404 during the first year of membership. How many visits did this person make to the club during that year?

3. Refer to Situation 2.4, in which a commercial airplane, after cruising at an altitude of 10,000 meters, glides to a landing at a rate of 450 meters per minute. Answer the following. For parts b through e, answer the question and record the calculation, equation, or inequality you used to find the answer.

     a. Write a rule for the altitude function A.

     b. What is the plane's altitude after 15 minutes of descent?

     c. How long does it take the plane to reach an altitude of 1000 meters?

     d. For what times is the plane's altitude more than 2500 meters?

     e. How long does the plane glide until it reaches the ground?

Exercises

By answering the following, you will gain experience in three very important parts of mathematical modeling:

–finding a rule that relates variables in the problem situation;

–writing calculations, equations, or inequalities that can be used to answer specific questions in the situation; and

–making estimates of the inputs or outputs likely to meet problem conditions.

As you have seen earlier in this course, technological tools are very useful in producing the calculations and solutions that you need. However, you need to plan and execute the work for those tools, and you should take some time to judge the reasonableness of the output.

1. **SITUATION** The cost of a long distance telephone call depends on the city you are calling, the time of day, and the length of the call in minutes. For a direct dial call from Hyattsville, MD, to State College, PA, during the daytime, one long-distance company charges $0.24 per minute.
   a. Write a rule giving cost in dollars as a function C of the length of a call t in minutes.
   b. Write a calculation that gives the cost of a 9-minute call.
   c. Write an equation for finding the length of a call costing $2.88. Estimate the root of that equation and test your estimate.
   d. Write an inequality that you would use to determine what calls (specified by length of call in minutes) cost under $10. Estimate and test three different inputs that satisfy the inequality.

2. **SITUATION** If you telephone long distance person-to-person, another company will charge $0.50 for operator service plus the charge per minute of calling time. For an evening call from Washington to Philadelphia, the charge per minute is $0.14. (For a call lasting a fractional part of a minute, assume that the phone company rounds the time up to the next minute.)
   a. Write a rule giving cost in dollars as a function C of the time taken, t minutes, for such a person-to-person call.
   b. Write a calculation that gives the cost of a 15-minute call.
   c. Write an equation for finding the length of such a call costing $11.75. Estimate the root of that equation and test your estimate.
   d. Write an inequality that you would use to determine which person-to-person calls cost less than $5. Estimate and test a value of t that *does not* satisfy this inequality.

3. **SITUATION** Submarines are best known as warships, but they are also used for research and engineering in deep ocean locations. Suppose such a submarine descends from the ocean surface at a rate of 45 meters per minute.
   a. Write a rule giving submarine location in meters as a function D of descent time t in minutes. Use negative numbers to indicate location below the ocean surface.

    b. Write a calculation that gives the submarine location after 35 minutes.

    c. Write an equation or inequality that can be used to answer the following question:

        When does the submarine reach a depth of 4410 meters, that is, location −4410?

    d. Write an equation or inequality that can be used to answer the following question:

        For what times is the submarine within 1000 meters of the ocean surface, that is, at a location above −1000?

        Identify and test at least one reasonable input value for t.

    e. What limitations, if any, would you put on the values for t?

4. SITUATION Suppose that you have a job selling peanuts at football games. For three hours of work you are to be paid $10, plus $0.15 for each bag of peanuts you sell.

    a. Write a rule giving your pay in dollars as a function P of the number of bags of peanuts sold n.

    b. Identify the input variable and the possible values it can have.

    c. Write an equation or inequality that can be used to answer the following question:

        What sales are needed if you are to earn at least $25?

        Estimate the solution(s) of that equation or inequality and test your estimate(s).

    d. Write a calculation that gives your pay if you sell 200 bags of peanuts.

5. SITUATION Suppose the concession manager sets the price of a bag of peanuts at $0.75.

    a. Write a rule giving the concessions revenue (number of dollars taken in) as a function R of the number of bags sold n.

    b. Write a calculation giving revenue if 1500 bags are sold.

    c. Write an equation or inequality that can be used to answer the following question:

        How many bags must be sold to give at least $1000 revenue?

    d. Estimate and test solution(s) to the equation or inequality in part c.

6. **SITUATION** Suppose the concession manager has the following costs for the business.

   Fixed: $150 for vendors; $125 permit to sell in the stadium
   Variable: $0.10 per bag to vendors; $0.25 per bag to buy peanuts

   Suppose also that the vendors sell the peanuts for $0.75 per bag.
   a. Write a rule that gives the concessions costs as a function of the number of bags of peanuts the vendors sell.
   b. Write a rule that gives the concessions revenue as a function of the number of bags of peanuts the vendors sell.
   c. Write a rule that gives the profit as a function of the number of bags of peanuts the vendors sell. Remember: Profit = Revenue – Costs.
   d. For each of the following, write a calculation, equation, or inequality that can be used to answer the question:
      i. What is the profit if 500 bags are sold?
      ii. What is the profit if 2000 bags are sold?
      iii. What number of bags of peanuts must be sold for concessions to break even?
      iv. What number of bags of peanuts must be sold for concessions to have at least $500 profit?
   e. On a given day, the manager starts the day with 2000 bags of peanuts. At the end of the day the manager notices that there are no bags of peanuts left. The profit, however, is different from your answer to part d.ii. Give several reasons why this could happen.

3 Tables of Linear Functions

When models of related variables are used, algebraic equations or inequalities often must be solved to find important or interesting questions. The search for solutions is often helped by use of technology-generated tables of data pairs, (input, output). At other times it is helpful to study a table of actual data to find a pattern relating changes in the variables.

The pattern of changes in tables of data is what distinguishes linear functions from nonlinear functions. As you work with the tables in this section, watch for patterns in how the related variables change.

SITUATION 3.1

In analyzing several business situations, you have seen cases where sales of a product depend on prices charged. Suppose market research has found that attendance at a concert will depend on price charged, and the function can be expressed using the rule

$$A(p) = 2500 - 175p.$$

To get a better picture of the relation between price and attendance for this concert, it is useful to produce a table of data pairs, (price, attendance), like the one below.

Price (dollars) p	Attendance $A(p)$	Price (dollars) p	Attendance $A(p)$
0	2500	8	1100
1	2325	9	925
2	2150	10	750
3	1975	11	575
4	1800	12	400
5	1625	13	225
6	1450	14	50
7	1275	15	-125

The function rule gives $A(15) = -125$, but that clearly can't happen in the actual situation! If we apply the function rule to large enough values of p, the expected attendance is negative. In this case the function rule describes the relation between price and attendance only for a limited range of input values. What do you think would happen in the actual situation if the output for a given price is negative?

The table of values, (price, attendance), gives answers to a variety of questions:

1. What is the attendance if ticket price is $10? *A(10) = 750. Answer: 750 people would attend.*

2. For what ticket price(s) is the attendance at least 1100? *2500 – 175p ≥ 1100 when p ≤ 8. Answer: Attendance is at least 1100 if the price is $8 or less.*

3. How many people will attend if admission is free? *A(0) = 2500. Answer: 2500 people will attend.*

You may have noticed that a linear function rule usually consists of two parts, a **constant term** (the number being added or subtracted) and what is called the **linear term** (here, the product of m and the variable x). If a function f has rule

$$f(x) = mx + b \text{ or, equivalently, } f(x) = b + mx$$

the constant term b always equals $f(0)$.

It is no coincidence that $A(0)$ has the same value as the constant term in the function formula $A(p) = 2500 - 175p$. This rule can be written in the form $A(p) = 2500 + (-175p)$, so the value of the linear term is $-175p$. This term is 0 when $p = 0$:

$$A(0) = 2500 - 175(0) = 2500 - 0 = 2500.$$

There are some interesting patterns in the table. For example, look at the differences between consecutive output values. Three of those differences have been calculated for you.

Price p (dollars)	Attendance $A(p)$	
0	2500	
1	2325	
2	2150	$2150 - 2325 = -175$
3	1975	
4	1800	$1800 - 1975 = -175$
5	1625	
6	1450	
7	1275	
8	1100	
9	925	
10	750	$750 - 925 = -175$
11	575	
12	400	
13	225	
14	50	
15	-125	

It looks as if each increase of $1 in price causes a drop of 175 in attendance. In other words, attendance drops at a rate of 175 per dollar increase in ticket price. You can check other entries in the table to test this observation.

Notice that –175, the rate of change in attendance, is the multiplier for p (called the **coefficient** of p) in the function rule $A(p) = 2500 - 175p$. Since –175 represents a rate of decrease (instead of a rate of increase), the coefficient of p in the function rule is negative (instead of positive).

To get a closer look at the relation between price and attendance, you can produce a refined table. For example, examine the table below.

Price	Attendance	Price	Attendance
4.00	1800.00	5.25	1581.25
4.25	1756.25	5.50	1537.50
4.50	1712.50	5.75	1493.75
4.75	1668.75	6.00	1450.00
5.00	1625.00		

Each increase of $0.25 in price predicts a drop of 43.75 in attendance. You know that attendance can occur only in whole numbers; the function rule is useful only as an estimate of what really happens.

From the information in these tables, would you expect a ticket price increase from $4 to $6 to cause the same drop in attendance as an increase from $6 to $8?

$$A(6) - A(4) = 1450 - 1800$$
$$= -350;$$

so the drop in attendance is 350.

$$A(8) - A(6) = 1100 - 1450$$
$$= -350;$$

so the drop in attendance is 350.

Is 350 the expected drop in attendance for *any* $2 increase in price? Checking a few more values confirms that pattern.

Attendance drops 175 for any $1 increase in ticket price and 350 for any $2 increase. The drop in attendance for a $2 increase is twice the drop for a $1 increase.

It is no accident that $350 \div 2 = 175$, the number that appears in the coefficient of p in the function rule, $A(p) = 2500 - 175p$.

When a rule is linear, it is easy to find the rate of change in output values by using a table. With a few easy calculations, the decrease of 175 persons per dollar increase in ticket price can be seen in many places in a table. As another example, the previous table shows that for every increase of $0.25 in ticket price the attendance drops by 43.75. Notice that

$$(43.75) \div (0.25) = 175.$$

This constant rate of increase or decrease is an important property of linear functions—a property that is not shared with other types of functions. Examine the table below for a nonlinear function.

x	$f(x) = 2 + 3x^2$
8	194
9	245
10	302
11	365
12	434

$302 - 245 = 57$

$434 - 365 = 69$

You can see that when x increases from 9 to 10, the increase in output value is 57, but when x increases from 11 to 12, the output increases by 69. *For a nonlinear function the rate of increase or decrease in output values is not constant.*

Exploration

As you work through the following situations, use a technological tool to produce tables that can help you answer the questions. Study the tables to see how the function rules relate to patterns you may observe.

SITUATION 3.2

The One Lap Around America is a 12,000-kilometer road rally in which cars try to travel at very precisely planned speeds. The object is not to go fastest, but to match a given elapsed time at various checkpoints all over the country. The driver controls the elapsed time by adjusting speed throughout the course. What matters most is the driver's average speed.

Suppose a driver starts out trying to average 85 kilometers per hour.

1. Copy and complete the following table giving desired distances at various times.

Time t (hr)	0	1	2	3	4	5	6	7	8	9
Distance $D(t)$ (km)	0	85								

2. At what rate does distance increase as time passes?

3. Write a rule that can be used to relate distance traveled $D(t)$ to time t.

4. Parts a through c refer to the following question:

   How long should the driver take to arrive at the rally checkpoint that is 335 kilometers from the start?
   a. Write an equation that can be used to answer this question.
   b. Use one or more tables to search for the solution to your equation. Locate a solution to the nearest 0.01 hour, then record three entries from the table that allowed you to locate the solution.
   c. Answer the question posed.

5. Parts a through c refer to the following question:

   At what time should the rally car arrive at a checkpoint 3500 kilometers from the start?
   a. Write an equation that can be used to answer this question.
   b. Use tables to search for the root of your equation. Locate a solution to the nearest 0.01 hour, then record three entries from the table that allowed you to locate the solution.
   c. Answer the question posed.

SITUATION 3.3

In Situation 3.1 you were given a rule relating attendance at a concert and the price charged for a ticket. For concerts in the winter, suppose that the rule

$$A_w(p) = 2700 - 225p$$

shows how attendance depends on ticket prices, and that the rule for concerts in the summer is

$$A_s(p) = 2700 - 150p.$$

Use tables to compare these two rules. Some technological tools allow you to make tables of several functions at the same time and display them side-by-side. If yours does not, you should record each table before producing another.

1. What summer attendance and winter attendance are expected if the ticket price is $7?

2. For what summer and winter ticket prices is the attendance zero?

3. For what ticket prices is summer attendance more than 1000?

4. For winter concerts, by how much does attendance change if the ticket price increases by $1? Compare that to the change in attendance if the ticket price increases $1 for summer concerts.

5. What are summer and winter attendances if tickets are free?

Exercises

Exercises 1, 2, and 3 give situations in which two variables are related. The relation is illustrated by data in tables. For each situation:

–determine whether the data fit a linear function;

–find a rule giving the relation between the variables, if you can; and

–write and answer three questions that might be of interest in the situation and can be answered using the table.

1. SITUATION Automobile advertising sometimes shows estimated annual costs of operating a car as a function of the distance driven. The table below gives typical data. Distance is in kilometers and cost in dollars.

Distance driven	0	1000	2000	3000	4000	10 000
Estimated cost	2000	2040	2080	2120	2160	2400

2. SITUATION As a submarine descends into the ocean, the pressure on its hull increases in roughly the following pattern. Pressure is measured in kilograms per square centimeter and depth in meters.

Depth	0	300	600	900	1200	1500
Pressure	0	32	64	96	128	160

3. SITUATION As the space shuttle climbs into the atmosphere, outside pressure on its hull decreases in roughly the following pattern. Altitude is in meters and pressure in millibars.

Altitude	0	3000	6000	9000	12 000
Pressure	1000	650	425	275	175

4. Create the following two tables so that one function is linear and the other is not.

   a.
Input	0	1	2	3	4	10	15	20
Output								

   b.
Input	0	1	2	3	4	100	200	300
Output								

5. Look back at the function rules used to describe the data in Situations 3.2 (distance traveled as a function of road rally time) and 3.3 (attendance at summer or winter concert as a function of the price of a ticket).

   a. How are all three function rules alike?
   b. How are these function rules different?
   c. How are the tables for these three function rules alike?
   d. How are the tables for these three function rules different?

4 Graphs of Linear Functions

For most students and users of algebra, a graph illustrates the relation between two variables more quickly and clearly than either a rule or table of values. In working with linear functions, it is helpful to be able to look at a function rule and draw a quick sketch of its graph or to look at the graph of a function and determine its rule. In this section, you will explore the relation between the graph for a linear function and its rule.

SITUATION 4.1

Two twins want to enter the summer lawn mowing business. They plan to buy a mower for $240. They hope to earn $12 per hour of work. Their profit in dollars $P(t)$ will be a function of time working in hours t with rule

$$P(t) = 12t - 240.$$

By now you know that the graph of data pairs, (input, output), for this function is a straight line. However, you have seen in earlier work that to draw by hand or to get technological help drawing a graph, several decisions must be made.

–Which variable should be represented by the horizontal axis and which by the vertical axis?

–What are reasonable values for the input and output variables?

–What scales on the axes are suitable in order to produce a clear picture of the interesting parts of the graph?

A graph of the twins' profit function is given at the right. A graphing tool calculates and plots each point shown. Since the function is linear, you need to locate only two points before drawing the expected line on a hand-drawn graph. However, it is a good idea to plot a few others to check your work.

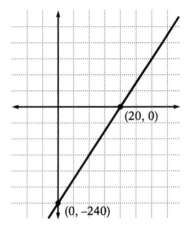

Notice that this graph includes data pairs, $(t, P(t))$, for some negative values of t. However, those points do not give any meaningful information about the problem situation—unless you can think of what negative working time might mean!

$P(t) = 12t - 240$

The horizontal (input) scale unit is 5 and the vertical (output) scale unit is 40.

Inspecting the graph answers a number of key business questions quickly. Some examples of these questions follow. You can apply the same kind of analysis that is used to answer these questions with the graph of any linear functions.

1. What is the twins' profit as they begin their business, before any lawns have been cut?

    The graph crosses the vertical axis at point $(0, -240)$. This shows that the twins start out $240 "in the hole", that is, their profit is -240, which commonly is reported as a *loss* of $240. This also can be found from the calculation:

$$P(0) = (12)(0) - 240$$
$$= -240.$$

Notice also that the point $(0, -240)$, at which the graph crosses the vertical axis, is related to the constant term, -240, in the function rule. You will find this fact useful later in this chapter when you are trying to determine a function rule by looking at its graph.

2. What is the twins' break-even point?

The graph crosses the horizontal axis at $(20, 0)$, which shows that the profit for the twins is $0 after 20 hours of work. This data pair also can be found by solving the equation

$$P(t) = 0 \text{ or } 12t - 240 = 0.$$

We can check the answer by evaluating $P(20)$.

$$P(20) = (12)(20) - 240$$
$$= 240 - 240$$
$$= 0.$$

3. For what numbers of hours worked will the twins be "in the hole", in other words, showing a loss?

The graph is below the horizontal axis for $t < 20$ and above the axis for $t > 20$. This means profit is negative until the twins work 20 hours. This corresponds to solving the inequality

$$P(t) < 0 \text{ or } 12t - 240 < 0.$$

4. How long must the twins work to have a profit of $300?

This question corresponds to the equation

$$P(t) = 300 \text{ or } 12t - 240 = 300.$$

Look on the diagram for the point where the profit graph reaches a height corresponding to $300. The value of t at that point is about 45. Verifying this estimate we find

$$P(45) = 12 \ (45) - 240$$
$$= 540 - 240$$
$$= 300.$$

5. At what rate does the twins' profit increase?

The profit line slants upward from left to right. This means that as time worked increases, profit increases too. More precisely, each increase of 5 hours in time worked produces an increase of $60 in profit. For example, using 35 and 30 as typical input values,

$$P(35) = 12(35) - 240$$
$$= 180 \text{ and}$$

$$P(30) = 12(30) - 240$$
$$= 120.$$

$$180 - 120 = 60$$

This means that profit increases at a rate of $60 per 5 hours, or 12 dollars per hour.

In the language of graphing, the rate of change in the function is called the **slope** of the line. Since slope is just rate of change, you can calculate slope by observing how much the output variable changes for any change of 1 in the input variable. In Example 5 above, profit increases at a rate of 12 dollars for every hour, so the slope is 12/1, or 12.

There is a visual way to calculate slope. The graph of the profit function is shown here.

To calculate the slope directly from the graph, first locate two points on the graph. Then determine the change in the value of the input variable (in this case the input variable increased by 5) and the change in the value of the output variable (in this case the output variable increased by 60).

The slope of this line is

$$\frac{\text{change in output variable}}{\text{change in input variable}} = \frac{60}{5}$$
$$= 12.$$

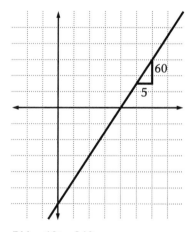

$P(t) = 12t - 240$

The horizontal (input) scale unit is 5 and the vertical (output) scale unit is 40.

This procedure for computing slope works for any two points on a line. Suppose, for example, you had chosen the points with t-values 35 and 50.

As shown in the diagram on the right, the value of the input variable increases by 15 and the value of the output variable increases by 180 (which is $P(50) - P(35) = 180$). The slope is

$$\frac{\text{change in output variable}}{\text{change in input variable}} = \frac{180}{15}$$
$$= 12.$$

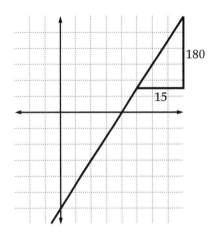

$P(t) = 12t - 240$

The horizontal (input) scale unit is 5 and the vertical (output) scale unit is 40.

In other words, an increase of $180 profit for any additional 15 hours worked is a rate of increase of 12 dollars per hour.

Exploration I

Using a graphing utility, it is easy to test the effects of a change in the profit function. In this exploration you are asked to consider several alternate options available to the two twins in their lawn-mowing project.

SITUATION 4.2

Suppose the twins decide to buy a used lawnmower that costs only $180.

1. The twins' profit in dollars is a function P_2 of t, the number of hours they work. Write a rule for $P_2(t)$.

    There is an easy way to find reasonable values for input and output variables in linear functions. To find reasonable values for the input variable in this case, ask yourself, "What is the least number of hours the twins work?" and "What is probably the greatest number of hours the twins work?" Then, to find reasonable values for the output variable, ask yourself, "For the least number of hours worked, what profit do the twins make?" and "For the greatest number of hours worked, what profit do the twins make?"

Estimate reasonable values for the input and output variables in this function. Using your estimates as a guide, produce a graph of this function.

For each of the following,
    a. write the calculation required or the equation or inequality to be solved;
    b. use the graph to find the answer, making estimates where necessary;
    c. where appropriate, give the coordinates of the point that provides the answer; and
    d. give the answer to the question.

2. What is the twins' profit before they start work?

3. What is the break-even point for this profit function?

4. How long do they work until they have a positive profit?

5. For what value of t is the profit $300?

6. Choose two points and calculate the slope of the graph of this profit function.

7. At what rate does the profit increase?

SITUATION 4.3

Next suppose that the twins get permission to use their parents' lawnmowers at no charge.

1. The profit in dollars is a function P_3 of t, the number of hours worked. Write a rule for P_3. Select a reasonable set of values for the input and output variables and graph the function. Use your graph to estimate answers for the following.

2. What is the twins' profit before they start work?

3. What is the break-even point for this profit function?

4. How long must the twins work in order to make enough money to purchase a stereo that costs $550?

5. What is the slope of the graph of this profit function?

6. What is the rate at which the profit increases?

SITUATION 4.4

Next suppose that the twins choose to buy a $270 lawnmower but keep their hourly charge at $12.

1. The profit in dollars is a function P_4 of t, the number of hours worked. Write a rule for P_4. Graph this function and use your graph to answer the following.

2. What is the twins' profit before they start work?

3. How many hours do the twins need to work in order to pay for the new lawnmower?

4. How long must the twins work in order to earn at least enough money for both of them to go to a summer sports camp (at $375 per person)?

5. What is the slope of the graph for this profit function?

   (Find the slope and check your answer by looking at two different pairs of points.)
   a. Write the coordinates for the first pair of points.
   b. Find the slope

   $$\left(\frac{\text{change in output variable}}{\text{change in input variable}} \right)$$

   using the first pair of points.
   c. Write the coordinates for the second pair of points.
   d. Find the slope using the second pair of points.
   e. Do your answers to parts b and d agree?

6. By how many dollars per hour does the profit increase?

SITUATION 4.5

Next suppose that the twins stick with their $270 lawnmower but raise their hourly charge to $15.

1. The profit in dollars is a function P_5 of t, the number of hours worked. Write a rule for P_5. Graph this function and use the graph to answer the following.

2. What is the twins' profit before they start work?

3. What is the break-even point for this profit function?

4. For what values of t is the profit positive?

5. What is the slope of this profit-function graph?

6. Compared with Situation 4.4, how many fewer hours do the twins have to work in order to earn enough money for their sports camp (at $375 per person)?

SITUATION 4.6

Next suppose that the twins stick with their $270 lawnmower but find it necessary to lower their hourly charge in order to attract business. Suppose they are willing to lower the rate as long as their break-even point does not require more than 30 hours of work.

1. Explore various choices to find one that meets this condition. Write your choice for a rule giving their profit P_6 as a function of time worked t.

Use a graph of $P_6(t)$ to answer these questions.

2. What is the profit for the twins before they start work?

3. For what values of t is the profit positive?

4. What is the slope of the graph of this profit function?

5. What is the rate at which the profit increases?

Making Connections

1. How are the function situations, rules, and graphs in Situations 4.2–4.6 similar? How are they different?

2. What connections do you see between the rules for these linear functions and the following aspects of their graphs?
   a. Slope
   b. Crossing point on vertical axis
   c. Crossing point on horizontal axis
   d. Rate of increase in the function

Exercises

1. SITUATION   Suppose the lawn-mowing twins buy a $400 lawnmower and charge $16 per hour of work.
    a. Write a rule giving profit as a function of time worked.
    b. Graph the function using scales of 5 and 100 on the respective axes.
    c. Label the startup, break-even, and $300 profit points with their coordinates.
    d. In addition to the price of the lawnmower and the number of hours worked, what other factors might affect the twins' profit?

2. SITUATION   Steve has a summer job painting houses for his mom. He is paid $7.25 per hour of work.
    a. Write a rule giving Steve's summer earnings as a function of the number of hours worked. Use t to represent time and $E(t)$ to represent his earnings.
    b. To graph this function, what decisions do you recommend on the following questions:
        i.   Are reasonable values for the input variable positive, negative, or both?
        ii. Are reasonable values for the output variable positive, negative, or both?
        iii. Which variable should be represented by the horizontal axis and which by the vertical axis?
        iv. What scale do you recommend on each axis, assuming 15 tic marks are available in any direction from the origin $(0, 0)$?
    c. Use your plan in part b to sketch a graph of $E(t)$.

3. SITUATION   Ingrid has a summer job as a waitress at the Terrapin Club. In a typical 8-hour day she earns $30 plus 15% of the

checks at her tables in tips. Her pay is a function P of the total checks c at her tables and can be expressed by the rule

$$P(c) = 0.15c + 30.$$

a. To graph this function, what decisions do you recommend on the following questions:

   i. Are the values of the input and output variables positive, negative, or both?

   ii. Which variable should be represented by the horizontal axis and which by the vertical axis?

   iii. What scale do you recommend on each axis, assuming 15 tic marks are available in any direction from the origin (0, 0)?

b. Use your plan in part a to sketch a graph of $P(c)$.

4. **SITUATION** The people in charge of hot dog sales at Nittany Lion football games have fixed costs for rental of space and equipment and for pay to vendors. They also have costs that depend on the number of hot dogs sold. Profit in dollars is a function P of number sold n with rule $P(n) = 0.50n - 450$.

Here is a graph of the profit function. For each of parts a through e, write a symbolic expression that can be used to answer the question, then use the given graph to find the answer.

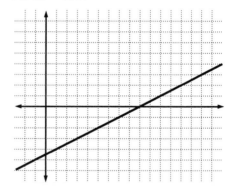

$P(n) = 0.5n - 450$

The horizontal (input) scale unit is 100 and the vertical (output) scale unit is 100.

a. How many hot dogs must be sold to break even?

b. How many hot dogs must be sold to reach $250 profit?

c. How many hot dogs are sold if the profit is at least $100?

d. What increase in profit occurs for an increase of 100 in the number of hot dogs sold?

e. What decrease in profit occurs corresponding to a decrease of 100 in the number of hot dogs sold?

f. What is the slope of this profit graph?

g. Which points on the graph tell something about the algebraic rule, but not about hot dog profits?

Exploration II

The examples in Exploration I all involved graphs of **increasing** linear functions, with positive slopes. Use a graphing utility to study the following examples of **decreasing** linear functions, which have negative slopes.

SITUATION 4.7

The basketball and wrestling coaches at a local high school are trying to increase interest in their games. They want good attendance, but must also bring in money from ticket sales to pay the bills. Suppose that the school statistics class did a market research survey that found the following information about interest in wrestling:

–If admission to the meet is free, attendance will average 750 people per meet.

–For each \$1 increase in the price of a ticket, average attendance probably will drop by 200 people per meet.

1. Wrestling attendance can be considered a function of ticket price. Using $W(p)$ to represent attendance when ticket price is p, write a rule for this function.

2. Graph this function in the first quadrant, using units of 0.50 and 150 on the horizontal and vertical axes, respectively. Make a rough sketch of the graph on your paper.

Use the graph to answer the following, estimating where necessary. In addition to the answer, include the calculation, equation, or inequality, as well as the graph points you use to find the answer.

3. If the wrestling ticket price is set at \$3, how many people attend?

4. For what wrestling ticket prices is average attendance at least 600 people?

5. What wrestling ticket price causes the attendance to be 0?

Suppose the statistics class finds that the demand function, B, relating attendance and ticket price for basketball games can be expressed as
$$B(p) = 1250 - 250p,$$
where $B(p)$ is the expected attendance at basketball games.

6. Graph this function on the same axes as that for wrestling. Use the graph to answer the questions that follow, estimating where necessary. Make a rough sketch of your graph for basketball attendance.

7. What basketball game ticket prices give an average attendance of at least 1000 people? Include the coordinates of any points you use to find your answers.

8. What basketball attendance occurs if admission is free? Include the coordinates of any points you use to find your answers.

9. What basketball ticket price causes the attendance to be 0? Include the coordinates of any points you use to find your answers.

Making Connections

Compare the two graphs of the demand functions for basketball and wrestling.

1. Which graph crosses the vertical axis at a higher point? What does that tell about demand for tickets?

2. Which graph crosses the horizontal axis farther to the right? What does that tell about demand for tickets?

3. Which graph is steeper? What does that tell about demand for tickets?

4. What connections do you see between the rules for these linear functions and the following aspects of their graphs?
   a. Slope
   b. Crossing point on vertical axis
   c. Crossing point on horizontal axis
   d. Rate of increase in the function

SITUATION 4.8

Suppose a submarine cruising 20 meters under the ocean surface dives at the rate of 25 meters per minute. Its position relative to the surface is a function of time in the dive. It is natural to indicate positions below sea level with negative numbers. When the submarine is at the surface, its position is 0; if it is 10 meters below the surface, its position is −10. Using this coordinate model, the submarine position at time t can be given by the rule

$$D_1(t) = -20 - 25t.$$

We use the subscript notation $D_1(t)$ to distinguish this function from similar functions you will use later.

Graph this function, choosing your viewing window so that it focuses on the fourth quadrant (positive input values along the horizontal axis and negative output values along the vertical axis) and choosing units of 1 and 50 on the horizontal and vertical axes, respectively. Use the graph to answer the following questions, estimating where necessary. Then do the following.

    a. Indicate any calculation, equation, or inequality you use.
    b. Give the data points that are important in finding the answer.
    c. Find the answer.

1. What is the submarine's position at the start of the dive?

2. When does the submarine reach a position of −200 meters?

3. For what values of t is the submarine more than 250 meters below the ocean surface?

4. Answer the following to help you find the slope of the graph for this submarine dive function.
    a. As t increases, does $D_1(t)$ increase or decreases?
    b. Each increase of 1 minute in time produces a change of how many meters in position?
    c. Position changes at a rate of how many meters per minute?
    d. What is the slope of the position graph?

SITUATION 4.9

Suppose another submarine begins a dive at the ocean surface and dives at the same rate, 25 meters per minute.

1. Let t represent minutes in the dive. Write a rule for the position of this submarine as a function D_2 of time.

Graph this function on the same fourth-quadrant diagram as the last function. Use the graph to do the following. You may need to estimate some of your answers.

    a. Give the coordinates for any points you need to answer the stated question.

    b. Answer the question.

2. What is the submarine's position at the start of the dive?

3. When does the submarine reach a position of –200 meters?

4. For what values of t is the submarine at a position below –250 meters?

5. What is the slope of this submarine dive graph?

SITUATION 4.10

Suppose a third submarine, cruising at 100 meters beneath the ocean surface, also begins a dive. Explore different graphs to find a rule for a dive rate that brings this submarine to a position of –200 meters after an eight-minute dive.

1. Let t represent minutes in the dive. Write a rule for the position function D_3 of meters.

2. What is the slope of the graph for this submarine dive function?

SITUATION 4.11

Suppose a fourth submarine, located 160 meters beneath the surface of the ocean, begins to rise toward the surface at the rate of 12 meters per minute.

1. **a.** Again using negative numbers to represent positions below sea level, write a rule expressing the submarine position $D_4(t)$ as a function of the time t since the sub began its rise.

    b. Graph this function.

2. Give the coordinates of the point where the graph intersects the output or vertical axis, and explain what this point tells about the motion of the submarine.

3. What is the position of the submarine eight minutes after it began to rise?

4. What is the slope of the graph?

5. When is the submarine 100 meters below the ocean surface?

6. Give the coordinates of the point where the graph crosses the input-axis, and explain what this point tells about the submarine's motion.

7. Find a positive number that is not an acceptable input for the function that models this submarine's motion.

Making Connections

Compare the lines that represent the graphs of the four functions of submarine positions.

1. Which line crosses the vertical axis at the highest point? What does that tell about the submarines?

2. Which line crosses the horizontal axis farthest to the right? What does that tell about the submarines?

3. Which of the four lines is steepest? What does that tell about the submarines?

4. What connections do you see between the rules for these linear functions and the following aspects of their graphs?
    a. Slope
    b. Crossing point on vertical axis
    c. Crossing point on horizontal axis
    d. Rate of increase in the function

Exercises

SITUATION

When the NASA space shuttle comes back to earth for landing, it descends in a long glide. Suppose you begin monitoring the shuttle altitude at a time when it is 100 kilometers (1 km = 1000 m) above the earth's surface and descending at a rate of 3500 meters per minute.

Altitude in meters is a function A of time t in minutes.

1. Write a rule for $A(t)$.

2. To graph this function, what decisions do you recommend on the following questions:
   a. In which quadrants do the points of interest lie?
   b. Which variable should be represented on the horizontal axis and which on the vertical axis?
   c. What unit do you recommend on each axis, assuming 15 tic marks are available in any direction from the origin, $(0, 0)$?

3. Use your plan in Exercise 2 to sketch a graph of $A(t)$.

4. For each of the following questions, write the question in symbolic form and then use your graph to find the answer.
   a. At what time does the shuttle reach an altitude of about 5000 meters?
   b. At what times is the shuttle within 3000 meters of the earth's surface?
   c. For each 5 minutes of time in flight, how much does the shuttle altitude decrease?
   d. When will the shuttle reach the ground? Why might this prediction not match the actual event exactly?

5. What is the slope of the graph of $A(t)$?

5 Solving Equations and Inequalities

In situations that involve linear functions, many questions call for the solution of equations or inequalities. You have seen how tables, graphs, and technological tools help—especially when there are complicated function rules involved or many related equations and inequalities to solve.

In some cases the situation involves only a single equation or inequality with only a few operations and manageable numbers. In those cases it is usually easiest to solve the equation or inequality mentally. Sometimes, you may prefer to use a calculator (without programming or graphing) for more complicated arithmetic and your paper and pencil to record your thinking.

SITUATION 5.1

High school students often have part time jobs at fast-food restaurants. For example, at the Pierside Crab Carryout students typically earn $6 per hour, after taxes have been taken out.

Suppose an employee wants to know how many hours it will take to earn $111 for a new portable stereo tape deck.

The pay function can be expressed by the rule $P(t) = 6t$, where t is the number of hours worked and $P(t)$ is in dollars. The question can be answered by solving the equation $111 = 6t$. To solve it you might reason as follows:

–To calculate earnings for a given number of hours worked, we know that we multiply this number of hours by 6, because 6 is the number of dollars earned per hour.

–To find how much time must be spent to get a specified $111, we must find the number of hours that, when multiplied by 6 dollars per hour, gives us $111. In this case, we can think of dividing $111 into 6-dollar parts. In other words, we need to "undo" the multiplication by dividing the 111-dollar earnings by the 6-dollars-per-hour rate.

–We can express the solution to this problem as: the time required to earn $111 is

$$111 \div 6 = 18\frac{1}{2} \text{ hours.}$$

Or, we might express the answer in another way: 18.5 hours is the time needed to earn $111.

–In the diagram shown here, we can see that the result we calculated by hand agrees with that found by using a graph for the pay function at the Pierside Crab Carryout. Do you see how $t = 18.5$ is graphically linked with $P(t) = 111$?

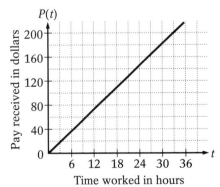

–If we examine a table of this pay function, like the table below, we should find agreement between the tabular entries and the hand-calculated value, namely, $P(t) = 111$ when $t = 18.5$. Can you see how our answer of 18.5 hours fits into this table?

t hours	$P(t)$ dollars	t hours	$P(t)$ dollars
0	0	12	72
2	12	14	84
4	24	16	96
6	36	18	108
8	48	20	120
10	60	22	132

SITUATION 5.2

At the Chicken Coop, a similar fast food job pays $7 per hour after taxes, but each worker must first spend $72.40 for a uniform and shoes.

How many hours must one work at this job to earn $111 for a tape deck?

The earnings function C can be expressed by the rule $C(t) = 7t - 72.40$, where t is the number of hours worked, and $C(t)$ is in dollars. The question can be answered by solving the equation $111 = 7t - 72.40$. To solve it you might reason as follows:

–To calculate earnings for given time, we multiply the number of hours by 7 dollars per hour and then subtract 72.40 dollars for the uniform and shoes.

–To calculate time worked from a specified earnings goal, you must reverse your thinking. To have $111 in total earnings, an employee must work long enough to have total earnings equal to $111 for the tape deck plus $72.40 for the uniform and shoes. Divide this total by the 7-dollars-per-hour pay rate to determine the number of hours to be worked.

–The process to undo the operations then must be: add 72.40 to 111 and then divide the result by 7.

–The time required to earn $111 at the Chicken Coop is

$$(111 + 72.40) \div 7 = 26\frac{1}{5}$$

$$= 26.2 \text{ hours.}$$

SITUATION 5.3

There are 50,000 seats in a college's football stadium available for season tickets and every year these seats are sold out. Besides the money from season tickets, the Athletic Office at the college takes in an average of 1.27 million dollars per year from other sources (mainly advertisers and radio and television networks).

If the college's Athletic Office took in $7,870,000 last year, what was the average cost of a season ticket? The total revenue is a function of the cost of a single season ticket t can be expressed by the rule

$$R(t) = 50\ 000\ t + 1\ 270\ 000.$$

To answer the question, we can solve the equation

$$R(t) = 7\ 870\ 000 \text{ or}$$
$$50\ 000\ t + 1\ 270\ 000 = 7\ 870\ 000.$$

–To calculate the total income, we multiply the cost of a season ticket by 50 000 and then add 1 270 000 to the result.

–To calculate the cost of a single season ticket when we are given the total income, we subtract 1 270 000, which is the number of dollars coming from sources other than season ticket sales; and then we divide the result by 50 000, which is the number of season tickets sold.

Briefly, to undo the operations in the function rule, we first subtract 1 270 000 and then divide by 50 000.

–The cost of a single season ticket is

$$(7\ 870\ 000 - 1\ 270\ 000) \div 50\ 000 = 132 \text{ dollars.}$$

SITUATION 5.4

The East Valley Soccer Club, which has a $22 annual membership fee, charges each member $3 per month to help pay the referees for the league. During the winter, practices are held in a local gym, which charges $5.75 for admittance to each practice.

Kathy wants to keep her total annual costs under $130. How many winter practices can she attend?

The total annual cost of membership in the club is a function C of the number of winter practices attended p and can be expressed by the rule

$$C(p) = 22 + (3)(12) + 5.75p \text{ or}$$
$$C(p) = 58 + 5.75p.$$

The question can be answered by solving the inequality

$$C(p) < 130 \text{ or}$$
$$58 + 5.75p < 130.$$

–In order to find the annual cost for a given number of winter practices, we multiply the number of practices attended by 5.75 and add the product to 58.

–In order to find the number of practices attended when the total annual cost of membership is known, we subtract 58, the number of dollars a member must pay, and then divide the difference by 5.75, the number of dollars charged per practice.

Briefly, we undo the operations shown in the function rule by first subtracting 58 and then dividing by 5.75.

–Since $(130 - 58) \div 5.75 \approx 12.5$, we see that Kathy could attend 12 winter practices and keep her total membership cost under \$130. Since a smaller number of practices would cost even less, the answer to the question is that Kathy can attend no more than 12 winter practices. That is, $p \leq 12$.

Remember: It is often easiest to solve an equation like those above mentally, using a calculator (but no programming or graphing) to do some of the computations. Using more advanced technology to draw a graph or list a table can take longer in these cases.

Exploration

In answering the following questions, you will be working with linear functions. As usual, you might want to have a calculator on hand to help with some of the arithmetic. For each of the following situations, follow these steps:

–Write a calculation, equation, or inequality that can be used to answer the given question.

–Solve the equation or inequality using reasoning like that illustrated in the previous examples.

–Write short sentences to briefly describe the steps in your thinking.

–Write your answer in a complete sentence.

When you have answered the questions, look for general procedures and patterns that seem to fit all the types of linear equations and inequalities. You may want to use technology to experiment with tables and graphs for each situation to see how these methods give results that are consistent with what you answered.

1. If Cedric earns $3.75 per hour for babysitting, how long does it take him to reach $50 in total earnings?

2. The Video Connection charges $25 for annual membership and $3.50 per day for video cassette rentals.
   a. What is the cost of renting cassettes for 30 days during one year?
   b. Suppose that this year you plan to spend $100 renting video cassettes. For how many days this year can you rent a cassette from Video Connection?

3. An airplane cruising at altitude of 5000 meters begins to descend at a rate of 250 meters per minute.
   a. What is the plane's altitude after 5 minutes?
   b. When does the plane reach an altitude of 1000 meters?

4. A parking lot at an ocean beach charges $6.75 for the first day and $3.50 for each day after the first.
   a. What is the cost of parking for 6 days?
   b. How long can a vacationing family park at this lot if they want to keep the parking cost under $50?

Exercises

1. In each of the following situations, write a sentence or phrase stating the sequence of arithmetic operations needed to calculate the quantity described in the situation.
   a. The total pay for a babysitter who charges $2.75 per hour.
      Sample answer: Multiply number of hours worked by 2.75.

b. The profit of students who enter the lawn-mowing business by spending $200 for a power mower and then earning $15 per lawn cut.

c. The earnings of a peanut vendor at football games when the vendor is paid $15 per game plus $0.25 per bag sold.

d. The position (where meters below sea level are reported as negative numbers) of a submarine that dives from the ocean surface at a constant rate of 25 meters per minute.

e. The position of a submarine that starts from a depth of 100 meters under the ocean surface and dives at a constant rate of 25 meters per minute.

2. What does the graph for the submarine situation described in Exercise 1e look like? What might the graph look like if the diving rate were not constantly 25 meters per minute but were a variable rate instead?

3. In each of the following situations write an equation or inequality that can be used to answer the given question. Then write a sequence of arithmetic operations that solve the equation or inequality and thus answer the given question.

a. If a babysitter charges $2.75 per hour, how long must she work to earn $50?

Sample answer: 50 = 2.75 h (h is the number of hours worked)

$$50 \div 2.75 \approx 18.2 \ hours$$

b. If two students enter the lawn-mowing business by spending $200 for a power mower and then earn $15 per lawn cut, how many lawns do they mow if their profit is at least $350?

c. If a peanut vendor is paid $15 per game plus $0.25 per bag sold, how many bags must be sold to earn at least $35?

d. If a submarine dives from the ocean surface at a rate of 25 meters per minute, how much time has passed when its position is –350 meters?

e. If a submarine dives from a starting depth of 100 meters below the ocean surface at a rate of 25 meters per minute, how much time passes before its position is –350 meters?

4. Solve each of the following equations. For at least two of the problems, write an explanation of your reasoning. (For part a, two sample responses are given.)

a. $5x + 11 = 41$
Sample answer:

$$(41 - 11) \div 5 = 30 \div 5$$
$$= 6.$$

If 41 is 11 more than 5x, then 5x must be equal to 30.
If the product of 5 and x is 30, then x must be 6.
or
The given equation states that first multiplying x by 5 and then
adding 11 gives the result 41. Reversing the operations, we
start with 41, subtract 11, and finally divide by 5.

b. $5x - 11 = 44$	c. $11 - 5x = 44$	d. $0.5x + 7 = 8$
e. $0.5x - 10 = 23$	f. $7x - 4 = 24$	g. $-3x + 5 = 20$
h. $21 - 2x = 11$	i. $x - 25 = 70$	

5. Solve each of the following inequalities. For at least two of the
inequalities, explain your reasoning.

a. $3x + 5 < 29$	b. $3x + 5 \geq 80$	c. $3x - 5 > 21$
d. $2x - 8 > 18$	e. $12 + 5t \leq 72$	f. $12 - 5t < 72$

6 Linear Patterns in Data

The linear functions and equations you have studied thus far in this
course have all appeared in an easy-to-use form. In some problems,
given facts led to a rule of the form $f(x) = mx + b$. In others, the function
rule could be found from a table or graph that fits a precise linear pattern. Once the rule was known, it was an easy task to answer questions
or make predictions about the relation.

In many practical situations, the starting point for study of a relation
between variables is data collected from a scientific experiment, a business market survey, records of a health survey, or some other set of data
pairs, (input, output). While patterns in these data often *suggest* a familiar type of rule, the individual data points seldom fit a simple rule exactly. However, it may still be possible to find a useful rule that matches
the pattern.

SITUATION 6.1

A skydiver who jumps from a plane gains speed very quickly while falling. However, when the parachute opens, speed slows for a gentle drift to landing. Consider the function which relates altitude to time after the parachute opens.

For one jump, altimeter readings were made every ten seconds with the initial reading taken when the parachute opened. The data are presented in the following table and graph:

Time (sec)	Altitude (m)
0	400
10	330
20	280
30	220
40	180
50	150
60	100
70	60
80	10

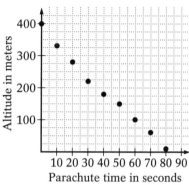

There are scientific principles that describe the way an *ideal* parachute flight might go, but the real world is more complex than the conditions for which those principles make predictions. In situations such as this, the recorded data often do not fit a linear rule exactly. The rate at which altitude changes over time is not constant, and the graph points do not lie on a straight line. However, it seems likely that a linear function rule can model the pattern in the data very well. The only problem is finding the rule that is the best fitting model.

6.1   Linear Function Rules to Model Data Patterns

There are at least three common and sensible strategies for finding a function rule to match linear data. Each helps find numbers m and b so that the data pairs, (input, output), for a linear function $f(x) = mx + b$ are very close to those of the experimental data.

Strategy I: Studying Numerical Patterns In the rule $f(x) = mx + b$ for a linear function, the coefficient m tells the steady rate at which outputs change as inputs increase. The constant term b tells the function output when input is 0. Looking again at the table of data, (time, altitude), in the parachuting situation, we get some strong clues to values of both m and b.

Time (sec)	Altitude (m)	Rate of change
0	400	$(330 - 400) \div 10 = -7$
10	330	$(280 - 330) \div 10 = -5$
20	280	$(220 - 280) \div 10 = -6$
30	220	$(180 - 220) \div 10 = -4$
40	180	$(150 - 180) \div 10 = -3$
50	150	$(100 - 150) \div 10 = -5$
60	100	$(60 - 100) \div 10 = -4$
70	60	$(10 - 60) \div 10 = -5$
80	10	

If we use the letter t to indicate the input variable time and $A(t)$ to indicate altitude at time t, the first line in the table shows that $A(0) = 400$. Thus it seems reasonable that the constant term b should equal 400.

The rate of change in the output variable, altitude, is not constant, but the tabular values, varying from –3 to –7, are moderately close to one another. The average rate-of-change is about –4.9 meters per second, which suggests using $m = -4.9$ or, rounding off for simplicity, $m = -5$. With this choice, the function rule is $A(t) = -5t + 400$.

Strategy II: Matching the Graphical Pattern A second practical strategy for finding the rule of a good-fitting linear function is to draw a line that represents the pattern you see in the data. Then find a rule describing the line you have drawn by observing the slope and the output-axis intercept of the line. This strategy is illustrated in the following sketch.

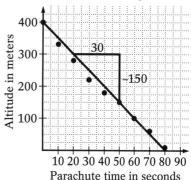

The line is drawn roughly along the pattern of data points—about half of the points are above the line and half below it. The line intersects the output-axis at 400. It has a slope of –5, since the points (20, 300) and (50, 150) appear to lie on the line shown in the sketch above and

$$\frac{(150 - 300)}{(50 - 20)} = \frac{-150}{30}$$
$$= -5.$$

Thus, again using t to represent time in seconds and $A(t)$ to represent altitude in meters, the rule for the function with that line as its graph is $A(t) = -5t + 400$.

Strategy III: Using Computer Function-Fitting Tools Finding function rules that model imperfect patterns in data is a very common task. Thus scientists and mathematicians have devised a variety of technological tools to help find the rules that best fit given data sets. For most such tools you only need to supply the data pairs and then enter a command or keystroke instructing the tool to create a linear function model. The tool then applies one of several possible procedures to find a good fit. Later in the section you will be given the opportunity to use this strategy—after you have gained more experience with Strategies I and II.

Exploration

Function-fitting techniques are often used in situations involving numerical data. The following exploration gives you the opportunity to find and use linear function models for patterns in data from a variety of problem situations. As you work on these situations, keep a sharp eye out for the limitations of prediction from such models.

SITUATION 6.2

Suppose another skydiver, using a different type of parachute from the one used in Situation 6.1, reports the following data, (time, altitude), from a jump.

Time (seconds)	0	10	20	30	40	50	60	70
Altitude (meters)	500	420	350	300	230	160	100	20

1. Plot these data points on a diagram like the one shown here. Then draw a line that you believe fits the pattern in the data well. Compare your line with the lines drawn by other students.

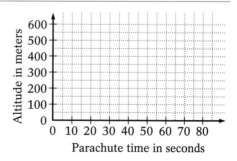

Parachute time in seconds

2. Using the given data, (time, altitude), or the line you drew on the graph, find the rule for a linear function A that you believe models the pattern in the data. Give your function rule in this form:

$$A(t) = \underline{\quad\quad} \, t + \underline{\quad\quad}.$$

Compare the rule you have chosen with the rules found by other students. Is it possible for two students to have difficult rules that are both acceptable?

3. According to your linear function rule, what is the altitude at which the parachute opens? What is the average speed for the rest of the jump?

4. Copy and complete this table comparing your model to actual data.

Time (sec)	Altitude (m)	Model's predicted altitude $A(t) = \underline{\quad}t + \underline{\quad}$	Actual rate of change	Model's rate of change
0	500			
10	420		$-80 \div 10 = -8$	
20	350			
30	300			
40	230			
50	160			
60	100			
70	20			

5. What are some factors that might cause the actual flight of a parachutist to differ from the linear function rule that you found to model the data pattern?

6. Predict the skydiver's altitude at 15 seconds.

7. Predict the skydiver's altitude at 35 seconds.

8. Predict the time when the skydiver reaches the ground. Explain your answer.

9. In Situation 6.1 the linear function model for a parachutist's jump after the chute opens predicts an average speed of about 5 meters per second. What could cause the actual speed to be different?

SITUATION 6.3

Ocean-going power boats often compete in long-distance races. One race of about 200 kilometers goes from Miami to Nassau.
Suppose that every 30 minutes the distance one boat has traveled is recorded by the driver. The data are given in the following table.

Time (minutes)	0	30	60	90	120	150
Distance (kilometers)	0	45	80	120	155	195

1. Plot these data on a diagram like the one shown here. Draw what you feel is the best fitting line for the data, then find a rule for that line.

2. Using the rule for your function, answer the following questions. For each question, show how you arrive at your answer by writing the calculation, equation, or inequality that you used.

   a. If this boat continues racing beyond the 150 minutes for which data are given, what distance probably would be reported when the time is 180 minutes?

   b. About how long would it take this boat to travel 300 kilometers?

   c. According to your rule for the best-fitting linear function, what is the average speed of this boat?

   d. Why do you suppose the boat does not report constant speed equal to your answer in part c?

We mentioned earlier in this section that there are technological tools to help in the analysis of patterns in data from experiments. The questions following the next situation give directions to help you use a function-fitting tool.

SITUATION 6.4

Summer weather brings heavy clouds and thunderstorms on many late afternoons. As warm, moist air rises, it cools. When the air has cooled to the condensation temperature, water drops form. The data below were recorded by a weather balloon released on a warm day.

Altitude (m)	Temperature (°Celsius)	Altitude (m)	Temperature (°Celsius)
0	32.0	2000	14.5
500	27.0	2500	9.0
1000	23.0	3000	3.5
1500	18.0	3500	–3.0

1. Enter the given data pairs, (0, 32), (500, 27), *etc.*, into a function-fitting tool. Then enter the appropriate commands or keystrokes to get a plot of the data. Try to see the pattern into which the data fall.

2. Now enter the proper commands or keystrokes so that a linear function is fitted to the data. Record the rule relating temperature $f(x)$ to altitude x, rounding the coefficient of x and the constant term to two decimal places.

   Explain what the slope of the graph and the constant term of the rule tell about the temperature as it is related to altitude.

3. Compare the plot of the data to the fitted function. How well does the rule match the experimental data?

4. Make the following predictions:
   a. The temperature at 5,000 meters altitude.
   b. The temperature at 10,000 meters altitude.
   c. The altitude at which temperature is 0°C, the freezing temperature of water.

5. Can you suggest reasons why the data, (altitude, temperature), do not exactly fit a linear model?

SITUATION 6.5

A student government association is considering sale of special T-shirts as part of a spirit-week campaign. They do some market research to see what price students would be willing to pay, and they make the following estimates.

Price per shirt (dollars)	5.00	7.50	10.00	12.50
Estimated sales	725	500	225	50

1. Plot these data on a diagram like the one shown here. Draw what you feel is the best fitting line for the data.

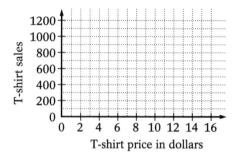

   a. Find a rule for the line that you have drawn.

   b. Find a linear function rule by using a function-fitting tool.

2. Using either rule you have found, predict sales for other possible prices. For each of the following questions, show how you arrive at your answer by writing the calculation, equation, or inequality that you used.

   a. What are sales likely to be if the price is set at $6 per shirt?

   b. What are sales likely to be if the price is set at $9?

   c. For what prices do you expect sales to be at least 400?

   d. What output corresponds to the input 0, and what does it tell about T-shirt sales prospects?

   e. What is the coefficient of the input variable in your function rule and what does it tell about the relation between the price and the sales of T-shirts?

   f. Why did you choose the function rule that you chose for answering the questions in parts a through e, instead of using the other that was available?

SITUATION 6.6

Inflation refers to the rate at which wages and prices go up each year. The salaries and prizes for athletes are an interesting measure of American inflation. The following tables show the winner's speed and the prizes at the Indianapolis 500 Auto Race for selected years since 1953.

Year	Winner	Winner's speed (mph)	Year	Prize (dollars)
1953	Bill Vukovich	129	1953	246 000
1957	Sam Hanks	136	1957	300 000
1961	A.J. Foyt	139	1961	400 000
1965	Jim Clark	151	1965	628 000
1969	Mario Andretti	157	1969	805 000
1973	Gordon Johncock	159	1973	1 012 000
1977	A.J. Foyt	161	1977	1 117 000
1981	Bobby Unser	139	1981	1 609 000
1985	Danny Sullivan	153	1985	3 271 025
1989	Emerson Fittipaldi	168	1989	5 723 725
1993	Emerson Fittipaldi	157	1993	7 681 300

Source: *The World Almanac and Book of Facts.* 1994.

Source: *The Indianapolis Hall of Fame Museum.*

Use these data to answer the following.

1. Use one of your strategies to find rules for a function, *s*, relating year to winning speed and for a function, *p*, relating year to winning prize money.

2. Now use those rules to make predictions for the following years.
   a. Predicted speed in 1987.
   b. Predicted prize in 1987.
   c. Predicted speed in 2000.
   d. Predicted prize in 2000.

3. In 1987, the winner's actual speed was 162 mph, and the actual prize was $4,490,375. What factors do you believe might cause the actual race time or race prize for any particular year to differ from the predictions modeled by your rule?

4. Compare your confidence in the predictions you made for 1987 with the predictions you made for the year 2000. Explain.

The situations you have worked on in this section are only a few of the cases in which it is helpful to fit a linear-function model to the pattern in data, (input, output), and to use the rule to answer questions about the relationship. Whenever you apply the rule, you must remember that it is merely a model of the actual situation. Predictions based on the rule may be reasonable and helpful—but not always. Be cautious!

Exercises

In each of the following situations:

–graph the given data;
–draw a line that you feel fits the data well;
–find a rule that fits the prediction line; and then
–answer the questions using your prediction line.

1. SITUATION Stores buy items for much lower prices than they charge their customers. Consider the following sample data from a shoe store.

Price paid by store x (dollars)	5	7	10	15
Price charged by store $f(x)$ (dollars)	15	19	28	42

Based on these data, predict answers to the following questions.
   a. What might one expect the store to charge if it pays $12 for an item?
   b. What might one expect the store to charge if it pays $20 for an item?
   c. If the store charges $50 for an item, what did it probably pay for that item?

2. SITUATION When a car's driver sees trouble ahead, the car travels some distance before the driver can even touch the brakes. The data below show the average "reaction distance" as a function of the car's speed.

Speed of car s (kilometers per hour)	15	50	80	110
Reaction distance $r(s)$ (meters)	3	10	17	25

Based on these experimental data, predict answers to the following questions.
   a. What is the reaction distance if speed is 25 km/hr?
   b. What is the reaction distance if speed is 70 km/hr?
   c. What is the reaction distance if speed is 100 km/hr?
   d. If the reaction distance is 30 meters, what is the speed?
   e. What is the rate at which reaction distance changes as speed increases?

3. **SITUATION** For many companies, the sales of their products are functions of the prices they charge. Suppose that one company, which makes satellite dish antennas for home TV reception, has a new product. According to the market, the following pairs, (price, sales), are likely.

Price charged (dollars)	995	600	450	380
Number sold	100	750	1500	2500

   a. If the price is $750, what are the expected sales?
   b. If the price is $500, what are the expected sales?
   c. What are the possible prices that can be charged if sales of at least 2000 are desired?

4. **SITUATION** The world's death rate is lower than the birth rate, which means that the world's population is growing. The following table shows estimated populations for various years.

Year	1975	1980	1983	1986	1991
World population (billions)	4.1	4.5	4.7	4.9	5.4

   Source: *World Population by Country and Region.* Washington, D. C. U. S. Government. 1993.

   Use a graph of these data to predict each of the following. It may be helpful if you choose the year 1975 as $t = 0$.
   a. World population in 2000
   b. World population in the year 2010
   c. The year when world population might reach 8.0 billion

5. **SITUATION** The Olympic Games is the most famous sports event in the world. As training methods and competition improves, the winning times in Olympic events improve also. Given below are data showing some winning times in the women's 100-meter race.

Year	Athlete	Time
1948	Francina Blankers-Koen, Netherlands	11.9
1952	Marjorie Jackson, Australia	11.5
1956	Betty Cuthbert, Australia	11.5
1964	Wyomia Tyus, United States	11.4
1968	Wyomia Tyus, United States	11.0
1972	Renate Stecher, East Germany	11.07
1976	Annegret Richter, West Germany	11.08
1980	Lyudmila Kondratyeva, USSR	⎯⎯
1984	Evelyn Ashford, United States	10.97
1988	Florence Griffith-Joyner, United States	10.54
1992	Gail Devers, United States	10.82

Source: *The World Almanac and Book of Facts.* 1994.

Use a graph of these data to solve the following problems. To get a suitably accurate graph you may wish to choose 1928 as year 0 and a unit of 4 years on the horizontal axis. A time scale from 10.0 to 12.5 in units of 0.1 seconds on the vertical axis also may be helpful.

a. Estimate the winning time for Wilma Rudolph in 1960. (The actual winning time was 11.0 seconds.)

b. Predict the winning time in the year 2000.

c. If the Olympics had been held in 1940 and 1944, what might have been the winning times? (Do you know why the Games were not held in those years?)

d. Estimate the winning time in 1980. (The actual winning time was 11.06 seconds.)

6. For each of the questions of Exercises 1 through 5, discuss several reasons why the predictions you made might be different from what really happens.

Making Connections

Suppose you run an experiment or collect data about the relation between two variables. Explain how you would find the answers to the following questions.

1. Does a linear function or equation seem right as a model of the relation?

2. What predictions can be safely made from the data or rule?

6.2 Evaluating the Goodness of Fit for a Model

Once you have found a rule for a function that fits the pattern in a list of data, it is tempting to forget that the rule is *only a model* of the real situation. Predictions made from the rule are likely to be good if the rule is a good model. In some cases, actual events turn out differently from the model's predictions.

It is a good idea to check the accuracy of fit for any model you find. This is especially true when you use a technological tool to find the rule, because most such tools give some rule—even if the "best possible" rule is not very good. For instance, one function-fitting tool found the rule $f(x) = -0.4x + 6.1$ as the best-fitting linear model for the following data.

Input	0	1	2	3
Output	5	9	2	6

If you compare these data pairs with the pairs $(x, f(x))$ from the rule, you will see that they make a rather poor match.

Two basic ways to decide on the accuracy of a function fit are given here.

Eyeball Test The first way is to graph the data points and the fitted function on the same diagram and to compare their patterns. In the case of the given data and $f(x) = -0.4x + 6.1$, the picture is given here. The graph of the linear function does not seem to fit the plotted points particularly closely, but there certainly are lots of other lines that are much worse. Indeed, the trend in the data—up, then down, then up again—leads us to wonder whether *any* linear model is suitable for these data.

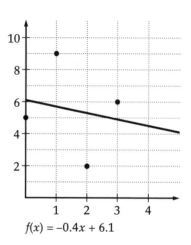

$f(x) = -0.4x + 6.1$

Measuring Goodness of Fit There are several ways to **measure** the accuracy of a model. The basic idea is to see how much the data outputs and the function outputs differ. The following table illustrates one way to compare the given data to the rule $f(x) = -0.4x + 6.1$.

Input	Output	Model's output $f(x) = -0.4x + 6.1$	Difference of outputs Model – Actual	Error |Model – Actual|
0	5	6.1	1.1	1.1
1	9	5.7	– 3.3	3.3
2	2	5.3	3.3	3.3
3	6	4.9	– 1.1	1.1

In this comparison, we are concerned with how different the model's outputs are from the actual data. However, we are *not* concerned with whether the model's output is greater or less than the actual data, so we've added the final column. A number in this column is the *absolute value* of the number in the previous column. That is, it is the difference of the outputs without the sign (+ or –). For instance, for the input 1, the data output is 9 and the modeling rule has output 5.7. The function rule is different from the data output by

$$\text{Model} - \text{Actual} = 5.7 - 9$$
$$= -3.3.$$

The negative sign shows that the model's output is less than the actual data, but we only need to know that the model missed the actual data by 3.3 units.

The differences without the sign are sometimes called **errors**. The average of all such errors for this model and data set is 2.2. That means that, on average, the rule output differs from the actual output by about 2.2 units.

When you are choosing a linear model to fit a given data set, you want the line to be *close* to the plotted points on the graph. The closer, the better! In other words, you want the differences or errors to be small numbers. Rather than looking at the individual differences, the above method for measuring goodness of fit looks at the average absolute value of the differences; we want the average to be *small*—the smaller, the better!

When you looked at the diagram in the first method, you may have noticed that not one of the plotted points lies on the line. Perhaps you believed that a different line might fit the points better. You might have chosen the line shown in this diagram, where the plotted points are the same as before.

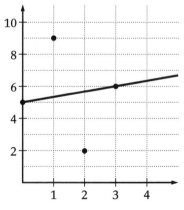

Notice that two plotted points lie on the new line and the other two points may be a little farther away from the new line than from the old line.

Which of these two lines is a better fit? Why do you think so? Discuss your ideas with your classmates.

Since the new line passes through the two points $(0, 5)$ and $(3, 6)$, its intercept on the output-axis is 5 and its slope is

$$\frac{6-5}{3-0} = \frac{1}{3}.$$

A function rule for the new line is

$$g(x) = \frac{1}{3}x + 5.$$

The calculations needed to measure goodness of fit are summarized in the following table, where numerical entries appear with two decimal places. (Here, we use the rule $g(x) = 0.33x + 5$ instead of using fractions.)

Input	Output	Model's output	|Model – Actual|
0	5	5.00	0.00
1	9	5.33	3.67
2	2	5.66	3.66
3	6	5.99	0.01

Average difference: 1.84

The average 1.84 obtained from the function g is considerably smaller than the average 2.2 from the function f. According to this method of measuring goodness of fit, the second choice with rule $g(x) = 0.33x + 5$ seems to be better than the first with rule $f(x) = -0.4x + 6.1$.

Why, then, did the function-fitting tool find $f(x)$ as the best fit for the data? As we mentioned earlier, there are several ways to measure how good a model is. Each of these methods has its advantages and its disadvantages. Two different methods may or may not agree on what model is best.

Most technological tools use a method that is preferred by statisticians. This procedure is more complicated than the method described above, and has several strong advantages in its favor. Because it involves more advanced ideas, we shall not study it in detail now. Of course, no matter what method has been programmed into your technological tool, you should free to use it as another way to find function rules.

The next exercises give you some practice in measuring goodness of fit by applying the second method described in this section. While it is not important that you become skillful at using this method, it is important to keep in mind that some models are better fits than others. Some models are not very good at all and perhaps should not be used!

Exercises

For each of the following tables:
    a. plot the given data points on a coordinate diagram, with suitable scales on the axes;
    b. draw a line that you believe fits the pattern in the data points as well as possible;
    c. calculate the average absolute value of the difference between the outputs for your model and the data; and
    d. write your opinion about how well the model fits the data.

1.

Input	0	1	2	3	4
Output	0	2.1	3.8	6.3	8.4

2.

Input	0	1	2	3	4
Output	0	1	4	9	16

3.

Input	−4	−1	0	2	5
Output	-3	1	2	3	5

4.

Input	−5	−2	1	4	7
Output	4.5	2	0.6	−0.9	−3

5.

Input	1	2	3	4
Output	7	5	4	4

6. To compare your function-fitting to a technological routine, enter each data set given in the above exercises and find the respective rule produced by your tool. Then describe and try to explain any differences between your results and those of the technological aid.

 ## Exploration

One of the most common and useful tools is the coil spring. From closing doors to powering clocks and cushioning cars, the elasticity of springs is an important, often unseen, convenience in our lives. In designing an apparatus that uses springs, the key factor is the relation between the force applied and the stretch (or compression) distance. The following experiment shows how that relation can be discovered and written with an algebraic formula. Although this experiment can be done on your own, you may find it helpful to work with a classmate or two.

For the experiment, you will need an apparatus such as the following. A spring is suspended from a hook with a tray for weights attached below the spring. A pointer shows stretching distance along a scale (the distance from the hook to the tray may be used instead). The following illustration shows one way this equipment may be assembled.

On your apparatus, note the pointer location (or the distance from the hook to the tray) before weights are added to the tray. Put a weight on the tray and measure the distance that the pointer (or tray) has moved downward from its original position. This distance is called the *displacement* of the spring.

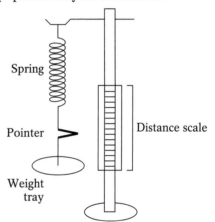

1. Copy and complete a table like the following, in which you will record the weight w and the displacement $D(w)$. Put another weight on the tray; record the total weight and the total displacement from the starting position. Repeat this process with several weights, at least four, each time recording the total weight w and the displacement $D(w)$.

Weight w	Displacement $D(w)$
0	___
___	___
___	___
___	___

2. What relation do you note between w and D?

3. Express this relation as a formula D showing how the displacement of this spring depends on the weight. Describe how you arrived at this rule.

Now select a second spring, either stronger or weaker than the original. Repeat the experiment for this spring and record the values for w and $D(w)$.

4. By studying patterns in the table, try to find a function D_2 that shows the relation between weight and displacement for this second spring. Describe how you arrived at this rule.

5. Use function-fitting technology to find rules for the displacement of each spring as a function of weight. Let $F_1(w)$ and $F_2(w)$ represent the rules for the first and second springs, respectively. How do your own rules compare to these new rules?

6. One use of rules for the functions relating weight and spring displacement is to predict the effect of situations other than those for which you have experimental data.

   For parts a through c, use the spring-displacement rule $D_1(w)$.
   a. Write a question that, to find an answer, requires only the calculation of a function output for a given input.
   b. Write a question that, to find an answer, requires the solution of an equation.
   c. Write a question that, to find an answer, requires the solution of an inequality.

   For parts d through f, use the spring-displacement rule $D_2(w)$.
   d. Write a question that, to find an answer, requires only the calculation of a function output for a given input.
   e. Write a question that, to find an answer, requires the solution of an equation.
   f. Write a question that, to find an answer, requires the solution of an inequality.

7. Suppose you conducted the spring displacement experiment for a third spring. Can you use your results from the first two springs to predict the rule for the third spring? Explain your answer.

7 Summary

In this chapter you have studied linear functions. By now you should be able to recognize linear relationships between quantities when you see tables, graphs, symbolic rules, or conditions in applied problems. These functions have the following key properties.

1. The outputs of a function are related to the inputs by a rule that can be written as

$$f(x) = mx + b,$$

   where the values of m and b are constants that depend on the situation being modeled.

2. In any table of pairs of values, (input, output), the outputs change at a constant rate as inputs change. That rate of change is the number m in the function rule $f(x) = mx + b$.

   Every increase of 1 in the value of x causes a change of m in $f(x)$. This means that for any input x, the output at the point 1 unit greater, namely at $x + 1$, can be obtained from the output at x by adding the number m. Expressed in functional notation,

$$f(x+1) = f(x) + m.$$

   In general, for any two pairs, $(x_1, f(x_1))$ and $(x_2, f(x_2))$,

$$m = \frac{\text{change in output}}{\text{change in input}}$$

$$= \frac{f(x_2) - f(x_1)}{x_2 - x_1}.$$

3. The constant b in any linear function rule $f(x) = mx + b$ is the output for the input 0, because

$$\begin{aligned} f(0) &= m(0) + b \\ &= 0 + b \\ &= b. \end{aligned}$$

4. The points of a graph for any linear function lie on a straight line. Furthermore, that line intersects the output-axis at the point $(0, b)$ and has slope m.

Because of these four properties, linear functions are very useful as models of relations among variables. The tables, graphs, and rules can be applied to answer many important questions that require calculation of outputs for given inputs, solution of equations, and solution of inequalities.

Review Exercises

1. SITUATION The printer of the school newspaper gives the following table showing estimated cost for sample numbers of copies.

Copies	500	1000	1500	2000
Cost (dollars)	100	150	200	250

   a. Is the relation between copies printed and cost linear? Show how you could check your answer using at least two different methods.

   b. If the relation between copies printed and cost is linear, find a rule for cost in the form $c(x) = $ _____ $x + $ _____. Explain what the constant term and the coefficient of the input variable x tell about cost.

2. SITUATION Suppose the fares on Metrorail are based on a formula: $0.75 base fare plus $0.075 per km. Give a rule for fare as a function of distance traveled.

3. SITUATION Suppose you are offered a job selling magazine subscriptions door to door. You buy $50.00 worth of samples and sell subscriptions for a commission of $0.25 for each subscription dollar sold.

   a. Write a rule giving your pay as a function p of the dollar value of subscriptions you sell.

   b. Write an equation or inequality that must be solved to answer each of the following questions. Do not actually solve the equations or inequalities.

   i. What dollar sales total makes your pay (after expenses) equal to $100?

   ii. How much must you sell so that your pay is at least $200?

   iii. What is your break-even point in sales?

   c. Solve part b.i by a guess-and-test method, recording your guesses and tests.

e. Solve part b.iii by reasoning with the symbolic rule for the pay function p as you did in Section 5 of this chapter.

f. Compare the three methods you used in parts c, d, and e. What are the advantages and limitations of each method?

g. What is the slope of the graph of the pay function? What does the slope tell you about the relation between your pay and the dollar value of the subscriptions you sell?

h. At what point does the graph of the pay function cross the vertical axis? What does the point of intersection tell you about the relation between your pay and the dollar value of the subscriptions you sell?

4. **SITUATION** The data below show a relation between monthly sales of a soft drink and money spent on television advertising.

Advertising (thousands of dollars)	125	175	250	100
Sales (thousands of dollars)	600	725	825	500

a. Create a graph of sales (in thousands of dollars) as a function of advertising (in thousands of dollars).

b. Sketch a line that seems to fit the data well. From your graph, determine a function rule of the form $s_1(a) = \underline{\quad} a + \underline{\quad}$.

c. Use a function-fitting program to try to find a linear rule that fits the data better (if a better fit is possible). Write the rule in the form $s_2(a) = \underline{\quad} a + \underline{\quad}$.

d. Use the technology-generated function rule from part c to make the following predictions.

　　i. Predict the effect on sales of $300,000 in advertising.

　　ii. Predict the effect on sales of $150,000 in advertising.

e. Write and solve an equation that tells how much advertising is needed to reach $1,000,000 in sales.

5. **SITUATION** Suppose you know that f is a linear function of x. Tell what special properties you expect to see in each of the following:

a. the function rule for f;

b. the graph of f; and

c. a table of function values for f.

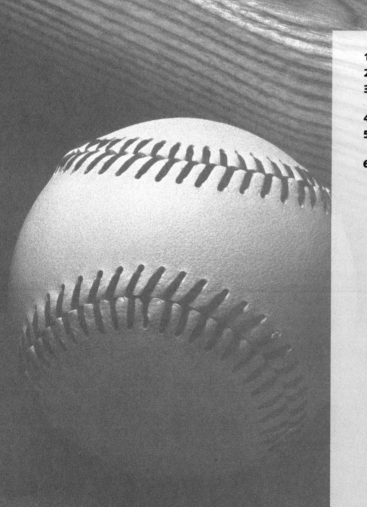

When a baseball player hits the ball, how high will it go and how long will it stay in the air? This is only one example of using mathematics to describe moving objects.

4 Quadratic Functions

The distance required to stop a moving car depends on many things, including the speed, the mass, the braking system of the car and the condition of the road on which the car is travelling. The path of a basketball or an artillery shell depends on things such as the initial velocity and elevation of the shot and the force of gravity. The profit of a business depends on the prices, sales, and costs of its products as well as more complicated issues.

In each of those situations, the important variables are often related by a rule of the form

$$f(x) = ax^2 + bx + c \ (a \neq 0),$$

where the numbers a, b, and c are constants determined by specific conditions in the situation that the function models. A function like this is called a **quadratic function**. By the end of this chapter, you should be good at recognizing situations in which quadratic functions occur and at using various methods for answering questions about those situations.

1 Recognizing Quadratic Functions

As you study any family of functions, it is important to find properties common to all functions in that family. For instance, in Chapter 3, you studied linear functions with rules of the form $f(x) = mx + b$. The values of the constants m and b have special importance in determining the patterns in the graph and table of data pairs, (input, output). The graph crosses the $f(x)$-axis at $(0, b)$. A linear function always has a graph with constant slope m. That is, the graph is a straight line, and the outputs change at a constant rate as the inputs change. For each increase of 1 in the value of x, the value of $f(x)$ increases by m.

A quadratic function, however, has the form

$$f(x) = ax^2 + bx + c,$$

where a, b, and c are constants and $a \neq 0$. (Why do you suppose a cannot be 0?) The addition of the **quadratic** or **square term** (x^2) in a quadratic function rule gives quite different patterns in the graph and tables of values. As you find properties of quadratic functions it will be helpful to compare them to properties of functions you have studied earlier.

As you did with linear functions, it is generally a good idea to start by generating a graph and a table of value pairs, $(x, f(x))$. Use the graph and table to answer specific questions of interest, but look also for connections between patterns in graphs and tables and in the constants a, b, and c that determine the function rule.

The examples in this first section illustrate the variety of situations modeled by quadratic functions, the types of graph and table patterns that can occur, and the kinds of questions that can be answered using those mathematical tools.

Exploration I

One of the most important questions facing the driver of an automobile, bus, or truck is "How long will it take me to stop if there is trouble ahead?" The answer depends on many variables.

SITUATION 1.1

For a car with mass of 1000 kilograms, traveling on a dry asphalt road over a flat surface, the rule

$$d(s) = 0.005s^2 + 0.14s$$

can be used to predict stopping distance $d(s)$ in meters from speed s in kilometers per hour.

For instance, if you are driving at 60 km/hr on a dry asphalt road when you spot an overturned truck blocking the road and you stop as quickly as possible, you can expect to travel about 26 meters from the time you start braking until the car stops, as the following calculations show.

$$d(60) = 0.005(60)^2 + 0.14(60)$$
$$= 0.005(3600) + 0.14(60)$$
$$= 18 + 8.4$$
$$= 26.4.$$

In this situation there are at least three kinds of questions that can be answered using the function rule.

1. What is the stopping distance for a given speed?

2. What speeds permit a stop within a given distance?

3. At what rate does stopping distance increase as speed increases?

1. For the stopping distance function rule given above, generate and record a table of value pairs, $(s, d(s))$. Use steps of 10 km/hr for speed. As you select starting and ending input values, remember to consider speeds that would be reasonable in the situation you are studying.

2. Next plot the data from your table to make a graph for $d(s)$ on a coordinate grid. Make sure you mark your scales on the grid. Check your graph by using a graphing utility.

3. Use the table and/or graph to answer the following questions.
   a. Describe the trend in stopping distances as the speed increases.
   b. Which properties of the relation between speed and stopping distance seem easier to learn from the table? Which seem easier to learn from the graph?
   c. How does the graph of this function seem to differ from those of linear functions you have studied?
   d. How does the table of values seem to differ from those of linear functions you have studied?
   e. Use the rule to calculate $d(-50)$ and $d(3000)$. What do these values mean for stopping distance?

Use technological tools and the function d to answer the following. Remember the input variable s is the speed of the vehicle in kilometers per hour and the output $d(s)$ is stopping distance in meters when the speed is s kilometers per hour.

For each of Exercises 4 through 6, do the following.
   a. Write an algebraic expression that can be used to answer the given question. This might be an equation or an inequality to be solved, or it might represent a calculation to be done. You may use function notation wherever you like.
   b. Describe a suitable method to solve the problem. This includes what tool (or tools) to use and how to use it (or them).
   c. Solve the problem. Then write a sentence stating your answer to the original question. Give your answer to the nearest meter (for stopping distance) or to the nearest 5 km/hr (for speed).

4. What is the stopping distance for a car traveling 50 kilometers per hour?

5. What speed requires exactly 35 meters stopping distance?

6. What is the stopping distance for a speed of 150 kilometers per hour?

For each of Exercises 7 through 10, answer the question and write a short report analyzing your solution. If you wish, you may follow the pattern used in Exercises 4 through 6 for a report, but some interesting questions can be of a different type, so you may choose your own style to describe the analysis. Be sure to end with a sentence giving your answer.

7. What is the stopping distance for a speed of 0 kilometers per hour?

8. What speed(s) permit a car to stop in less than 90 meters?

9. A car stopped in 125 meters. How fast was it going?

10. What are the stopping distances for speeds greater than 100 kilometers per hour?

Exploration II

One of the most important discoveries of science was the description of the way that gravity affects objects rising from or falling to the earth's surface. In 1638 Galileo claimed that the height of such an object is a quadratic function of its time in the air.

SITUATION 1.2

For example, suppose that a baseball player hits a high pop-up straight above home plate. If the bat meets the ball 1.5 meters above the ground and sends it up at a velocity of 30 meters per second, then the height of the ball, in meters, t seconds later is predicted well by the rule

$$h(t) = -4.9t^2 + 30t + 1.5.$$

In general, if any projectile has **initial height** h_0 meters and **initial vertical velocity** v_0 meters per second, then its height in meters after t seconds can be predicted by a rule of the form

$$h(t) = -4.9t^2 + v_0t + h_0.$$

This rule fits real world data quite well, as long as air resistance is not a significant factor.

1. Make a table of values and a graph for the function $h(t) = -4.9t^2 + 30t + 1.5$ with an appropriate graphing utility. Remember t is time in seconds and $h(t)$ is height in meters. Make a record of your table and graph.

2. Use the table and graph to answer the following questions.
    a. Describe the trend in height values as the time increases.
    b. Which properties of the relation between time and height seem easier to learn from the graph and which from the table?
    c. How does the graph of this function seem to differ from those of linear functions you have studied?
    d. How does the table of values seem to differ from those of linear functions you have studied?
    e. Use the rule to calculate $h(-5)$ and $h(13)$. What do these results mean in this situation?

3. In any projectile motion situation there are several common questions of interest. For example, what is the projectile height when time is 3 seconds? List at least 3 different types of questions that could be answered using the height function for a fly ball.

Use the rule $h(t) = -4.9t^2 + 30t + 1.5$ to answer Exercises 4 through 9. For each, write a report analyzing your solution. This should include the following.

a. An equation, inequality, or calculation that can be used to answer the given question.

b. A description of a suitable method for solving the problem. Tell what tool (or tools) to use and how to use it (or them).

c. A sentence giving your answer to the original question. Express an answer for time to the nearest second, or an answer for height to the nearest meter.

4. What is the height of the ball when the time is 2 seconds?

5. When is the ball more than 40 meters above ground?

6. When is the ball exactly 35 meters above the ground? (Be careful!)

7. What is the height of the ball when the time is zero seconds?

8. What is the height of the ball when the time is 4.5 seconds?

9. When is the ball less than 50 meters above the ground?

10. Sketch the graph of $h(t) = -4.9t^2 + 30t + 1.5$ on a coordinate grid such as the one shown here. Choose your scales so that you can answer the questions that follow. Be sure to indicate time and height scales along the axes.

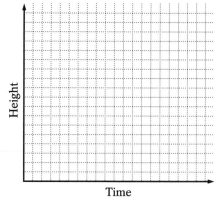

a. What point shows the starting height of the ball? Label this point A and give its coordinates.

b. What point shows height of the ball after 2 seconds? Label this point B and give its coordinates.

c. What point(s) show a height of 20 meters? Label it (or them) C and give the coordinates.

d. What point shows where the ball reaches its maximum height? Label this point D and give its coordinates.

e. What point shows when the ball hits the ground? Label this point E and give its coordinates.

f.   Trace, with a different colored pen or pencil, the part of the graph that shows when the ball is at least 30 meters in the air.

g.   What does the part of the graph between A and D show about the flight of the ball?

h.   What does the part of the graph between D and E show about the flight of the ball?

Making Connections

1. How are the tables of the projectile motion and stopping distance functions similar to, and different from, each other?

2. How are the graphs of the two functions similar to, and different from, each other?

Exercises

1. A Porsche 944 has a mass of about 1250 kg. For this car, operating on a dry asphalt road over a flat surface, a rule relating speed and approximate stopping distance is $d(s) = 0.006s^2 + 0.14s$.

    a.   Copy and complete the table shown here. Sketch a graph of speed versus distance on a coordinate grid.

s	0	20	40	60	80	100	120	140	160	180	200
$d(s)$	0	5.2		30.0		74.0					

    b.   Use the table, the graph, or a technological tool to solve each of the following. Then write sentences explaining what each answer tells about speed and stopping distance for this car. Note that for some questions you might have to make estimates.

    i.   Find $d(s)$ when $s = 60$.

    ii.   Find $d(s)$ when $s = 120$.

    iii.   Find $d(180)$.

    iv.   Find $d(150)$.

    v.   Find s when $d(s) = 74$.

    vi.   Find s when $d(s) = 10$.

    vii.   Find s when $d(s) \leq 22$.

    viii.   Find s when $d(s) > 100$.

2. Recall from Chapter 3 that as input values change in a linear function, output values change at a constant rate. To discover whether rate of change is also constant for quadratic functions, use the rule $d(s) = 0.005s^2 + 0.14s$ and the table and graph you created in Exploration I for the following.
   a. Copy and complete the table below.

Stopping distance	Difference
$d(10) = 1.9$	
$d(20) = 4.8$	$d(20) - d(10) = 4.8 - 1.9 = 2.9$
$d(30) =$	$d(30) - d(20) =$
$d(40) =$	
$d(50) =$	
$d(60) =$	
$d(70) =$	
$d(80) =$	

   b. Does this function have a constant rate of change? Explain your reasoning.
   c. Does the rate of change increase, decrease, or stay the same as speed increases?
   d. What do your answers to parts b and c mean in terms of the relation between speed and stopping distance?

3. An approximate rule for judging stopping distance is to allow one car length, about 6 m, for each 15 km/hr of speed.
   a. Copy and complete the following table using this relation between speed and stopping distance.

Speed (km/hr)	15	30	45	60	75	90	105	120
Distance (meters)	6							

   b. Write an algebraic rule relating speed s and distance $f(s)$.
   c. Sketch a graph of this relation.
   d. How are the predictions of this rule similar to, and different from, the predictions of the quadratic rule for stopping distance, $d(s) = 0.005s^2 + 0.14s$, that you studied in Exploration I?

4. In the quadratic rule for stopping distance, $d(s) = 0.005s^2 + 0.14s$, the term $0.14s$ accounts for average reaction time before braking. If you were traveling at 30 km/hr and suddenly needed to stop, you would travel $0.14(30)$ meters further before you applied your brakes.

a. Make a table and a graph for $r(s) = 0.14s$.

b. Write and answer three questions involving this function rule.

c. If a function of this type were to describe reaction distance while a person is driving under the influence of fatigue, certain cold medicines, or alcohol, how might its rule be different from the rule for $r(s)$?

d. If a function like this were to describe reaction distance after a person drinks five cups of coffee, how might it be different?

5. When a projectile is hit or thrown or shot into the air, its height as a function of time depends on its initial height and vertical velocity. Write a height function for each of the following situations and then write two questions that your height function could help answer. (Remember, the general equation is $h(t) = -4.9t^2 + v_0t + h_0$.)

a. A ball is thrown into the air from atop a 50-meter tower with initial vertical velocity of 40 meters per second.

b. A football is punted into the air with initial upward velocity of 20 meters per second. It leaves the punter's foot at an initial height of 1.25 meters.

c. A tennis ball is lobbed into the air with upward velocity 24 meters per second. It leaves the racquet 1.5 meters above ground level.

6. The Washington Monument is about 170 meters tall. Suppose that a metal ball is dropped (initial velocity is 0 meters per second) from the top.

a. Write the quadratic function rule that relates height of the ball above ground to time in flight. (Remember, the rule has the form $h(t) = -4.9t^2 + v_0t + h_0$.)

b. Copy and complete the following table for your height function.

Time (sec)	Height (m)	Time (sec)	Height (m)
0.0		3.5	110
0.5	169	4.0	
1.0		4.5	71
1.5	159	5.0	
2.0		5.5	
2.5	139	6.0	
3.0			

c. Use the table to sketch a graph of the height function.

d. Use the table or graph to estimate the time the ball hits the ground.

e. The average speed of a projectile is the change in distance divided by the change in time. For example, if a projectile travels 24 meters in 3 seconds, its average speed is 8 meters per second. If it travels 12 meters in 1 second, its average speed is 12 meters per second. Use the table data to calculate the average speed of the ball from:

    i.   0 seconds to 1 second

    ii. 1 second to 2 seconds

    iii. 4 seconds to 5 seconds

f. How fast do you think the ball is moving as it hits the ground?

2 Properties of Quadratic Functions

The "stopping distance" and "projectile motion" functions you studied in Section 1 are only two particular examples in the quadratic family. Tables, graphs, and symbol manipulations help in answering questions about such functions. However, to use those tools efficiently and wisely, it helps to know some of the basic properties of quadratic functions,

$$f(x) = ax^2 + bx + c \ \ (a \neq 0)$$

and to know how the values of the coefficients, a, b, and c, affect the tables and graphs in individual cases.

There are some properties of quadratic functions that you may have noticed already from examples in Section 1 of this chapter and in earlier chapters. For instance, the graph of a quadratic function is a **symmetric** curve like each of the ones shown here.

Each of the curves is called a **parabola**. Notice that each curve is a mirror image of itself along the **axis of symmetry**. The

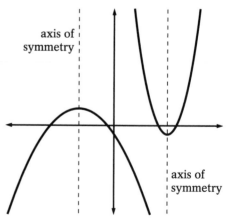

patterns in tables of values, (input, output), for quadratic functions match the trends and symmetry in those graphic patterns.

The explorations that follow outline a study of quadratic functions. Your task is to learn how the patterns in tables and graphs of such functions can be predicted from the coefficients, a, b, and c.

You will learn that usually the most important factor is the coefficient of x^2 in the **quadratic term**, ax^2. Thus the first examples focus on it. The effect of the **constant term** c will be fairly easy to detect. But the role of the coefficient of x in the **linear term** bx, is more subtle and is better addressed in later study.

Exploration

One way to discover the effect of each coefficient in a quadratic function rule is to study a series of examples in which two of the three coefficients are fixed while the third is varied. The observed differences in the table or graphs are then due to the varied coefficient.

Experiment Series 1: *The Effect of* a *in* ax² + bx + c

What does the value of a, the coefficient of x^2, tell you about the data pairs, (input, output), and the graph of the quadratic function? The rules in each of the following sets of quadratic functions differ only in that factor. For each set, use a technological tool to produce a table and a graph of each function. Then study the results to answer the given questions about patterns observed.

In the first cases we suggest you make tables of values for even numbers from $x = -10$ to $x = 10$ and graphs from $x = -6$ to $x = 6$. These should help you recognize important patterns. However, you should also keep in mind the fact that each given rule produces an output value for every possible input value—for $x < -10$, for $x > 10$, and for fractions as well as integers.

Experiment 1.1. What does the **sign** (+ or –) of the value for a tell about a table or graph of the quadratic function?

Complete tables and graphs for the following sets of functions and look for similarities and differences that might be due to change in the sign of the coefficient of x^2. When using tables to look for differences in the output values, it may be helpful to use three columns as shown.

Set 1. Compare $f(x) = 1x^2 - 4x + 3$ and
$$g(x) = -1x^2 - 4x + 3.$$

Record the tables and graphs you use. A sample table follows.

x	$f(x)$ $1x^2 - 4x + 3$	$g(x)$ $-1x^2 - 4x + 3$
–10		
–8		
–6		
–4		
–2		
0		
2		
4		
6		
8		
10		

Patterns observed:

a. How are the tables of values, (input, output), for the two functions alike and how are they different?

i. Are there any input values for which both functions give the same output value? If so, identify each of them.

ii. What do the patterns in the tables of values suggest about possible maximum or minimum outputs for each of these functions?

iii. Do the output values of these two functions change at constant rates as x increases like those of linear functions? Give some numerical examples to support your answer.

b. How are the graphs for the two functions alike and how are they different?

i. Do they appear to have any points in common? If so, give coordinates.

ii. What do the patterns or symmetry in the graphs of these functions suggest about possible maximum or minimum points? List approximate coordinates for any such points.

iii. Do the patterns in these graphs suggest that the output values of the functions change at constant rates as x increases? Explain how the diagrams support your conclusion.

c. In what other interesting ways are the tables and graphs of these two functions alike or different from each other?

Set 2. Compare $f(x) = 0.5x^2 - 2x - 2.5$ and
$$g(x) = -0.5x^2 - 2x - 2.5.$$

Record the tables and graphs you use. A sample table follows.

	$f(x)$	$g(x)$
x	$0.5x^2 - 2x - 2.5$	$-0.5x^2 - 2x - 2.5$
−10		
−8		
−6		
−4		
−2		
0		
2		
4		
6		
8		
10		

Patterns observed:

a. How are the tables of values, (input, output), for the two functions alike and how are they different?

   i. Is there any input for which both functions give the same output? If so, list each of the inputs.

   ii. What do patterns in the tables of values of these two functions suggest about possible maximum or minimum outputs?

   iii. Do the output values of these two functions change at constant rates as x increases? Give some numerical examples to support your answer.

b. How are the graphs for the two functions alike and how are they different?

   i. Do they appear to have any points in common? If so, give coordinates for these points.

   ii. What do the patterns or symmetry in the graphs of these functions suggest about possible maximum or minimum points? List approximate coordinates for any such points.

   iii. Do the patterns in the graphs suggest that the output values of these functions change at constant rates as x increases? Explain how the graphs support your conclusion.

c. In what other interesting ways are the tables and graphs of these two functions alike or different from each other?

Set 3. Write rules for four other quadratic functions of your own design—two that have maximum outputs and two that have minimum outputs. Test your ideas with computer tables and graphs before recording your answers. Indicate which two have maximum outputs and which two have minimum outputs.

Making Connections

List your conclusions about the sign of the coefficient a in a quadratic function with rule $f(x) = ax^2 + bx + c$.

1. What properties does the function have if $a > 0$?

2. What properties does the function have if $a < 0$?

Experiment 1.2. What is the effect of changing the **absolute value** of a?
     Complete tables and graphs for the following sets of quadratic functions. Then study the results to find connections between the size of the coefficient for x^2 and the patterns in those tables and graphs. Tables with four columns, such as the ones shown, may be helpful.

Set 1.   Compare $f(x) = 2x^2 - 5x + 4$,
$$g(x) = x^2 - 5x + 4, \text{ and}$$
$$h(x) = 0.5x^2 - 5x + 4.$$

Record the tables you use. On one coordinate system, sketch and label the graphs of all three functions. A sample table follows.

x	$f(x) = 2x^2 - 5x + 4$	$g(x) = x^2 - 5x + 4$	$h(x) = 0.5x^2 - 5x + 4$
-5	79	54	41.5
-4			
-3			
-2			
-1			
0			
1			
2			
3			
4			
5			

Patterns observed:

    a. How are the tables of data pairs, (input, output) for these three functions alike and how are they different?

    b. How are the graphs of these functions alike and how are they different? How do the patterns in the graphs match the patterns you noticed in the tables of data pairs?

Set 2. Compare $f(x) = -3x^2 + x + 6,$

$$g(x) = -x^2 + x + 6, \text{ and}$$
$$h(x) = -0.3x^2 + x + 6.$$

After you have made tables and graphs for these three functions, answer the following questions.

Patterns observed:

    a. How are the tables of values, (input, output), for these three functions alike and how are they different?

    b. How are the graphs of these functions alike and how are they different? How do the patterns in the graphs match the patterns you noticed in the tables of values?

Set 3. Write two additional pairs of quadratic functions that differ only in the absolute value of the coefficient a. (The coefficient should have the same sign for both.)

    1. For the first pair, choose quadratic functions f and g that both have positive values for a.

    2. For the second pair, choose quadratic functions h and j that both have negative values for a.

    3. Compare tables of values and graphs for each pair to test the ideas you have formed about the effect of the coefficient a.

Making Connections

Discuss your ideas about the effect of changing the magnitude of a, the coefficient of x^2, with classmates and your teacher. Then, write your conclusions about the differences between two quadratic functions, $f(x) = a_1x^2 + bx + c$ and $g(x) = a_2x^2 + bx + c$, where $|a_1| > |a_2|$.

Experiment Series 2: *The Effect of c in* $ax^2 + bx + c$

Now that you have some ideas about how the coefficient of the quadratic term x^2 affects the patterns in tables of values and the graph for a quadratic function, it seems natural to study the effect of the other terms as well.

You probably have some hunches about what can be learned from the **constant term** c in such a rule. To check your ideas, create and study tables and graphs for the following sets of quadratic functions. In each set the rules differ only in the term c.

Experiment 2.1. Compare $f(x) = 0.8x^2 - 3.2x + 3$,
$$g(x) = 0.8x^2 - 3.2x + 0, \text{ and}$$
$$h(x) = 0.8x^2 - 3.2x - 2.$$

Patterns observed:
    a. How are the tables of data pairs, (input, output), for these functions alike and how are they different?
    b. How are the graphs of these functions alike and how are they different?

Experiment 2.2. Compare $f(x) = -1x^2 - 3x - 2$,
$$g(x) = -1x^2 - 3x + 0, \text{ and}$$
$$h(x) = -1x^2 - 3x + 3.$$

Patterns observed:
    a. How are the tables of values, (input, output), for these functions alike and how are they different?
    b. How are the graphs of these functions alike and how are they different?
    c. Now test your ideas about the effect of the constant term c in a quadratic function on several examples of your own design before you go ahead to the summary that follows.

Making Connections

For any two quadratic functions with rules of the form $f(x) = ax^2 + bx + c_1$ and $g(x) = ax^2 + bx + c_2$, what conclusions can you draw about:

1. the value tables for these functions?

2. the graphs for these functions?

Experiment Series 3: *The Effect of* b *in* ax² + bx + c

As we mentioned at the beginning of this section, in a quadratic function rule the role of the **linear term** bx with its coefficient b is more difficult to discover and describe. Given below are two sets of quadratic functions in which b is varied, while the other coefficients are held fixed. Generate tables of values and graphs for those sets of functions and look for similarities and differences that you can relate to changes in the coefficient of the linear term.

Experiment 3.1.   Compare $f(x) = 0.75x^2 + 6x - 2$,
$g(x) = 0.75x^2 + 0x - 2$, and
$h(x) = 0.75x^2 - 6x - 2$.

Experiment 3.2.   Compare $f(x) = -1x^2 + 4x + 3$,
$g(x) = -1x^2 + 0x + 3$, and
$h(x) = -1x^2 - 4x + 3$.

Now try some series of quadratic functions of your own design, changing only the coefficient b at each step. Look at the graphs and tables to try to determine how changes in the linear term affect the overall pattern of the function values and graphs.

Patterns observed:
   a.   How does the value of the linear term coefficient seem to affect the tables of values for quadratic functions?
   b.   How does the value of the linear term coefficient seem to affect the graphs of quadratic functions?

Exercises

1. Consider the five function rules and six graphs that follow.

   Match each rule to the graph that best represents it. Explain your reasoning in each choice. Horizontal (input) and vertical (output) scales are both 2 units per tic mark.

   $f(x) = 2x^2 - 8x - 5$ $\quad$ $g(x) = 2x^2 - 8x + 3$ $\quad$ $h(x) = 0.5x^2 - 2x + 3$

   $j(x) = 2x - 5$ $\qquad\qquad$ $k(x) = -0.5x^2 - 2x + 3$

   a.

   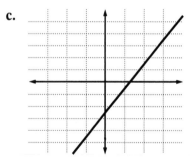

   b.

   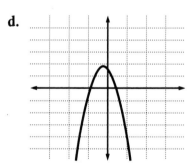

   c.

   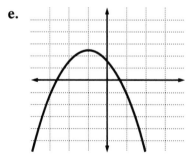

   d.

   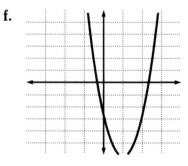

   e.

   f.

2. Consider the five function rules and six graphs that follow.

Match each rule to the graph that best represents it. Explain your reasoning in each choice. Horizontal and vertical scales are both 2 units per tic mark.

$f(x) = -x^2 + 4x + 5$ $\qquad$ $g(x) = -x^2 - 5x + 5$ $\qquad$ $h(x) = -5x + 5$

$j(x) = 0.2x^2 + 1.6x - 1$ $\qquad$ $k(x) = x^2 - 5x - 1$

a.

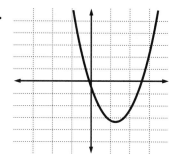

b.

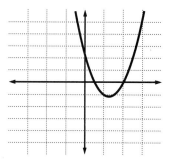

c.

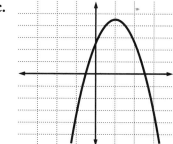

d.

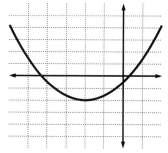

e.

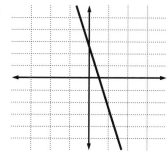

f.

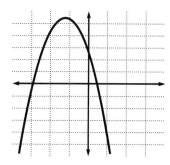

3. Each of the following is a table of data pairs, (input, output), for a linear or quadratic function with rule of the form $f(x) = mx + b$ or $g(x) = ax^2 + bx + c$.

For each table answer the following.
  i. Is the function linear or quadratic?
  ii. If it is linear, what are the numbers m and b? If it is quadratic, what can be determined about the values of a, b, or c?

a. x	$r(x)$		b. x	$s(x)$		c. x	$t(x)$
−2	−1		−2	−3		−2	−6
−1	−3		−1	−1		−1	0
0	−1		0	1		0	4
1	5		1	3		1	6
2	15		2	5		2	6
3	29		3	7		3	4

d. x	$u(x)$		e. x	$v(x)$		f. x	$w(x)$
−2	−6		−2	4		−2	4
−1	−1		−1	1		−1	1
0	2		0	−2		0	0
1	3		1	−5		1	1
2	2		2	−8		2	4
3	-1		3	−11		3	9

4. For each of the following function rules, make a sketch that shows the expected shape and position of the graph if you were to get a computer plot. You need not calculate and plot exact points—just give the general shape and location. For example, $f(x) = 0.6x^2 + 3.6x - 2$ gives a graph like the one shown here.

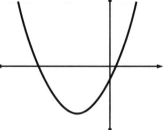

  a. $g(x) = -0.6x^2 + 3.6x + 2$
  b. $h(x) = x^2 - 5$
  c. $j(x) = x - 5$
  d. $k(x) = 35x^2 - 700x - 1700$

5. Copy and complete each of the following tables of values, (input, output), assuming that each function is quadratic. For each function, determine what you can about the coefficients a, b, and c in the rule.

a.	x	$r(x)$	b.	x	$s(x)$	c.	x	$t(x)$
	−2	4		−5			11	385
	−1	−1		−4			15	
	0	−4		−3	−0.5		19	513
	1	−5		−2	−1		23	529
	2	−4		−1	−0.5		27	513
	3			0	1		31	465
	4			1	3.5		35	

6. Each of the following quadratic function rules is a special case of the rule pattern $f(x) = ax^2 + bx + c$.

   a. $g(x) = x^2$ b. $h(x) = x^2 + 2$ c. $j(x) = x^2 - 3$
   d. $k(x) = 4x^2$ e. $m(x) = -3x^2$ f. $p(x) = -x^2$

   (Reminder: To calculate $-x^2$ you must square the value of x first.)

   For each rule do the following.

      i.   Identify the values of a, b, and c. For example, if $g(x) = x^2$, then $a = 1$, $b = 0$, and $c = 0$.

      ii. Determine whether the function has a maximum or a minimum output.

      iii. Graph the function. Find the point where the graph crosses the vertical axis and record that point's coordinates.

      iv. Complete a table of pairs, (input, output), for $x = -10$ to $x = 10$ in steps of 2.

Use of a spreadsheet may be helpful for Exercises 7 and 8.

7. Consider the quadratic function $f(x) = 3x^2 - 6x + 4$. Copy and complete the following table showing how each term in the rule contributes to the value of $f(x)$.

x	$3x^2$	$-6x$	$3x^2 - 6x + 4$
−100	30 000	600	30 604
−50			
−20			
0			
1			
5			
10			
20			
50			

8. Generate tables like that in Exercise 7 for each of the following functions.
   a. $g(x) = x^2 - 10x + 2$
   b. $h(x) = -2x^2 + 25x + 1$
   c. $j(x) = 2x^2 + 25x + 1$
   d. $m(x) = -x^2 - 10x + 2$

9. Look at the results of your work in Exercises 7 and 8 and try to find an explanation for the fact that a, the coefficient of x^2, determines whether the quadratic function has a maximum or a minimum value and whether its graph "opens up" or "opens down."

10. When a function rule describes the relation between two variables in a situation, specific information about the situation can often be discovered from the coefficients in the rule. For each of the following, explain what you can learn about the given situation by studying the coefficients of the function.
    a. The height of a certain projectile, in meters, is related to its time in flight, in seconds. Height, $h(t)$, can be estimated by the quadratic rule $h(t) = -4.9t^2 + 5t + 10$, where t is the time in flight.
    b. The profit a particular company earns, in dollars, is related to the price it charges, in dollars, for its main product. This relation can be modeled by the quadratic rule Profit $= -300p^2 + 3500p - 1500$, where p is the price.

3 Solving Quadratic Equations and Inequalities

When quadratic functions are encountered in problem situations, many of the most important questions are answered by solving equations or inequalities related to the functions. For instance, in the Talent Show simulation, the function relating ticket price x in dollars and profit in dollars might be

$$P(x) = -100x^2 + 800x - 700.$$

By now you should know quadratic functions well enough to look at that rule and predict the pattern in a table of values or a graph for this function. Do you also know what to expect as you search for answers to questions like the following?

For what ticket prices is the profit $500?

What ticket prices give a situation that does not lose money, in other words, give a profit at least equal to $0?

In symbolic form, these questions are

Solve: $P(x) = 500$ or $-100x^2 + 800x -700 = 500$, and
Solve: $P(x) \geq 0$ or $-100x^2 + 800x - 700 \geq 0$.

You have used function tables, graphs, and computer-algebra programs to solve such equations and inequalities before. Again, to use those tools efficiently and wisely, it helps to know what kinds of answers can be expected to occur. In the explorations that follow you will study a series of equations and inequalities. Look for patterns suggesting methods and strategies to help you find solutions for quadratic equations and inequalities.

Exploration I

In the experiments that follow, use the given function tables and graphs to estimate solutions to the given equations and inequalities. Then find more precise solutions using addtional tables or technology with equation-solving capabilities.

Experiment 1. Shown here is a graph of the quadratic function $f(x) = 0.2x^2 - 0.8x - 1$. Create a table of data pairs, $(x, f(x))$, using input values from $x = -7$ to $x = 10$ and a step of 1 (so your input values will be $-7, -6, -5, ..., 8, 9, 10$). Technological tools may be useful to help you find entries for the table and to help you answer the questions that follow.

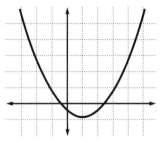

$f(x) = 0.2x^2 - 0.8x - 1$

The horizontal (input) scale unit is 2 and the vertical (output) scale unit is 2.

a. On the graph, locate any points for which $f(x) = 0$ and give their coordinates. What do these points tell about solutions for the equation $0.2x^2 - 0.8x - 1 = 0$?

b. Give the coordinates of any points for which $f(x) = 8$. What do these points tell about roots of the equation $0.2x^2 - 0.8x - 1 = 8$?

c. Give the coordinates of any points where $f(x) = -1.8$. What do these points tell about roots of $0.2x^2 - 0.8x - 1 = -1.8$?

d. Give the coordinates of any points where $f(x) = -3$. What do these points tell about solutions for $0.2x^2 - 0.8x - 1 = -3$?

For parts e through j, describe all values of x that satisfy the given inequality.

e. $0.2x^2 - 0.8x - 1 \le 0$.

f. $0.2x^2 - 0.8x - 1 < 8$.

g. $0.2x^2 - 0.8x - 1 < -1$.

h. $0.2x^2 - 0.8x - 1 \le -3$.

i. $0.2x^2 - 0.8x - 1 > 0$.

j. $0.2x^2 - 0.8x - 1 > -10$.

Experiment 2. All of the following equations and inequalities involve questions about the same quadratic function and can be solved by using a graph, a table of values, or symbol manipulation applied to that function. Examine the equations and inequalities to find the function involved.

i. $-x^2 - 2x + 8 = 0$		**v.** $-x^2 - 2x + 8 = 9$
ii. $-x^2 - 2x + 8 = 8$		**vi.** $-x^2 - 2x + 8 = 10$
iii. $-x^2 - 2x + 8 \ge 0$		**vii.** $-x^2 - 2x + 8 < 0$
iv. $-x^2 - 2x + 8 > 8$		**viii.** $-x^2 - 2x + 8 \ge 10$

Now answer the following.

a. What is the function rule?

b. Complete a table of values, using input values from $x = -8$ to $x = 8$ and a step of 1, and sketch the graph of the function. Label each point where the graph crosses the output axis or the input axis.

c. Solve each of the equations and inequalities. That is, find *all* values of x that satisfy the given condition.

Making Connections

In the two experiments of Exploration I, you have found solutions to many equations and inequalities involving quadratic functions. Examine your results of Exploration I, then answer the following questions and try to find general patterns. If you have an idea about an answer, but are unsure, try several other specific examples to see what may happen. Write an equation or inequality of your choice, and use a graph, a table of data pairs, (input, output), or a technological tool to see what the solution possibilities are.

1. For a quadratic equation of the form $ax^2 + bx + c = d$, how many different roots can there be?

2. How is the solution of an inequality of the form $ax^2 + bx + c \leq d$ different from the solution of an equation of the form $ax^2 + bx + c = d$?

Exercises

1. For each of the following equations or inequalities, write the rule for an associated function $f(x)$ and rewrite the problem using function notation.
   a. $x^2 + 2x - 7 = 0$
   b. $x^2 + 2x - 1 = 2$
   c. $-x^2 + 7x - 3 = 9$
   d. $3x^2 - 2.5x + 7.2 = 125$
   e. $4x^2 - 2x + 3.1 \leq 84$
   f. $-7x^2 + 12x \geq 23$

2. This diagram shows the graphs of $f(x) = x^2 + 2x - 7$ and the graph of $g(x) = -x^2 + 7x - 3$. Use these graphs to estimate solutions for the following equations and inequalities.
   a. $x^2 + 2x - 7 = -4$
   b. $x^2 + 2x - 7 < -4$
   c. $x^2 + 2x - 7 = -8$
   d. $x^2 + 2x - 7 \leq -8$
   e. $f(x) = -10$
   f. $f(x) > -10$

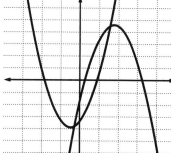

The horizontal (input) scale unit is 2 and the vertical (output) scale unit is 2.

g. $f(x) = 5$ **h.** $f(x) \geq 5$

i. $-x^2 + 7x - 3 = 9$ **j.** $-x^2 + 7x - 3 \geq 9$

k. $-x^2 + 7x - 3 = 12$ **l.** $-x^2 + 7x - 3 \leq 12$

m. $g(x) = 0$ **n.** $g(x) \leq 0$

o. $x^2 + 2x = 3$ **p.** $-x^2 + 7x = 12$

3. The tables shown here display a sample of data pairs for the functions $f(x) = 2x^2 + 3x - 5$ and $g(x) = -0.5x^2 + 8$, respectively. Use those tables to estimate solutions for the following equations and inequalities.

a. $2x^2 + 3x - 5 = 0$ **b.** $2x^2 + 3x - 5 = 10$

c. $f(x) = 100$ **d.** $2x^2 + 3x - 5 \leq 100$

e. $f(x) = 50$ **f.** $f(x) \geq 50$

g. $-0.5x^2 + 8 = 0$ **h.** $-0.5x^2 + 8 > 0$

i. $-0.5x^2 + 8 = 8$ **j.** $-0.5x^2 + 8 \geq 8$

k. $g(x) = 10$ **l.** $g(x) \leq 10$

m. $g(x) = -20$ **n.** $g(x) > -20$

o. $2x^2 + 3x = 55$ **p.** $-0.5x^2 = -28$

x	$f(x) = 2x^2 + 3x - 5$	x	$g(x) = 0.5x^2 + 8$
−10	165	−10	−42.0
−9	130	−9	−32.5
−8	99	−8	−24.0
−7	72	−7	−16.5
−6	49	−6	−10.0
−5	30	−5	−4.5
−4	15	−4	0.0
−3	4	−3	3.5
−2	−3	−2	6.0
−1	−6	−1	7.5
0	−5	0	8.0
1	0	1	7.5
2	9	2	6.0
3	22	3	3.5
4	39	4	0.0
5	60	5	−4.5
6	85	6	−10.0
7	114	7	−16.5
8	147	8	−24.0
9	184	9	−32.5
10	225	10	−42.0

4. Write a brief plan for use of a technological tool, graph, table, or guess-and-test search to solve each of the following equations or inequalities. You should include a choice of tool (different problems might be best solved by different tools) and steps you will follow in using that tool. Then follow the plan to find the requested solution.

   a. $x^2 - 5x - 24 = 0$ b. $-4.9x^2 + 98x = 490$
   c. $x^2 - 5x - 24 \leq 0$ d. $-4.9x^2 + 98x \geq 490$
   e. $3.4x^2 + 4.5x + 23.5 = 50$ f. $3.4x^2 + 4.5x + 23.5 < 50$
   g. $3.4x^2 + 4.5x + 23.5 = 0$ h. $-4x^2 + 7x + 15 = 0$
   i. $-4x^2 + 7x + 15 = 20$ j. $-4x^2 + 7x + 15 > 0$

Exploration II

The following sets of equations and inequalities involve simple quadratic functions that can be solved with a little thought and perhaps some calculator arithmetic.

1. Shown here is a graph of $f(x) = x^2$. It is a quadratic function in which the coefficient a is 1, the coefficient b is 0, and the term c is also 0.

   Solve each of the following equations using mental or calculator arithmetic. Copy the graph of $f(x) = x^2$ and label the points corresponding to each solution with the matching letter and coordinates.

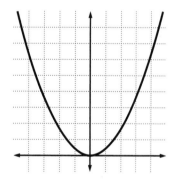

$f(x) = x^2$

The horizontal (input) scale unit is 2 and the vertical (output) scale unit is 10.

   a. $x^2 = 16$ b. $x^2 = 64$ c. $x^2 = 42.25$
   d. $x^2 = 0$ e. $x^2 = -20$

   Describe the solutions for each of the following inequalities. Again, it may be helpful to refer to the graph of $f(x) = x^2$ in determining the full solution set.

   f. $x^2 \leq 49$ g. $x^2 < 16$ h. $x^2 \geq 16$
   i. $x^2 \leq -10$ j. $x^2 \leq 20.25$ k. $x^2 > 0.25$

2. One of the best ways to solve a new kind of mathematics problem is to show how the new problem can be reduced to familiar pieces. For example,

the equation $x^2 + 5 = 41$
has the same solution as $x^2 = 41 - 5$,
which is equivalent to $x^2 = 36$,
which has the roots $x = 6$ and $x = -6$.

Use this strategy to solve each of the following equations, recording the steps in your thinking.

a. $x^2 + 8 = 44$ b. $x^2 + 8 = 72$
c. $x^2 - 23 = 77$ d. $x^2 - 23 = 26$
e. $x^2 + 25 = 16$ f. $x^2 - 45 = 0$
g. $25 + x^2 = 41$

3. You may have discovered from the preceding examples that if you can reduce a quadratic equation to the form $x^2 = a$, the solution requires only one finding the square root of a. This may involve little more than locating and pressing a square root button on your technological tool. The next examples can be reduced in a similar way, using the same kind of reasoning you have used with linear equations—that is, by reasoning about the arithmetic involved. Solve each of the following equations, recording the steps in your reasoning.

a. $3x = 48$ b. $3x^2 = 48$
c. $7x = 63$ d. $7x^2 = 63$
e. $23x^2 = -69$ f. $5x + 12 = 92$
g. $5x^2 + 12 = 92$ h. $13.5x^2 = 486$
i. $12x^2 - 8 = 116$ j. $7x^2 - 32 = 31$
k. $8x^2 + 28x = -24$

As you can see, some quadratic equations and inequalities are much easier to solve with technological assistance. However, you should now be able to recognize and solve many simple cases that are easily done with only mental arithmetic or a calculator.

Exercises

1. Solve each of the following equations.
   a. $x^2 = 64$ b. $x^2 = 6.25$
   c. $49 = x^2$ d. $x^2 = -9$
   e. $x + 12 = 93$ f. $x^2 + 12 = 93$
   g. $41 = x^2 - 8$ h. $12 - x^2 = -69$
   i. $13x - 15 = 37$ j. $13x^2 - 15 = 37$
   k. $5x^2 + 23 = 103$ l. $49 = 15 + x^2$
   m. $16 - x^2 = 25$ n. $x^2 + 25 = 25$

2. The area of a circle is a function of its radius r with rule $A = \pi r^2$. We sometimes approximate π by 3.14; in this problem, use the approximate rule $A(r) = 3.14r^2$.

   Write and solve an equation to answer each of the following questions.
   a. What is the area of a circle with radius 20 m?
   b. What radius gives a circle with area 314 m²?
   c. What radius gives a circle with area 1000 m²?

3. A square with side length s has area $A(s) = s^2$. Answer each of the following and explain what your answers tell about squares.
   a. What is the value of $A(10)$?
   b. What is the value of $A(1.414)$?
   c. Find s so that $A(s) = 169$.
   d. Find s so that $A(s) = 3$.
   e. Find $A(11) - A(10)$ and $A(101) - A(100)$.

4. A steel ball dropped from the top of the Washington Monument has initial velocity $v_0 = 0$ meters per second and initial height $h_0 = 170$ meters. (Remember, this type of projectile motion was introduced in Section 1.)
   a. Write a rule giving height of the ball $h(t)$ in meters as a function of the time in flight t in seconds.
   b. Write and solve equations to answer each of the following questions.
      i. At what time does the ball reach the ground?
      ii. At what time is the ball halfway to the ground?

    c. Write and solve inequalities to answer each of the following questions.

        i. At what times is the ball at least 100 meters above the ground?

        ii. At what times is the ball less than 50 meters from the ground?

5. Consider the quadratic functions $f(x) = x^2$ and $g(x) = x^2 + 3$.

    a. For each function make a table of values with x from -5 to 5 in steps of 0.5.

    b. Use the tables to plot graphs of the two functions on a single diagram.

    c. Explain how the tables or the graphs you have made illustrate the procedure used for solving equations of the form $x^2 + a = b$.

4 Applying Quadratic Functions

Using functions to solve real problems involves three kinds of skills:

1. finding an algebraic rule that models a relation among variables of the situation;

2. studying the rule, graph, and tables for the function to answer questions involving equations and inequalities, maximum or minimum values, and rates of change in the variables; and

3. interpreting your calculations in the situation from which the questions arose.

In Section 1, you used given rules for quadratic functions and tables, graphs, or symbol manipulations to solve equations or inequalities that you wrote to answer questions about stopping distances for cars and heights of projectiles in motion. In many situations, however, the relation between two variables does not easily follow from some well-known formula. Instead, it makes sense to search for a rule that fits experimental data. In the example of this section, you can put your knowledge of quadratic functions to work in the process of finding rules from data and answering questions about the situation.

Exploration I

Biologists studying insect populations often make measurements at many different temperatures. Based on their data, these scientists frequently make the conclusion that the number of insects in a population is a function of temperature. It is often helpful to describe the pattern in the measurements with a mathematical rule relating temperature and population.

SITUATION 4.1

Suppose the following data relate the temperature and the number of insects found in samples of water.

Temperature (in degrees Celsius)	Insect Population
0	20
10	620
20	950
30	920
40	670
50	75

Can we find an algebraic rule that fits these data well? If so, what predictions about other temperatures can you make from the rule?

For most scientists, the first step in looking for a pattern in experimental data is to construct a graph. On a coordinate grid like the one shown here, plot the given data pairs, (temperature, population).

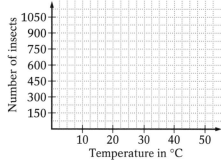

1. Do the data suggest a linear or a quadratic relation between temperature and population? Explain your answer.

2. There are many technological tools that help you to find rules for quadratic functions by searching for coefficients to match your data.
    a. Use a technological tool to find a quadratic rule matching the experimental data. Record your rule.

b. Copy and complete the following table to test the fit.

Temperature	Population	Prediction from the rule
0	20	
10	620	
20	950	
30	920	
40	670	
50	75	

3. The rule you discovered may be similar to

$$P(t) = -1.5t^2 + 75t + 20.$$

Using this function rule, calculate and explain the meaning of the following.

a. $P(0) =$
b. $P(25) =$
c. $P(35) =$

Answer each of the following using guess-and-test, a table of values, a graph, or symbol manipulation. Show a record of your work and give your answers to the nearest 0.1.

d. Find t so that $P(t) = 800$.
e. Find t so that $P(t) \leq 400$.
f. Find the value of t that gives the maximum predicted population and find the size of that population.
g. The size of an insect population will be influenced by factors other than temperature. List several other factors that you think might be significant.

Exercises

1. SITUATION The cost of operating a car is a function of the average speed at which the car is driven. Suppose the data in the table below are results of fuel economy tests.

Average speed s (km/hr)	Cost $C(s)$ (dollars/km)
30	0.20
60	0.18
90	0.15
120	0.17
150	0.26

a. Graph these data points.
b. Use the graph of part a to estimate the average speed that is likely to make the operating cost a minimum.
c. Use function notation to write calculations, equations, or inequalities that can be used to answer the following questions. Then use your graph to estimate answers.
   i. What cost per kilometer results from an average speed of 75 km/hr?
   ii. What average speeds give a cost of $0.17 per km?
   iii. What average speeds keep the cost under $0.18 per km?
d. If a quadratic function rule were chosen to fit these data, do you expect the coefficient of s^2 to be positive or negative? What about the constant term? Explain your reasoning.

Exploration II

In any business one of the most important questions is the relation between prices and sales. Raising the price of a product often will lower the number sold, but if company officials know the *rate* at which the number sold will decline as price increases, they can choose a price that will maximize revenue.

SITUATION 4.2

Suppose a vending machine company that operates juice machines experiments with price and finds the following data.

Price p (dollars)	Daily sales s (bottles)
0.40	147
0.50	127
0.60	87

There are several natural questions to ask about this situation:

1. What daily sales are likely for other prices?
2. What revenues result from different prices?
3. What price yields the maximum revenue?
4. What price yields the maximum profit?

Rules for the sales, revenue, cost, and profit functions can help answer these questions.

1. Plot the given data pairs, (price, sales), on a coordinate grid.
   a. Draw a line that seems to fit the three points well and find (and record) the rule for that linear function.
   b. Use a technological tool to find (and record) a rule that fits the data. Graph this "best-fitting" linear function on the same grid.

2. Your rules from Exercise 1 relate the price of a product to its sales. They are examples of **demand functions**. Each of the rules should be close to $s(p) = -300p + 270$, where $s(p)$ is the sales when the price is p.

   Use this function $s(p)$ and your technological tools to calculate the following values, then explain what each result means.
   a. $s(0.30)$
   b. $s(0.80)$
   c. $s(0.55)$
   d. Find p so that $s(p) = 60$.
   e. In the rule for $s(p)$ what can be learned about juice sales from the coefficient -300? What can be learned from the term 270?

3. The revenue of a business is the money brought in by sales. For example, if 90 juices are sold each day at $0.60 per bottle, the daily revenue is $90 \times 0.60 = 54.00$ dollars. In general,

   $$\text{Daily revenue} = \text{Number sold} \times \text{Price per item.}$$

   a. Find the rule for the juice vendor's revenue, using the same demand function $s(p) = -300p + 270$ and $R(p)$ to represent revenue when the price per item is p.
   b. Calculate and explain the meaning of each of the following.
      i. $R(0)$
      ii. $R(0.40)$
      iii. $R(0.55)$
      iv. $R(0.80)$
   c. On a coordinate grid, graph the revenue function
      $$R(p) = -300p^2 + 270p.$$

   Be sure to indicate scales used for price p and revenue $R(p)$.
   d. Mark your diagram according to the following instructions.
      i. Locate the point that shows revenue when price is $0.75 per drink. Label the point A and give its coordinates.

      ii. Locate any points that show revenue of $100 per day. Label the points B and give their coordinates.

      iii. Locate any points where the revenue has its maximum. Label the points C and give their coordinates.

      iv. Trace with a different color pen or pencil the part of the graph that shows revenue at least $50 per day.

  e. How many juices do you expect to be sold at the price that gives maximum revenue?

    Hint: Use the demand rule $s(p) = -300p + 270$.

  f. What limitations can you see in the analysis of business projections such as the one you have just done?

4. Profit for a business depends on a combination of revenue and cost. Suppose that a juice vendor must pay 20 cents per bottle to buy the drinks. Total cost per day can be predicted from unit cost as follows:

$$\text{Total daily cost} = \text{Number sold} \times \text{Unit cost}$$
$$= s(p) \times 0.20.$$

  a. Write a rule for daily cost as a function C of price per bottle p.

  b. The daily profit is also a function of price per bottle.

$$\text{Profit} = \text{Revenue} - \text{Cost}$$
$$= R(p) - c(p).$$

  Write a function rule for profit $PR(p)$ at price p.

  c. Use a technological tool to answer each of the following questions.

    i. What is the maximum possible daily profit?

    ii. What price gives the maximum profit?

    iii. For what prices is the profit greater than $0?

In this Exploration, we suggested that you find a *linear* demand function. When you try to model situations, other types of functions are possible if the linear fit does not seem good. Any conclusions you make from your functions rules are only as good as the assumptions from which you start. Such assumptions include which types of functions are appropriate for a particular situation.

Exercises

1. SITUATION The owners of a concession stand at the Capital Centre have been experimenting with the price for hot dogs. With the price set at $1 they sell 2500 hot dogs. With the price raised to $1.25 they sell 1500. What is the best price to charge?

   For a linear demand function based on these data, the rule

   $$s(p) = -4000p + 6500$$

   relates price p in dollars and sales $s(p)$.
   a. Use this demand function to predict:
       i.   sales if the price is $1.10;
       ii. sales if the price is $1.50; and
       iii. the price that gives sales of 2000 hot dogs.
   b. Write a rule for the revenue function R implied by the demand function. Remember: Revenue = Sales × Price.
   c. Make a table of values with three columns: Price p ranging from $0.60 to $2.00 in steps of $0.20, Sales $s(p)$, and Revenue $R(p)$.
   d. From your table of values, estimate the price that gives maximum revenue.
   e. Suppose that each hot dog costs an average of $0.40 for the concession stand staff to make. Write a rule for total cost as a function of hot dog price. Remember, total cost = unit cost × number sold.
   f. Write a rule for concession stand profit on hot dogs as a function of hot dog price.
   g. Explain how to find the price that leads to maximum profit.
   h. List any important factors in this situation that have not been considered in deciding on the best price.

2. SITUATION When a high school charged $2.50 for admission to its basketball games, the gym was usually full with attendance of 2000 people. When the price was raised to $3.50, attendance fell to an average of 1500.
   a. Find a linear demand function giving attendance as a function a of the admission price p in dollars.
   b. Using your demand function, find a rule for revenue as a function R of admission price. Remember that the revenue is the money received from ticket sales. Write the rule using the form $R(p) = ap^2 + bp + c$.

    c. Make a table of values for $R(p)$ with p ranging from \$1 to \$7 in steps of \$0.50.

    d. Use the table to estimate the price that gives maximum revenue.

    e. Use the table to estimate:

        i. prices that give revenue at least \$4000; and

        ii. prices that give revenue at least \$5000.

3. SITUATION The operator of Speedy Car Wash has studied the records of the business and made several projections.

    a. Explain why each of the following projections might or might not be reasonable. Hint: For each prediction function, consider the shape of the graph to help you explain your decision.

        i. Average daily revenue is a function of the price charged to wash a car and can be expressed by the rule

$$R(p) = -20p^2 + 200p.$$

        ii. The average cost of washing a car depends on how many cars are washed each day. This relation can be expressed using the rule

$$c(x) = 0.001x^2 - 0.31x + 26.$$

        iii. The average daily profit depends on how many cars are washed in a day. This relation can be expressed using the rule

$$PR(x) = -0.027x^2 + 8x - 280.$$

        iv. The average daily profit depends on the price charged for each car. This relation can be expressed using the rule

$$PR(p) = -12p^2 + 125p - 108.$$

    b. List some factors other than price that may affect revenue and profit for the car wash company.

    c. List ten questions that would be reasonable to ask about the price, revenue, cost, sales, and profit of the car wash business. For each question write the equation or inequality that might be solved to answer it, or explain any other calculations that would be needed.

5 Polynomial Functions of Degree Greater Than Two

So far your study of algebra has emphasized only linear and quadratic functions. These two types of functions are often useful in solving practical problems. They are simple examples of a large family of functions called **polynomials**. The functions with rules

$$f(x) = 4x^3 + 5x^2 + 6x - 1,$$
$$g(x) = 9x^4 - 3x + 8, \text{ and}$$
$$h(x) = 23x^{45} - 11x^{17} + 3x$$

are also called polynomial functions. In fact, any function with rule of the form

$$p(x) = a_n x^n + a_{n-1} x^{n-1} + \ldots + a_1 x + a_0$$

(where the coefficients a_0, a_1, a_2, a_3, ... , a_n are specific numbers for any particular function, and $a_n \neq 0$) is called a **polynomial function**. The **degree** of a polynomial is the largest exponent occurring when the function rule is written in this form. For instance, the polynomial function

$$g(x) = 9x^4 - 3x + 8$$

has degree 4. Using the general subscript notation as in the rule above, the coefficients of this particular polynomial are $a_4 = 9$, $a_3 = 0$, $a_2 = 0$, $a_1 = -3$, and $a_0 = 8$.

The basic reason for being interested in polynomial functions of degree higher than one or two (the linear or quadratic cases) is that some relations between quantitative variables do not fit the graphic or numerical patterns modeled well by the simple linear or quadratic rules. Polynomial functions of higher degree may have graphs and tables of data pairs, (input, output), with different kinds of symmetry, different rates of change, different maximum/minimum patterns, and different types of solutions for related equations and inequalities.

Exploration

The following examples illustrate a few of the numerical and graphic patterns that can occur with polynomial functions of higher degrees.

Experiment 1. *Polynomial functions of degree 3*

A polynomial of degree 3 is called a **cubic polynomial**. For each of the following cubic function rules, use a technological tool to make a graph and a table of values, (input, output). Make notes of interesting patterns in the results for each individual function and for all examples taken together as a family. Record any sketches and other interesting observations you make.

1. $f(x) = 0.5x^3 - 6x$

2. $g(x) = -0.5x^3 + 6x$

3. $h(x) = x^3 + 5x^2 - 29x - 105$

   (Hint: Study this function for $-10 \le x \le 10$, that is, input values from $x = -10$ to $x = 10$ inclusive.)

4. $k(x) = x^3 + 2$ and $m(x) = x^3 - 2$

5. Make up rules for several additional cubic polynomial functions. Record those rules and sketches of their graphs, then list any observations you make.

Making Connections

As you compare the graphs and tables for various cubic polynomials to those of linear and quadratic functions that have been studied earlier, what similarities and differences stand out?

1. Compare the shapes of the graphs of cubics.

2. Compare the tables of values, (input, output), of cubics.

Experiment 2. *Polynomial functions of degree 4*

A polynomial of degree 4 is called a **quartic polynomial**. For each of the following quartic function rules, use a technological tool to make a graph and a table of values, (input, output). Make notes of interesting patterns in the results for each individual function and for all examples taken together as a family. Record any sketches and other interesting observations you make.

1. $f(x) = x^4 + x^3 - 20x^2 + 15$

2. $g(x) = -x^4 - x^3 + 20x^2 - 15$

3. $h(x) = 0.25x^4 - x^3 + 5$

4. $k(x) = 0.2x^4 - 6$ and $m(x) = 0.2x^4 + 2$

5. Make up rules for several additional fourth-degree (or quartic) polynomial functions. Record those rules and sketches of their graphs, then list any observations you make.

Making Connections

As you compare the graphs and tables for various quartic polynomials to those of linear, quadratic, and cubic polynomial functions studied earlier, what similarities and differences stand out?

1. Compare the shapes of the graphs of quartics.

2. Compare the tables of values, (input, output), of quartics.

Exercises

1. A partial graph for the cubic polynomial function $p(x) = x^3 - 9x$ is shown here. Use the graph to estimate roots for the following equations.

   a. $x^3 - 9x = 0$ b. $x^3 - 9x = 4$
   c. $p(x) = 10$ d. $p(x) = -10$

2. Use the graph of $p(x) = x^3 - 9x$ to find solutions for the following inequalities.

   a. $x^3 - 9x < 0$ b. $p(x) \geq 0$
   c. $x^3 - 9x \leq 10$ d. $p(x) > 10$

3. Estimate the coordinates of each maximum point and each minimum point shown on the graph of $p(x)$ below.

4. A partial graph for the quartic polynomial function $q(x) = 0.5x^4 - x^3 - 10x^2 + x + 30$ is shown below. Use the graph to estimate roots for each of the following equations.
   a. $q(x) = 0$ b. $q(x) = 10$
   c. $q(x) = 50$ d. $q(x) = -30$

5. Identify any maximum or minimum values that the function $q(x)$ appears to have.

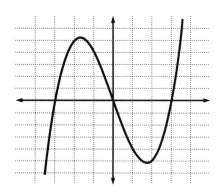

$p(x) = x^3 - 9x$

The horizontal (input) scale unit is 1 and the vertical (output) scale unit is 2.

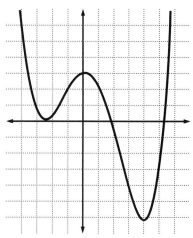

$q(x) = 0.5x^4 - x^3 - 10x^2 + x + 30$

The horizontal (input) scale unit is 1 and the vertical (output) scale unit is 10.

6 Summary

In situations that you know are modeled well by quadratic functions (of the form $f(x) = ax^2 + bx + c$), finding the answers to the important questions almost always requires one of the following.

1. Find the coefficients a, b, and c that give a rule relating the two variables of interest.

2. Evaluate $f(x)$ for given values of x.

3. Solve equations and inequalities involving $f(x)$, such as $f(x) = d$, $f(x) \leq d$, and $f(x) \geq d$.

4. Find the maximum or minimum value of $f(x)$.

5. Describe the rate at which $f(x)$ changes as x is changed.

Once you decide that a quadratic rule can be helpful in a situation, you have a variety of mental and technology-aided strategies for putting that rule to work. As you use those tools, remember these general facts about quadratic functions.

1. The coefficient of x^2 determines whether the function has a maximum or a minimum output and (correspondingly) whether the graph has a highest or a lowest point. If the coefficient is negative, the function has a maximum; if it is positive, the function has a minimum.

2. The coefficient of x combines with the coefficient of x^2 to determine the location of the maximum or minimum point and the line of symmetry for the graph of a quadratic function.

3. The constant term in a quadratic rule sets the point $(0, c)$, where the graph crosses the vertical axis.

4. Output values of a quadratic function do not change at a constant rate as input values change. In general, the rate of change increases as distance from the axis of symmetry increases. The coefficient of x^2 is the key factor in determining this rate of change.

5. A quadratic equation has 0, 1, or 2 real roots, depending on the function rule involved. Quadratic inequalities usually have solutions described by intervals of the form $e \le x \le f$ or "tails" of the form $x < e$ and $x > f$.

6. Polynomial functions of degree higher than 2 usually have quite different patterns in their graphs and tables of values, (input, output).

Review Exercises

1. The graphs of four quadratic functions are shown here. Match each graph with a rule from the following list.
   a. $0.4x^2 + 3x - 7$
   b. $-0.4x^2 - 3x + 7$
   c. $x^2 + 3x - 7$
   d. $-2x^2 + 5$

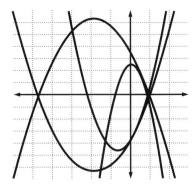

The horizontal (input) scale unit is 2 and the vertical (output) scale unit is 2.

2. Use the graphs from Exercise 1 to estimate the solution for each of the following equations and inequalities.

   a. $-2x^2 + 5 = -3$ b. $0.4x^2 + 3x - 7 = 0$

   c. $x^2 + 3x - 7 < 0$ d. $-0.4x^2 - 3x + 7 \leq 0$

   e. $-2x^2 + 5 > 3$ f. $-2x^2 + 5 = 13$

3. Make a table of values, (input, output), for the quadratic function $g(t) = -t^2 + 4t + 5$ for $t = -5$ to $t = 10$ in steps of 1.

4. Use the table produced in Exercise 3 to estimate solutions for each of the following equations and inequalities.

   a. $-t^2 + 4t + 5 = 0$ b. $-t^2 + 4t + 5 = 5$

   c. $-t^2 + 4t + 5 = 9$ d. $-t^2 + 4t + 5 \geq 5$

   e. $g(t) = 10$ f. $g(t) < 0$

   g. $-t^2 + 4t + 5 = -23$ h. $-t^2 + 4t = 3$

5. Solve each of the following simple quadratics using only mental or calculator arithmetic.

   a. $x^2 = 0.49$ b. $3x^2 + 8 = 56$

   c. $3x^2 + 12 = 10$ d. $x^2 - 4 = -4$

   e. $43 - x^2 = 18$ f. $475 = 55 + x^2$

   g. $x^2 < 64$ h. $x^2 \geq 20$

6. Gravity on the moon is approximately 0.16 of that on the earth. Further, the moon has no atmosphere to give air resistance to flying objects. One rule giving height (in meters) of a projectile as a function H of time in flight t (in seconds) is

$$H(t) = -0.83t^2 + v_0 t + h_0$$

where v_0 is the initial velocity of the projectile and h_0 is the projectile's initial height.

Suppose an arrow is shot off the moon's surface at 40 m/s.

   a. Write the rule expressing the height of the arrow as a function of its time in flight.

   b. Explain briefly how you would answer each of the following questions about your rule. If an equation or inequality is involved, write it in symbolic form. Then describe the tool (including mental or calculator arithmetic) you would choose and how you would use it.

   i. What is the maximum height of the arrow? When does it reach that height?

ii.   When does the arrow return to the moon's surface?

iii. What is the average speed of the arrow from $t = 0$ to $t = 1$ and from $t = 2$ to $t = 3$?

iv.   When does the arrow reach a height of 100 m?

Review Project

Part I. One of the first explorations in Chapter 1 was a simulation of business decision–making. You made choices about the planning of a talent show involving setting ticket and concession prices, deciding whether to hire a celebrity master of ceremonies, planning advertising, and ordering concession supplies. Then you were given a prediction of the attendance, costs, revenues, and profit from the show if your decisions were actually followed. A program of rules relating the input and output variables was used for these predictions—rules chosen by the authors, based on their judgment of how the variables would probably be related in real life. If you know some algebra and the relations built into the program, you can choose more easily a decision strategy to yield maximum profit.

For example, consider the case in which you choose to have a celebrity come to emcee the show:

1. The TALENT SHOW program has (in this case) a built-in demand function a relating ticket price p in dollars and attendance $a(p)$ by the rule

$$a(p) = -50p + 800.$$

Find the following values and explain the meaning of each.

a. $a(3)$ b. $a(10)$

c. p so that $a(p) = 600$ d. p so that $a(p) > 700$

2. The ticket revenue is calculated by multiplying the price per ticket, p dollars, by the attendance $a(p)$. Find the rule giving ticket revenue $R(p)$ as a function of ticket price, and use that rule to find each of the following values.

a. $R(5)$ b. $R(10)$

c. p so that $R(p) = 2000$ d. p so that $R(p) > 500$

e. p so that $R(p)$ is as large as possible

3. The demand function for the talent show without a celebrity as MC (emcee) is

$$d(p) = -100p + 800.$$

Write the revenue function in this case and find the ticket price for which the revenue is greatest.

Part II. The TALENT SHOW program includes other function rules relating concession prices to concession sales. Each is based on assumptions the program authors thought made sense. However, the relations among input and output variables in such a business–planning situation are usually discovered by some sort of market research. The following activities outline the steps that are typically involved. Use what you have learned about finding and analyzing functions to study the given situation.

S I T U A T I O N

Each year Northwestern has a special fund–raising campaign in which t-shirts are sold. Suppose that you have the task of ordering a supply of shirts and setting the price. Your goal is to arrange plans to maximize profit for the project.

1. Estimate the demand function for the t-shirts as follows. Survey your class and find out how many shirts would be purchased at each of the following prices. Then use the fact that there are a total of about 2000 students to estimate the number of shirts that would be sold at each price. Copy and complete the following table with this data.

Price in dollars	Students in your class who would buy at this price	Fraction of your class who would buy at this price	Expected total number of students who would buy at this price
4			
8			
12			
16			
20			

Find a demand function $s(p)$ that predicts sales of t-shirts as a function of price p.

2. Use your demand function $s(p)$ to find a rule that predicts revenue as a function R of price. Find the price that gives maximum revenue, then calculate the number of shirts you would expect to sell at this price.

3. Suppose that the company that will make the t-shirts quotes the following cost data.

Number of shirts	Total cost in dollars
100	500
250	1100
500	2000
750	2750
1000	3450
1500	4950
2000	6000

   Find a rule that gives total cost as a function of the number of shirts ordered.

   Then find a rule for the cost as a function of the selling price per shirt by using the rule for $s(p)$.

4. Find a rule giving the profit PR as a function of the selling price per shirt. Remember that profit equals revenue minus cost.

   Find the price for which your functions predict maximum profit.

   Find the number of shirts that will be sold, the revenue, and the cost at that price level.

   Find the price that will guarantee a break-even result (profit is 0), and find the predicted sales at that price level.

5. Write a report of your results. Discuss any factors other than price that you believe are important in the decision process.

Most cars are worth less each year than they were the year before. This decline in value is called depreciation, and can be modeled by an exponential function. What do you think a new car will be worth in five years?.

5 Exponential Functions

When a doctor gives a drug to a patient, the concentration of that drug in the patient's blood is a function of the time since the injection. When a *sky-diver* jumps from an airplane, speed in freefall is a function of time since the start of the jump. When a *disease epidemic* like flu or AIDS starts, the number of people with the disease is often simplified as a function of time since the start of the epidemic. When a mountain climber goes to higher altitudes, the *atmospheric pressure* decreases as a function of altitude. When you deposit money in an interest-bearing savings account, your *bank balance* increases as a function of time since the initial deposit.

Each of the very different situations just described can be modeled by a type of function different from the ones you have studied so far. The key variables in each situation are often expressed by a function rule having the form

$$f(x) = a^x,$$

where a is a constant determined by the situation being modeled. A function of this form is called an **exponential function** because the input variable x is used as an **exponent** in determining the value of the output variable.

By the end of this chapter you should be good at recognizing situations in which exponential functions of this type occur and at using properties of their tables, graphs, and rules to solve problems.

1 Recognizing Exponential Functions

With the previous types of functions you studied, we noted many patterns common to members of each type. A linear function, with rule of the form $f(x) = mx + b$, has a straight-line graph with constant slope m and y-intercept b. As the input variable of a linear function changes, the output variable always changes at a constant rate. On the other hand, a quadratic function, with rule of the form $f(x) = ax^2 + bx + c$, has a graph that is a symmetric curve called a parabola. The graph has y-intercept c, has either a maximum point or a minimum point, and has outputs which change at varying rates. The output variable in a quadratic function does not change at a constant rate as the input variable changes.

As you study the examples of exponential functions in this section, look for patterns in the various graphs and tables and the relation between those patterns and the rules of the form $f(x) = a^x$.

SITUATION 1.1

The winner in the Sweepstakes America has two options for payoff: $100,000 paid immediately, or a special savings bond with initial value of $1 that doubles each year until it pays off 20 years later. Which payoff would you prefer?

There are many factors to consider in making a decision like this—some can be described mathematically and some are related to your own personal desires. However, the obvious unknown is "What would that savings bond be worth in 20 years?" One way to find this is to construct a table of values for the relation between time and bond value. A start on this process is given here.

Time (years)	Bond value (dollars)
0	1
1	$1 \times 2 = 2$
2	$2 \times 2 = 4$
3	$4 \times 2 = 8$
4	$8 \times 2 = 16$
5	$16 \times 2 = 32$
6	$32 \times 2 = 64$
7	$64 \times 2 = 128$

This table clearly shows the doubling pattern from one year to the next. You could continue the table in this manner, since you are interested in the value of the bond after 20 years. This would require many more lines in the table!

Because this doubling pattern occurs in many situations, mathematicians have devised a more efficient way to write and to calculate output values for a rule of this type. The next table shows some of the required calculations in a different way.

Time (years)	Bond value (dollars)
0	1
1	$1 \times 2 = (1) \times 2 = 2$
2	$2 \times 2 = (1 \times 2) \times 2 = 4$
3	$4 \times 2 = (1 \times 2 \times 2) \times 2 = 8$
4	$8 \times 2 = (1 \times 2 \times 2 \times 2) \times 2 = 16$
5	$16 \times 2 = (1 \times 2 \times 2 \times 2 \times 2) \times 2 = 32$

This partial table suggests that to get the value of the bond after 20 years we should multiply the initial value \$1 by 20 factors, each of which is 2. The repeated multiplication by 2 can be written in very brief mathematical notation using the **exponential form** 2^{20}; that is

$$2^{20} = 2 \times 2 \times 2 \times 2 \times 2 \times 2 \times 2 \times 2 \times 2 \times 2 \times 2 \times 2 \times 2 \times 2 \times 2 \times 2 \times 2 \times 2 \times 2 \times 2.$$

Here the number 2 is called the **base** and the number 20 is called the **exponent.**

Of course, the exponential form can be used as shorthand for any number of repeated factors. For instance, $2^3 = 2 \times 2 \times 2 = 8$ and $2^5 = 2 \times 2 \times 2 \times 2 \times 2 = 32$. The expression 2^3 is usually read "2 to the third power" and 2^5 is read "2 to the fifth power." In general, 2^n is shorthand for a product of n **factors**, each of which is **2**:

$$2^n = \underbrace{2 \times 2 \times 2 \times \cdots \times 2 \times 2}_{n}$$

The bracket and n shown above are used to indicate that there are n factors.

In a similar way, the notation "5^3" stands for "5 times 5 times 5" or "5 to the third power." For any base a and any *positive* integer exponent n,

$$a^n = \underbrace{a \times a \times a \times \cdots \times a \times a}_{(n \text{ factors})}$$

and we read a^n as "a **to the nth power**".

Because calculations like 2^5 and 5^3 occur so often, most technological tools are equipped to compute values for exponential expressions with only a few key strokes. For instance, many scientific calculators have "y to the xth power" keys $\boxed{y^x}$. To use this exponent key for a calculation like 2^5, one usually proceeds as follows.

1. Enter 2 (the base).
2. Press the key $\boxed{y^x}$.
3. Enter 5 (the exponent).
4. Press the key $\boxed{=}$.

The numeral 32 would then show in the display.

Find the proper method to compute exponential values on your technological aid, then evaluate 2^{20}. Are you surprised at how large this number is? Using this information, which do you think might be the better option for the Sweepstakes America winner: to wait 20 years for a large sum of money, or to receive $100,000 today?

As you will see in the explorations that follow, exponential expressions of the form y^x and some simple variations are useful in modeling a wide variety of relations among quantitative variables. Further, while you have just seen how to use positive integral exponents, there are natural ways to calculate and to interpret the values of exponential expressions with negative or fractional exponents as well.

Exercises

In Exercises 1 through 8, write each of the calculations in shorter form using exponential notation.

1. $2 \times 2 \times 2 \times 2$

2. $3 \times 3 \times 3 \times 3 \times 3 \times 3$

3. $21 \times 21 \times 21$

4. $(0.5) \times (0.5) \times (0.5)$

5. $a \times a \times a \times a$

6. $500 \times 4 \times 4 \times 4 \times 4 \times 4$

7. $2 \times 2 \times 2 \times 2 \times 21 \times 21 \times 21$

8. $21 \times 21 \times 21 \times a \times a \times a \times a$

In Exercises 9 through 15 write the indicated calculations in equivalent form without exponents, then use technology to find the value in standard whole–number or decimal form. In Exercises 13 through 15, the caret (^) is used to signify the exponent operation, as is done with some technology.

9. 4^3

10. 25^2

11. $(1.5)^3$

12. 10^6

13. (1000×2) ^ 3

14. 1000×2^ 3

15. $1000 \times (2$ ^ $3)$

Exploration I

The example in this exploration illustrates an important application of exponential functions—modeling of growth in populations of people and many other kinds of living things. As you study the situation, notice particularly the way that the population at any time depends directly on the population at an immediately preceding time.

SITUATION 1.2

Public health departments monitor the cleanliness of restaurants, grocery stores, swimming pools, and other public facilities, because harmful bacteria can grow very quickly in untreated conditions.

For instance, data in the following table show a typical pattern of bacterial growth in water of a swimming pool if no filtration or chlorine is used.

Time (days)	Number of bacteria per cubic centimeter
0	1000
1	2000
2	4000
3	8000

Copy and complete the table for input values through 10 in a pattern that matches the entries in the first four rows.

The pattern in this table is easy to extend and to describe. From each day to the next, the bacteria population density *doubles*:

$$2000 = 1000 \times 2,$$
$$4000 = 2000 \times 2,$$
$$8000 = 4000 \times 2,$$

and so on. If d represents the time in days and $N(d)$ represents the number of bacteria per cubic centimeter on day d, then the above pattern can be written as follows:

$$N(2) = N(1) \times 2,$$
$$N(3) = N(2) \times 2,$$
$$N(4) = N(3) \times 2,$$

and so on. This pattern can be summarized by the rule

$$N(d+1) = N(d) \times 2.$$

This means that the number on any given day equals twice the number on the previous day. We also might say doubling the number on any given day produces the number on the next day.

While the doubling pattern is simple, using it to calculate the bacteria population density on the 10th day or the 15th day seems to require calculation of the population densities for all the days from 0 through the desired day. Exponential notation gives another way to write a rule for $N(d)$. Then technology can be used to find the output value for any input.

Copy and complete the following table showing calculations used to find the outputs corresponding to inputs 1, 2, 3, 4, 5, 6, and 7.

Time (days)	Number of bacteria per cubic centimeter
0	1000
1	$1000 \times 2 = 1000 \times 2^1 = 2000$
2	$2000 \times 2 = 1000 \times 2 \times 2 = 1000 \times 2^2 =$ ____
3	___ $\times$ ___ $= 1000 \times$ ____ $= 1000 \times 2\text{-} =$ ____

This table suggests that to get the bacteria count on day d we should multiply the initial count 1000 by d factors, each equal to 2. Using exponential notation, the function rule can be written

$$N(d) = 1000 \times 2^d.$$

Remember: In calculating output values for the function with rule $N(d) = 1000 \times 2^d$, the first step is to evaluate 2^d and the second step is to multiply that result by 1000. Order of operation rules prescribe *exponentiation before any multiplication, division, addition or subtraction.* For example,

$$N(5) = 1000 \times 2^5$$
$$= 1000 \times 32$$
$$= 32\,000.$$

Notice what would happen if you multiplied *before* evaluating the exponent:

$$(1000 \times 2)^5 = 2000^5 = 32\ 000\ 000\ 000\ 000\ 000.$$

SITUATION 1.3

Suppose that a sample of water from a swimming pool shows 1500 bacteria per cubic centimeter at 8 a.m. on a Monday morning and that the density of bacteria in that pool doubles every day thereafter.

1. Copy and complete the following table showing number of bacteria per cubic centimeter as a function, N, of time from Monday (0 days) to the following Monday (7 days). Then plot the data pairs, (time, density), using a unit of 5000 on the output axis.

Time (days)	Number of bacteria per cubic centimeter
1	
2	
3	
4	
5	
6	
7	

Use the table and graph to answer the following.

2. Write a rule giving number of bacteria per cubic centimeter as a function, N, of time in days.

3. Describe the pattern in bacteria population density as time passes.

4. How is the table of values for this function similar to, and how is it different from, tables for linear and quadratic functions?

5. How is the graph of this function similar to, and how is it different from, graphs of linear and quadratic functions?

6. For each of the following questions, write a report describing your solution of the problem. In detail:
   –Use the function rule to write a calculation, equation, or inequality that can be used to answer the question.

−Decide on the best tool or tools to solve the problem. You might choose a calculator, a computer and its table, graph, or symbol manipulation program, or the power of your own brain with some paper-and-pencil aid in recording results.

−Solve the problem and write a sentence stating your answer to the original question.

a. What is the bacteria count per cubic centimeter on the Saturday after the test?

   i. Record the calculation, equation, or inequality you will use.

   ii. Indicate which tool or tools you feel are best to solve the problem.

   iii. Answer the original question.

b. If health officials set 200,000 per cubic centimeter as the maximum bacteria count for pool water that is safe to use, how long will this pool stay in the safe range?

   i. Record the calculation, equation, or inequality you will use.

   ii. Indicate which tool or tools you feel are best to solve the problem.

   iii. Answer the original question.

7. The number of bacteria on Monday, at the start of the test counting, is 1500 per cubic centimeter. The rule for the number of bacteria $N(d)$ after d days have passed says that this value should be $N(0) = 1500 \times 2^0$. What value for the expression "2^0" makes sense?

8. Use the rule for $N(d)$ and a technology-generated table or graph to find the values of the following outputs. Then explain what each tells about the bacteria count in the pool. For example, $N(0) = 1500$. This means the bacteria count in the pool on the day of the test (day 0) was 1500 per cubic centimeter.

a. $N(-1)$

b. $N(-2)$

c. $N(-3)$

d. Compare your answers to parts a through c to the original value of 1500. As you did in Exercise 7, decide what values makes sense for 2^{-1}, 2^{-2}, and 2^{-3}.

9. Calculate each of the following, and explain what information the results give about the pool-bacteria situation.

a. $N(1) - N(0)$

b. $N(2) - N(1)$

    c. $N(5) - N(4)$

    d. What do these results say about the *rate of change* in the bacteria count as time passes?

10. If you ask for technological help in graphing $N(d) = 1500 \times 2^d$, you will find that most graphing utilities give a curve something like the one shown here.

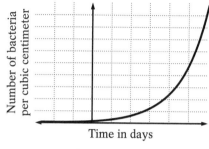

Use the graph to estimate the value for each of the following calculations. Then explain what each expression might tell about the bacteria population density.

$N(d) = 1500(2^d)$

The horizontal (input) scale unit is 1 and the vertical (output) scale unit is 10,000.

    a. $N(2.5)$

    b. $N(4.5)$

    c. $N(-1.5)$

Exploration II

The bacteria count problems in Exploration I show one important application of exponential functions and the tables and graphs that can occur for rules in the form $f(x) = Ca^x$. The example in Exploration II shows another situation in which an exponential rule helps to model relations among variables.

SITUATION 1.4

Nearly every year the prices of new cars go up. Once a car is purchased, however, its value often begins decreasing. In fact, for many cars the decline in value (called the depreciation) is about 30% per year. In other words, the value of a one-year-old car is only 70% of its purchase price.

1. With a car that sells for $12,000, this process of depreciation leads to the sequence of values begun in the following table. Copy and complete the following table to show how the car's value each year is derived from its value the year before.

Age of car (years)	Value of car (dollars)
0	12 000
1	12 000 × 0.70 = 8400
2	8400 × 0.70 = _____
3	_____ × 0.70 = _____
4	_____ × 0.70 = _____

2. The value of a car is a function V (in dollars) of the age of the car y (in years). Copy and complete the following examples showing the pattern of year-to-year changes in value.
   a. $V(2) = V(1) \times$ _____ .
   b. $V(5) = V(4) \times$ _____ .
   c. $V(7) = V(\_) \times$ _____ .
   d. The general pattern these examples show is $V(y + 1) =$ _____ .

3. The entries in the table you completed for Exercise 1 can be adapted to show that the relation is expressible in exponential form. Copy and complete the next table, using the pattern begun in the rows shown.

Age of car (years)	Value of car (dollars)
0	12 000
1	12 000 × 0.70 = (12 000) × 0.70
	= 12 000 × 0.70^1 = 8400
2	8400 × 0.70 = (12 000 × 0.70^1) × 0.70
	= 12 000 × 0.70^2 = 5880
3	5880 × 0.70 = (12 000 × 0.70^2) × 0.70 = _____

4. Using exponential notation, write a rule for $V(y)$, giving dollar value of the car at age y.

Notice that the patterns in this situation and in the earlier doubling situations show two ways to evaluate outputs for an exponential function $f(x) = Ca^x$. One way is to construct a table: start with the output C associated with the input 0; then, as the inputs increase with step size 1, multiply any output by the base a in order to find the next output. The other way is to evaluate the power a^x directly and then multiply it by the factor C. The equivalence of the two methods is an important feature of exponential functions.

5. Graph the car value function $V(y)$ on a coordinate grid such as the one shown here. Then use the tables, rule, and graph to complete Exercises 6 and 7.

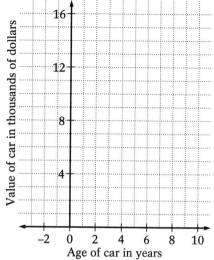

6. How are the table and graph of this depreciation function similar to and different from those of the bacteria growth example in Exploration I?

7. Use the function rule and a technological aid to answer the following questions.

   a. What are the values and meanings of the following?
      i. $V(9)$
      ii. $V(10)$
      iii. $V(7.5)$

   b. What is the solution to the following equation, and what does it tell about the car value?

$$V(y) = 325 \text{ or}$$
$$12\,000 \times 0.70^y = 325.$$

   c. The purchase price of the car is $12,000. The rule for $V(y)$ calculates this as $V(0) = 12\,000 \times (0.7)^0$. What value makes sense for the expression $(0.7)^0$?

   d. Suppose that your $12,000 was a used car, already several years old. Let $y = 0$ indicate when you bought the car. What would each of the following values be, and what would it tell about the car's value?
      i. $V(-1)$
      ii. $V(-2)$
      iii. $V(-3)$

   e. What is the result of each of the following calculations, and what does it tell about the car-value function?
      i. $V(1) - V(0)$
      ii. $V(2) - V(1)$
      iii. $V(7) - V(6)$
      iv. What do all three calculations, taken together, show about the rate at which car value changes as time passes? How does

this answer compare to the similar question about rate of change for bacterial growth in Exploration I?

f.   Write three questions of your own concerning the car depreciation situation so that one requires calculation of a function value, one requires solution of an equation, and one requires solution of an inequality. Then write each question in symbolic form, and write the equation, inequality, or calculation required. Answer the questions using technological tools as needed.

Exploration III

What do bacteria in a swimming pool have in common with money in a bank savings account? You will see in this exploration that both have similar growth patterns.

SITUATION 1.5

To help prepare for college costs, many families set aside money in college funds long before their children are of college age. Suppose a family puts $7500 in a savings account that pays interest at a rate of 8% per year. How will the balance in that account grow over the years?

To start with a simple case, let us assume that interest is paid at the end of each year, and that the interest is added to the previous account balance, but that no other deposits are made. Then the balance grows in the pattern illustrated by the following table.

Year	Account balance
0	7500.00
1	$7500.00 + (0.08 \times 7500.00) = 8100.00$
2	$8100.00 + (0.08 \times 8100.00) = 8748.00$
3	$8748.00 + (0.08 \times 8748.00) = 9447.84$
4	$9447.84 + (0.08 \times 9447.84) = 10\ 203.67$

The basic pattern used in calculating the next year's balance is

$$\text{Old balance} + \text{Interest earned} = \text{New balance.}$$

Notice that since the interest is added to the balance, it also draws interest over the next year. This is called **compound interest**.

It seems again that finding the balance after 10 or 15 years will require all the calculations in between. However, clever bankers and mathematicians once again have found a shortcut by making use of two basic properties of the number system as follows. See if you can explain why each expression is equal to the one before it.

$$\begin{aligned}
7500.00 + (0.08 \times 7500.00) &= (1 \times 7500.00) + (0.08 \times 7500.00) \\
&= (1 + 0.08) \times 7500.00 \\
&= 1.08 \times 7500.00 \\
&= 8100.00.
\end{aligned}$$

In exactly the same way, you can see that

$$\begin{aligned}
8100.00 + (0.08 \times 8100.00) &= (1 \times 8100.00) + (0.08 \times 8100.00) \\
&= (1.08) \times 8100.00 \\
&= 8748.00,
\end{aligned}$$

and so on.

This observation allows us to rewrite the calculations required in the college account table as follows:

Year	Account balance
0	7500.00
1	$1.08 \times 7500.00 = 1.08 \times (7500.00) = 8100.00$
2	$1.08 \times 8100.00 = 1.08 \times (1.08 \times 7500.00) = 8748.00$
3	$1.08 \times 8748.00 = 1.08 \times (1.08 \times 1.08 \times 7500.00)$ $= 9447.84$
4	$1.08 \times 9447.84 = 1.08 \times (1.08 \times 1.08 \times 1.08 \times 7500.00)$ $= 10\,203.67$

You can see the pattern that is developing. If we use n to indicate the number of years the account has been earning interest and $B(n)$ to indicate the balance in dollars after n years, one compact rule for this function is

$$B(n) = 1.08^n \times 7500.00 \text{ or}$$
$$B(n) = 7500.00 \times 1.08^n.$$

Use this rule and technological tools to answer the following questions about the growth of this college savings account. In each case, write the calculation, equation or inequality that matches the given question and write your answer, explaining the method and tools you use.

1. If a family makes the $7500 deposit when their first child is born, and no other deposits are made, what will the balance be 18 years later when the child is ready for college?

2. How long will it take the account to reach at least $15,000?

3. Copy and complete the following table and then graph $B(n)$.

Year n	Balance $B(n)$	Year n	Balance $B(n)$
0	7500.00	8	
1	8100.00	9	
2	8748.00	10	
3	9447.84	11	
4	10 203.67	12	
5		13	
6		14	
7		15	

4. How are the table and graph for this bank account function similar to and different from those of the bacteria growth and car depreciation functions in Explorations I and II?

5. Calculate each of the following differences, and explain what each tells about the college bank account.
   a. $B(1) - B(0)$
   b. $B(5) - B(4)$
   c. $B(15) - B(14)$
   d. What do the three differences in parts a, b, and c show about the rate (in dollars per year) at which such a bank account grows? Explain why this pattern occurs.

Making Connections

Each situation in this section involves variables related by a rule of the form $f(x) = Ca^x$, where C and a are constants. The most common case occurs when $C = 1$, that is, when $f(x) = a^x$. The facts of each situation determine the numbers C and a.

How are the tables and graphs of these exponential functions similar to, and how are they different from, those of linear and quadratic functions?

Exercises

1. SITUATION The growth of bacteria populations often depends on temperature as well as time. Suppose that in another study of swimming-pool water, the bacteria count is 500 per cubic centimeter at the start, but, because of warmer weather, the count triples every day.
   a. Make a table showing the number of bacteria per cubic centimeter as a function of time from day 0 to day 5. Then plot the data pairs, (time, density), on a coordinate grid with unit 1 on the time axis and unit 5000 on the bacteria-count axis.
   b. Write a rule giving number of bacteria per cubic centimeter in this pool as a function N of time d since the start of the study.
   c. Use your rule, table, or graph and a technological tool to answer the following questions.
      i. In what ways are the tables of values, (input, output), for this situation and for the doubling function similar, and in what ways are they different?
      ii. In what ways are the graphs of this function and of the doubling function similar and in what ways are they different?
      iii. How long will the bacteria count in this pool stay in the safe range, below 200,000 per cubic centimeter? Explain how you found your answer.
   d. Use a technological tool and your rule for bacteria count as a function of time in days to find the following values. Explain how each expression might be interpreted in terms of bacteria counts.
      i. $N(0.5)$
      ii. $N(2.5)$
      iii. $N(1.75)$
      iv. $N(-1)$
      v. $N(-2)$
   e. Write an equation that shows the relation between $N(d+1)$ and $N(d)$ in this situation.

2. SITUATION Some expensive luxury cars do not depreciate in value as quickly as other cars do. Suppose one such $60,000 car loses only 9% of its value each year.
   a. Write a rule relating time and value for this car, then use that rule to complete a table and graph for the value function for years 0 to 10.

250 Concepts in Algebra

    b. How are the rule, table, and graph for this function similar to, and how are they different from, those for the car with initial value $12,000 and depreciation rate 30% per year?

    c. How are both depreciation tables and graphs similar to or different from those of linear and quadratic functions?

    d. For each of the following questions, use function notation or the algebraic rule for your function from part a to write an expression, equation or inequality that can be used to answer the question. Then write a sentence giving your answer for the question.

        i. What is the value of the $60,000 car after 8 years?

        ii. When does the value of that car decrease to only $30,000?

        iii. How much value does the car lose during the first 5 years?

        iv. What is the value of the car after 1.5 years?

    e. How well does your rule describe the value of cars that are more than 25 years old?

3. SITUATION Suppose $100,000 is deposited in a bank savings account that pays interest of 6% at the end of each year.

    a. Calculate the balance in this account at the end of each of the first five years, assuming no other deposits are made.

    b. Copy and complete the following rules relating time in years y and balance in dollars B in this bank account.

$$B(y + 1) = \underline{} \times B(y) \text{ and}$$
$$B(y) = \underline{} \, .$$

    c. For each of the following, use a rule, a table, or a graph to find the indicated value(s), then write a sentence explaining what your answer tells about the bank account.

        i. $B(8)$

        ii. $B(20)$

        iii. Values of y for which $B(y) \geq 200\,000$

        iv. $B(2) - B(1)$

        v. $B(10) - B(9)$

    d. Recall that in Situation 1.1, the value of the savings bond after n years is 2^n dollars. How does this plan compare with $B(y)$ after 5 years? After 10 years? After 15 years? After 20 years?

4. SITUATION Due to varying rates of inflation, the rate at which the value of a car actually depreciates may vary from year to year. Suppose a $15,000 car depreciates at the following rates over a five-year period.

Year	1	2	3	4	5
Rate of depreciation	36%	29%	35%	27%	23%

    a. Use these rates and a technological tool to find the value of the car at the end of each of the five years. You will need to proceed one year at a time.

    b. Find the average of the five given depreciation rates.

    c. Suppose that the average rate is used in an exponential rule for depreciation of the $15,000 car. Write the rule giving value of the car as a function of time in years.

    d. Use your rule from part c to calculate the value of the car after each of the 5 years and compare those values to the results with variable depreciation rates.

5. For any *exponential* function $f(n) = a^n$ with positive base a, when the input changes from any integer n to $n + 1$, the output changes according to the rule $f(n+1) = a \times f(n)$. Recall that for a linear function the slope of its graph is closely connected with how the output varies as the input changes from x to $x + 1$. Review this important connection by answering the following questions.

Each of the following is a table of values for a linear function. For each function: (i) find the rule in the form $g(x) = mx + b$ (ii) identify the slope of the graph, and (iii) try to find a relation between $g(x+1)$ and $g(x)$.

a. Input	Output		b. Input	Output		c. Input	Output
0	12		0	−7		−5	23
1	14		1	−4		−4	16
2	16		2	−1		−3	9
3	18		3	2		−2	2
4	20		4	5		−1	−5

Finally write a statement telling how the outputs $g(x)$ and $g(x+1)$ are related for any linear function with rule $g(x) = mx + b$ and any input x.

2 Properties of Exponential Functions

In the explorations of Section 1 you saw several different examples of the situations in which exponential functions $f(x) = a^x$ and $f(x) = Ca^x$ are good models for relations between variables. Several basic facts about exponents were probably quite clear; several others were suggested but not fully developed.

The starting point for use of exponential functions is the basic definition:

1. **For any number a and any positive integer n,**

    $$a^n = a \times a \times a \times a \times ... \times a \times a \ (n \text{ factors}).$$

Several situations suggested extending this definition to the case of 0 as an exponent. In Situation 1.3, the original number of swimming pool bacteria was 1500 per cubic centimeter. That is, $N(0) = 1500$. The function rule in that case is $N(t) = 1500 \times 2^t$. So $N(0) = 1500 \times 2^0$. It follows that $1500 = 1500 \times 2^0$. So $2^0 = 1$.

2. **For any value of a (except $a = 0$), $a^0 = 1$.**

There is also a natural way to define a^{-n} when n is an integer.

3. **For any number a (except $a = 0$) and any integer n,**

    $$a^{-n} = \frac{1}{a^n}.$$

This third property of exponents is probably more difficult to understand than the first two, but remember the following examples of bacteria growth and automobile depreciation where it was mentioned briefly.

Example 1. In a pool the bacterial population was 1500 per cubic centimeter at 8 a.m. on Monday and doubling every day. This led to the model

$$N(d) = 1500 \times 2^d.$$

It seems reasonable that the population density was 750 one day *before* that Monday [$N(-1) = 750$], 375 two days *before* that Monday [$N(-2) = 375)$], about 188 three days *before* that Monday [$N(-3) \approx 188$], and so on. It seems reasonable to conclude the following.

a. $N(-1) = 750 \Rightarrow$

$1500 \times 2^{-1} = 750 \Rightarrow$

$2^{-1} = \dfrac{750}{1500} \Rightarrow$

$2^{-1} = \dfrac{1}{2} \Rightarrow$

$2^{-1} = \dfrac{1}{2^1}.$

b. $N(-2) = 375 \Rightarrow$

$1500 \times 2^{-2} = 375 \Rightarrow$

$2^{-2} = \dfrac{375}{1500} \Rightarrow$

$2^{-2} = \dfrac{1}{4} \Rightarrow$

$2^{-2} = \dfrac{1}{2^2}.$

c. $N(-3) = 187.5 \Rightarrow$

$1500 \times 2^{-3} = 187.5 \Rightarrow$

$2^{-3} = \dfrac{187.5}{1500} \Rightarrow$

$2^{-3} = \dfrac{1}{8} \Rightarrow$

$2^{-3} = \dfrac{1}{2^3}.$

Example 2. If a car is bought used for \$12,000 and loses about 30% of its value every year, the value y years later can be predicted from the model $V(y) = 12\,000 \times 0.70^y$.

It seems reasonable to infer that the value in the years before the used car purchase can be calculated as follows.

a.   For the value one year earlier

$$V(-1) \times 0.70 = V(0) \text{ or}$$

$$V(-1) = \frac{V(0)}{0.70}.$$

If the model holds for one year earlier, then $V(-1) = 12\,000 \times (0.70)^{-1}$. We know that $V(0) = 12\,000$. We substitute for each, giving

$$12\,000 \times (0.70)^{-1} = \frac{12\,000}{(0.70)^1} = 12\,000 \times \frac{1}{(0.70)^1}.$$

So it seems that multiplying by $(0.70)^{-1}$ is the same as dividing by $(0.70)^1$, or multiplying by $\frac{1}{(0.70)^1}$. This implies

$$(0.70)^{-1} = \frac{1}{(0.70)^1}.$$

b. Similarly, for the value two years earlier, we have the following.

$$V(-2) \times (0.70)^2 = V(0) \Rightarrow$$

$$V(-2) = V(0) / (0.70)^2 \Rightarrow$$

$$12000 \times (0.70)^{-2} = 12\,000 / (0.70)^2.$$

So it seems that multiplying by $(0.70)^{-2}$ is the same as dividing by $(0.70)^2$ or

$$(0.70)^{-2} = 1 / (0.70)^2.$$

Finally, the fact that exponents are used to indicate repeated multiplication leads to a fourth useful property of exponents, which should make sense as you look at the following examples of its use.

Example 3. $\quad 2^3 \times 2^5 = (2 \times 2 \times 2) \times (2 \times 2 \times 2 \times 2 \times 2)$
$$= (2 \times 2 \times 2 \times 2 \times 2 \times 2 \times 2 \times 2)$$
$$= 2^8.$$

In words, when a product of three factors is multiplied by a product of five factors, there are altogether $3 + 5 = 8$ factors.

Example 4. $\quad a^3 \times a^2 = (a \times a \times a) \times (a \times a)$
$$= (a \times a \times a \times a \times a)$$
$$= a^5.$$

The general pattern suggested by these two examples is:

4. **For any positive number a and any integers m and n,**

$$a^m \times a^n = a^{m+n}.$$

The following exercises give practice in applying all of these properties of exponents.

Exercises

In Exercises 1 through 10, write each of the expressions in standard decimal form.

1. 4^3 **2.** $(1.5)^2$ **3.** 10^2 **4.** 10^{-2}

5. 2^{-2} **6.** $(0.5)^2$ **7.** 10^3 **8.** 10^{-3}

9. $\dfrac{1}{3^0}$ **10.** $\dfrac{1}{7^{-2}}$

In Exercises 11 through 16, solve each equation.

11. $2^x = 16$ **12.** $3^x = 9$ **13.** $50 \times 2^x = 400$

14. $2^x = 0.25$ **15.** $10^x = 1\,000\,000$ **16.** $1.85^x = 1$

In Exercises 17 and 18, you should apply the patterns relating exponents, multiplication, and addition.

17. By the definition of exponents, $2^8 = 2 \times 2 \times 2 \times 2 \times 2 \times 2 \times 2 \times 2$. To find the standard decimal name for this result, 256, you can use the fourth property of exponents we introduced to arrange the problem in more familiar pieces. For example,

$$2^8 = 2^{3+5}$$
$$= 2^3 \times 2^5$$
$$= 8 \times 32$$
$$= 256.$$

  a. There are 6 other ways to write 2^8 as a product of two groups of factors like this. Write each of them in the style shown above.

  b. There are two ways to write 2^8 as a product of two exponential factors using 0 as one of the exponents. Write both of them.

  c. There are many ways to write 2^8 as a product of two exponential factors if negative exponents are allowed. Write at least six examples, beginning with completion of the following:

$$2^8 = (2^{-3}) \times (2^-).$$

  See how many more ways you can write 2^8 as a product of two exponential factors.

18. Follow the pattern illustrated in Exercise 17 for this exercise.

  a. Find at least five ways to write 2^6 as a product of three exponential expressions using positive numbers or 0 as exponents.

    b. Find five different ways to write 2^6 as a product of three exponential expressions, each product involving at least one negative exponent.

    c. Find all possible ways to write s^7 as a product of two exponential expressions using only positive integers or 0 as exponents.

    d. Find five ways to write s^7 as a product of two exponential expressions using at least one negative exponent.

 # Exploration I

Evaluating a^x for x Between Integers In the Explorations of Section 1 you studied several situations in which function rules of the form $f(x) = Ca^x$ help model relations among variables. For integer values of x it was usually clear what the output of those functions said about the situations and how those output values could be calculated. For example, if the swimming pool bacteria count is 1500 per cubic centimeter on Monday morning at 8 a.m. and doubles each day thereafter, the count three days later is given by

$$
\begin{aligned}
N(3) &= 1500 \times 2^3 \\
&= 1500 \times (2 \times 2 \times 2) \\
&= 1500 \times 8 \\
&= 12\,000.
\end{aligned}
$$

However, the bacteria certainly do not wait *exactly* 24 hours and then double instantaneously, then wait another 24 hours and double instantaneously, and so on. Some of the bacteria are splitting into two new bacteria at times throughout the day. If we are interested in the quality of the water for swimming, we may need to ask questions such as: what is the bacteria count at 8 p.m. on Monday (0.5 days after the initial count), and what is the bacteria count at 2 p.m. on Thursday (3.25 days after the initial count)?

These two questions correspond to the following calculations:

$$
\begin{aligned}
N(0.5) &= 1500 \times 2^{0.5} \text{ and} \\
N(3.25) &= 1500 \times 2^{3.25}.
\end{aligned}
$$

We now have a problem. How do we calculate $2^{0.5}$ and $2^{3.25}$? The repeated multiplication process used to evaluate $N(3) = 1500 \times 2^3$ does not work when the exponent is not an integer. However, it seems reasonable that $2^{0.5}$ should be between $2^0 = 1$ and $2^1 = 2$. Similarly, $2^{3.25}$ should be between $2^3 = 8$ and $2^4 = 16$.

Using a technological tool to find these values, you may get outputs suggesting that $2^{0.5} \approx 1.4142136$ and $2^{3.25} \approx 9.5136569$, which support our hunches about what those values ought to be.

How are the expressions with decimal exponents evaluated? How are other powers with fractional exponents calculated? How do these calculations relate to exponential calculations with integral exponents?

There is a reasonable, although somewhat complicated, way to define powers with fractional exponents. Those exponents give very useful information if interpreted with common sense. As a practical matter, calculation of the output a^x when the input x is not an integer is usually done using technology.

1. In order to gain a better understanding of fractional exponents, calculate the outputs in the following three tables. Make suitable use of a technological tool, as needed.

a. x	$f(x) = 81^x$	b. x	$g(x) = 16^x$	c. x	$h(x) = (0.36)^x$
0		0		0	
0.25		0.2		0.05	
0.50		0.4		0.10	
0.75		0.6		0.90	
1		0.8		0.95	
2		1		1	
2.25		3		4	
2.50		3.1		4.4	
2.75		3.9		4.6	
3		4		5	

2. Look over the results you calculated for the three tables. For an exponential function $f(x) = a^x$ (where the base a is positive but different from 1), if an input x is between two numbers p and q, how is the output $f(x)$ related to $f(p)$ and $f(q)$? You may want to test your ideas by considering additional examples with different inputs or different bases.

In many cases, when an exponential function models a situation, fractional exponents are highly useful. In other cases, they may not be particularly meaningful—it depends upon the situation! You should be sure that any fraction you use as an exponent in an application is sensible.

There is one family of powers, those involving the fractional exponent 0.5 or 1/2, whose calculation can be explained easily. For any $a >$ 0, $\sqrt{a} \times \sqrt{a} = a$. However, using the properties of exponents, $a^{0.5} \times a^{0.5} = a^{0.5 + 0.5} = a^1 = a$. It seems reasonable, then, that $a^{0.5} = \sqrt{a}$. As it turns out, this is the case.

Because square roots occur in calculations required to solve many practical problems, mathematics reference books often contain tables of square roots. Technology has eliminated most of the need for such tables, but those values, and basic properties of exponents, still can be applied to evaluate a variety of exponential expressions.

The basic idea is to break an exponential calculation into two factors, one involving an integer exponent and the other a square root. Some examples are shown here.

$$2^{8.5} = 2^{(8 + 0.5)}$$
$$= (2^8)(2^{0.5})$$
$$= 256\sqrt{2}$$
$$\approx 256(1.414)$$
$$\approx 362.04.$$

$$5^{3.5} = 5^{(3 + 0.5)}$$
$$= (5^3)(5^{0.5})$$
$$= 125\sqrt{5}$$
$$\approx 125(2.236)$$
$$\approx 279.5.$$

$$(0.6)^{2.5} = (0.6)^{(2 + 0.5)}$$
$$= (0.6)^2(0.6)^{0.5}$$
$$= 0.36\sqrt{0.6}$$
$$\approx 0.36(0.775)$$
$$\approx 0.279.$$

As you can see, technology makes life with square roots and other fractional exponents much easier than working by hand and looking up square roots in printed tables.

Exercises

In Exercises 1 through 8, evaluate the given exponential expressions. Use a technological tool as needed.

1. $49^{0.5}$ 2. $32^{0.2}$ 3. $(0.64)^{1.5}$ 4. $27^{1/3}$

5. $7^{2.5}$ 6. $(0.6)^{-2.5}$ 7. $0^{23.1}$ 8. 23.1^0

In Exercises 9 through 12, calculate the first and third values without technology. Then use that information to *estimate* the remaining values.

9. 2^3, $2^{3.5}$, 2^4 10. 3^1, $3^{1.4}$, 3^2

11. $(4.5)^0$, $(4.5)^{0.6}$, $(4.5)^1$ 12. 6^2, $6^{2.3}$, 6^3

13. **S I T U A T I O N** Suppose a carton of milk contains 500 bacteria per cubic centimeter at the time it was bought, and that the number of bacteria doubles every day. The number of bacteria in the carton is then a function N of the number of days, *d*, since the milk was bought.

a. Explain what each of the following means for this situation.
   i. *N*(7)   **ii.** *N*(3.5)   **iii.** *N*(2.25)

b. Write a rule for the number of bacteria as a function of the number of days since the milk was bought.

c. Use the rule you wrote in part b to calculate the following values.
   i. *N*(3.5)   **ii.** *N*(0)   **iii.** *N*(14)

d. Suppose that it is not safe to drink the milk when the bacteria count is more than 20,000 per cubic centimeter. For how many days after you buy it can you safely drink the milk? Explain how you get your answer.

Exploration II *Properties of Tables and Graphs for* $f(x) = Ca^x$

Earlier in this chapter you investigated situations in which exponential functions with rules of the form $f(x) = Ca^x$ occur as models relating variables. You calculated inputs and outputs and related your results to situations such as population growth, automobile depreciation, and savings account interest.

These situations suggest questions such as "What if the number of bacteria quadrupled each day, instead of doubling or tripling?" or "What will my savings account be worth in 10 years if I start with $10,000 instead of $7500?" Maybe you are curious to know what the world population will be in the year 2525. A shrewd car buyer might want to know which would be more valuable in three years: A $15,000 car which depreciates 30% every year or a $12,000 car with a 20% depreciation rate. The buyer might want to compare the functions with rules $V_1(t) = 15\,000(0.7^t)$ and $V_2(t) = 12\,000(0.8^t)$. Exponential functions of the form $f(x) = Ca^x$ can be used to answer most of these questions by changing the values of C and a to fit each given situation.

As you might recall, for a linear function with rule $f(x) = mx + b$ you can use the values of m and b to determine rate of change in a table or the slope and intercept of a graph. For a quadratic function with rule $g(x) = ax^2 + bx + c$ the values of a, b, and c affect rates of change, the line of symmetry, and the location of the maximum or minimum point on the graph.

The goal of the experiments in this section is to help you find clues to the behavior of exponential functions. Since the conditions of a model can affect its accuracy, concentrate on what can be learned about the table and graph of $f(x) = Ca^x$ from the values of C and a.

Experiment 1

Given below is a list of six exponential functions, each with a rule of the form $f(x) = a^x$. To help you answer the questions that follow, use technology to display, for each function, a table of values (perhaps from $x = -8$ to $x = 8$ in steps of 1) and a graph (perhaps from $x = -6$ to $x = 6$).

The test functions: 2^x, 0.5^x, 1.2^x, 0.2^x, 0.8^x, and 3.5^x.

If you wish, you may include other test functions of the same type. It may help in your study of these functions to display several graphs on the same coordinate system or to display several tables at the same time.

1. Examine your tables and graphs for trends in the relation between inputs and outputs.
   a. In which of the tables of values or graphs do the outputs *increase* as the input values increase?
   b. Among those increasing functions, which has the greatest rate of increase? Which has the least?
   c. Find a pattern based on your answers to parts a and b. Why is the pattern you noted reasonable to expect?
   d. In which of the tables and graphs do the outputs *decrease* as the inputs increase?
   e. For which of those rules do the output values decrease most quickly? For which do the outputs decrease most slowly?
   f. Find a pattern based on your answers to parts d and e. Why is the pattern you noted reasonable to expect?

2. Is there any input x for which at least two of the six functions have the same output? If so, can you explain why?

3. For an exponential function with rule $f(x) = a^x$, how does the base a affect the table of data pairs, (input, output)? Hint: Consider the cases $a > 1$ and $a < 1$ separately.

Experiment 2

Given below is a list of six exponential functions with rules of the form $f(x) = Ca^x$. For each of the functions in the list and for any other similar functions of your own choosing, use technology to make a table of values and a graph. In each table you might pick values of x from –7 to 7 in steps of 1; use the tabular information as a guide in choosing scales for your graph.

The test functions: 0.6^x, $4(0.6^x)$, $8(0.6^x)$, 2.3^x, $4(2.3^x)$, and $8(2.3^x)$.

Identify the values of C and a in each of the test functions.

1. Examine your tables and graphs for trends in the relation between inputs and outputs.
   a. For which of these functions do output values always *increase* as input values increase?
   b. For which of these functions do outputs always *decrease* as input values increase?
   c. What common property of the rules in part a makes it reasonable that they should be *increasing,* and what property of the rules in part b makes it reasonable that they should be *decreasing?*

2. Are there any input values for which at least two of these functions give the same output? If so, why does that happen?

3. Describe the overall effect of C on the tables of values for functions of the form $f(x) = Ca^x$.

4. Is your conclusion about the effect of the number a in an exponential function a^x changed in the case Ca^x? If so, how?

Experiment 3

In the examples of Section 1 of this chapter, the graph of each exponential function of the form $f(x) = Ca^x$ (where C and a are positive numbers) fits one of the following patterns.

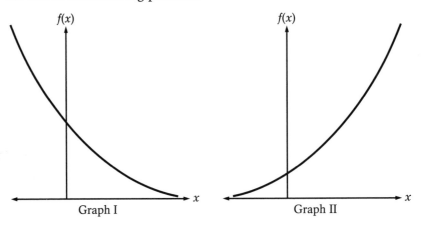

Graph I   Graph II

The following questions concern the relation between exponential function rules, $f(x) = Ca^x$, and the patterns of their graphs. Before trying to answer the questions, use a graphing utility to display the graphs of a wide variety of such functions with many different positive values of C and a.

In the sketches of Graph I and Graph II given above, notice that no grid points or scale tic marks are given, so that you must focus on the overall pattern of the graph—not on specific points.

1. Which values of a and C give graphs like Graph I above, and which give graphs like Graph II? Explain why your answer is reasonable.

2. What is the effect of the value of a on the graph? Why do you think this happens?

3. What is the effect of the value of C on the graph? Explain why your answer is reasonable.

Experiment 4

The following sketch includes graphs of three different exponential functions. No grid points or scale tic marks are given, so you should focus only on the overall relation of the three graphs.

1. Through experimentation with various exponential rules, find three rules that are related as those in the diagram. Call your rules $f(x)$, $g(x)$, and $h(x)$, as indicated in each corresponding graph.

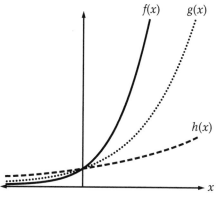

2. Next find three more exponential function rules with graphs related as those in the sketch at the right. Again, call your rules $f(x), g(x)$, and $h(x)$ to reflect the labels on the sketch.

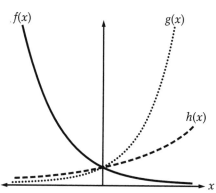

Making Connections

Consider the two exponential functions $f(x) = a^x$ and $g(x) = b^x$ with different positive bases a and b, respectively.

1. If a > b and each of the bases a and b is greater than 1, then which function (f or g) increases faster as the input x increases? Consider only positive values for x.

2. If a > b and each of the positive bases a and b is less than 1, then which function (f or g) decreases more rapidly as x increases? Consider only negative values of x.

3. Now consider two functions of exponential type, $h(x) = Ca^x$ and $j(x) = Da^x$, which have the same positive base a and which have different multipliers C and D. If C > D and each of the numbers C and D is positive, what can be said about the relationship between the graphs of the two functions h and j? You may wish to look back at graphs you have made such as those in Experiment 2. Explain how you arrived at your answer.

Exercises

1. Given below are sketches of the graphs of four exponential functions:

$$f(x) = 10\,(3)^x,$$
$$g(x) = 10\,(1.8)^x,$$
$$h(x) = 10\,(0.4)^x, \text{ and}$$
$$j(x) = 20\,(0.3)^x.$$

The same scales were used in sketching all the graphs, but the scales and tic marks have been removed from the drawings. Match each sketch with the rule it fits best.

a.

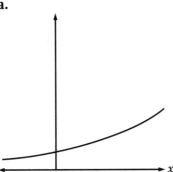

b.

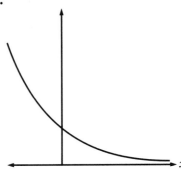

c.

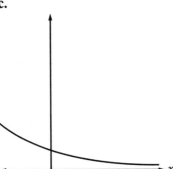

d.

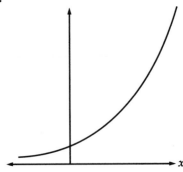

2. For each of the following four functions, sketch its graph on coordinate axes as shown in the sample response. On each graph, label three points with approximate coordinates. Select one with negative x-value, one with $x = 0$, and one with positive x-value.

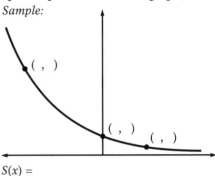

Sample:

(,)

(,) (,)

$S(x) =$

Rules:
a. $f(x) = 5\,(4^x)$
b. $g(x) = 5\,(0.4^x)$
c. $h(x) = 5\,(1.5^x)$
d. $j(x) = 12\,(0.3^x)$

3. Given below are partial tables of values for the following exponential functions:

$$f(x) = 8(2^x),$$
$$g(x) = 8(0.4^x),$$
$$h(x) = 9\,(1.5^x),\text{ and}$$
$$j(x) = 9\,(0.3^x).$$

Match the rules and tables by studying the patterns in the given data pairs, (input, output).

a.
Input	Output
-2	4
-1	6
0	9
1	13.5
2	20.25

b.
Input	Output
-2	50
-1	20
0	8
1	3.2
2	1.28

c.
Input	Output
-2	100
-1	30
0	9
1	2.7
2	0.81

d.
Input	Output
-2	2
-1	4
0	8
1	16
2	32

4. Consider the exponential function with rule $f(x) = 1^x$.
a. Make a table of pairs, $(x, f(x))$, for $x = -10$ to $x = 10$.
b. Sketch a graph of this function for the same x values.
c. How are the tables and graphs for all functions with rules $f(x) = C(1^x)$ similar, and how are they different?

Making Connections

The experiments in this section test the effects of changing values of C and a in rules for exponential functions of the form $f(x) = Ca^x$. Summarize your conclusions by answering the following questions.

1. How does the value of a affect the function tables and graphs?

2. How does the value of C affect the function tables and graphs?

3 Applying Exponential Functions

In previous sections of this chapter, you studied situations in which exponential functions are good models for relating variables and the patterns that occur in tables and graphs of exponential rules of the form $f(x) = Ca^x$.

In many cases the choice of an exponential model is dictated by problem conditions: the rule for depreciating value of a car or for paying interest on a bank account. In many other situations, such as the ones that follow, you will be faced with experimental data *suggesting* an exponential function, but not the values of C or a that make a rule fitting the data well.

SITUATION 3.1

Acquired Immune Deficiency Syndrome (AIDS) is probably the most serious public health concern in the United States today. In fact, medical researchers around the world are searching for causes of and cures for this serious disease.

One aspect of research on AIDS focuses on finding the rate at which the disease is spreading. The data below show the total number of *reported cases* in the United States through 1991. Note that these data give only the accumulated number of cases that have been reported with no subtractions for deaths.

Year	Total number of cases diagnosed	Year	Total number of cases diagnosed
Pre-1981	87	1986	40 638
1981	397	1987	68 433
1982	1 519	1988	102 501
1983	4 529	1989	141 753
1984	10 595	1990	182 761
1985	22 049	1991	225 233

Source: *The Universal Almanac.* NY: Andrews andMcMeel. 1993.

Is there an algebraic rule that fits these data well? If so, what does the rule predict about spread of the AIDS epidemic? To help answer these questions it is important to look more closely at the data for patterns relating time and number of cases.

Plot the given data pairs, (year, cases), on a coordinate grid. It simplifies work if you treat "Pre-1981" as year 0 and continue with 1981 as year 1, 1982 as year 2, and so on.

If you plot the data carefully, you should notice that the graph appears to have the shape of an exponential function. In fact, scientists have found that many diseases seem to spread exponentially as a function of time. Use this observation to answer the following questions.

1. Use function-fitting technology in order to find an exponential rule $N(y)$ that matches the given data. When entering the data points, remember to use 0 for the Pre-1981, 1 for 1981, and so on. Record the rule you found.

    Before going ahead, compare the graph of the data and the graph of the fitted function to judge the accuracy of the function model. Then copy and complete the following table to compare the function outputs with the data of reported AIDS cases. Extend the table to y=11.

Year y	Actual number of cases	Predicted cases calculated from $N(y)$
0	87	
1	397	
2	1 519	
3	4 529	
4	10 595	
5	22 049	
6	40 638	

2. Calculate and explain the meaning of each of the following function values.
   a. $N(-1)$ d. $N(4.5)$
   b. $N(0)$ e. $N(10)$
   c. $N(7)$ f. $N(20)$

3. Why would calculations such as the one you made for Exercise 2 worry public health officials?

4. Data indicate that by 1992 the number of AIDS cases reported in the United States reached a total of 253,448.
   a. What is the number of cases predicted by your rule for $N(y)$ for 1992?
   b. List some reasons why the two numbers are not the same.

5. Use your rule for $N(y)$ to write and solve equations or inequalities that match the following questions.
   a. The population of the United States in 1992 was approximately 252 million and was growing rather slowly. If the number of reported AIDS cases continues to grow according to the rule for $N(y)$, when will that number equal 252,000,000? Would that mean that every citizen of the U. S. had a reported case of AIDS?
   b. The population of China is now approximateiy 1,170,000,000. If the number of reported AIDS cases continues to grow according to the rule for $N(y)$, when will that number equal 1,170,000,000?
   Note: For numbers such as 1,170,000,000, many people (and technology) use **scientific notation** to write or display the number. In this notation, the number usually is written with only one digit to the left of a decimal point. Because multiplying or dividing a number by ten moves the decimal point one place, the true position of the decimal is shown as an exponent of ten. For example, 1,170,000,000 is often written as 1.17×10^9 or is often displayed by technology as 1.17E9 or 1.17E09. You may need to use this notation when entering 1,170,000,000 into your technological tool.
   c. World population in 1992 was approximately 5.42 billion (written 5,420,000,000; 5.42E9; or 5.42×10^9). According to your rule for $N(y)$, when will the predicted number of reported AIDS cases equal this number?

d. Based on your rule for $N(y)$, your answer to part c, and your own thoughts, will everyone eventually have AIDS? Why or why not? In what ways is the rule for $N(y)$ limited in predicting the future?

6. In March, 1985 an official of the National Institute of Allergy and Infectious Diseases estimated that the number of cases of AIDS was doubling every 9 months (0.75 year).
   a. Copy and complete five rows of the following table that relates time and AIDS cases in a way that fits the "doubling every nine months" pattern.

Years after 1980	Number of cases
5.00	22 049
5.75	44 098
6.50	

   Use function-fitting technology and these data to find a predicting rule $P(y)$ that follows this pattern.
   b. Use your rule for $P(y)$ to complete the following table. Compare its predictions to actual reported data.

Year y	Actual number of cases	Predicted cases $P(y)$	Year y	Actual number of cases	Predicted cases $P(y)$
0	87		7	68 433	
1	397		8	102 501	
2	1 519		9	141 753	
3	4 529		10	182 761	
4	10 595		11	225 223	
5	22 049		12	253 448	
6	40 638				

   c. What are some factors that may have caused this prediction to become inaccurate after 1986?

SITUATION 3.2

Storage of radioactive waste materials is an important environmental problem. Many hazardous radioactive waste products have been stored in various containers underground and in the ocean. As time passes, the radioactive substances decay into other, less harmful materials as they emit their radiation. The following are data showing a typical pattern for decay for cobalt, a radioactive metal.

Years in storage	Radioactive cobalt (grams)
0	100
2	75
4	56
6	42
8	30
10	24
12	20

One of the most important questions about such a relation is when will the radioactive material have decayed to an amount so small it is assumed to be no health threat. For instance, if a safe level of the substance for which data are given is 1 gram (1% of the original amount), we need a good prediction of the year in which that level will be reached. We must find a rule that fits the behavior of the substance.

1. Plot the seven data points on a coordinate grid to get some clues to the type of function rule suitable in this case.

If you plotted the data carefully you should notice that the pattern suggests an exponential function $f(x) = Ca^x$ for which the base a is less than 1. In fact, scientists have determined that such an exponential model fits the decay pattern for any radioactive substance. Thus, to find the rule for any particular substance, the problem is reduced to determining the decay constant a and the correct value for C.

2. a. Use your technological tools to find a rule that matches the data given above. Let $A(y)$ be the predicted grams of cobalt remaining after the substance has been in storage for y years.

   b. Study a diagram that shows both the data points and the graph of $A(y)$. How well do the two match?

   c. Copy and complete the following table comparing the actual data and the outputs predicted by your rule for $A(y)$.

Years in storage y	Radioactive cobalt (grams)	Predicted cobalt (grams) $A(y)$
0	100	
2	75	
4	56	
6	42	
8	30	
10	24	
12	20	

3. Using the rule for $A(y)$, calculate the following values and explain what each means in terms of Situation 3.2.
   a. $A(0)$
   b. $A(3.8)$
   c. $A(20)$
   d. $A(-2)$

4. Write and solve equations or inequalities that match the following questions.
   a. Scientists call the length of time it takes for half of a radioactive substance to decay the **half-life** of the substance. Different elements have different half-lives. Some are as short as 3 minutes (Polonium-218), and some are as long as 4,500,000,000 years (Uranium-238)! What is the half-life of radioactive cobalt? (Hint: How much is stored at time 0?)
   b. If 1 gram of radioactive cobalt is considered "safe", how long will it take for 100 grams to decay to a "safe amount"?

Each example in this chapter involves a function whose rule has the form $f(x) = Ca^x$ where the numbers C and a depend on the situation. The basic exponential form also appears in other function rules, as the next example shows.

SITUATION 3.3

After a sky-diver jumps out of an airplane, the diver's speed increases until the rip-cord is pulled and the parachute opens. However, the speed is affected by air resistance. Because of this, the quadratic model you studied in Chapter 4 does not make accurate predictions of speed as a function of time in free-fall. A rule involving an exponential function may provide a better fit.

If t represents the time of free-fall in seconds and $S(t)$ stands for the speed of the sky-diver in meters per second, a typical rule looks like

$$S(t) = 20 (1 - 0.6^t).$$

1. To understand the relation between time and speed that this rule implies, produce a table of values and a graph for values of t from 0 to 15 seconds.

2. Describe the pattern in the relation between time and speed that the table and graph show.

3. How are the table and graph of this function similar to, and how are they different from, those of other exponential, linear, and quadratic functions?

4. In this situation, $S(0) = 0$. Use the formula for $S(t)$ to show why that happens.

5. Calculate $S(t)$ for $t = 10, 15, 20,$ and 25. Then describe the pattern that appears when t is quite large.

   Copy and complete the following statements that should help explain the pattern you have observed.
   a. As t increases, 0.6^t....
   b. Thus as t increases, $(1 - 0.6^t)$....
   c. Thus as t increases, $20 (1 - 0.6^t)$....
   d. Combine your observations to explain why, in this situation, 20 meters per second is called the **terminal velocity**.

Exercises

1. Calculate each of the following values of the function $f(t) = 5 (1.02^t)$.
   a. $f(0)$ b. $f(1)$ c. $f(2)$ d. $f(10)$ e. $f(2.5)$

   Note: The input variable is labeled t in this rule. That is a common convention whenever time is the input variable.

2. Calculate each of the following values of the function $A(t) = 10 (0.20^t)$.
   a. $A(0)$ b. $A(1)$ c. $A(4)$ d. $A(9)$ e. $A(6.4)$

3. **SITUATION** The world population is approximately 5.292 billion and is growing at a rate of approximately 1.7% per year. This pattern is modelled well by the function with rule $P(t) = 5.292 (1.017^t)$ where t is the time in years from the present, and $P(t)$ is the number of billions of people at time t. Use this rule, your knowledge of exponential functions, and a technological tool to solve the following problems.
   a. Sketch a graph of $P(t)$. You need not plot specific points. Show only the trend in the relation between time and population.
   b. What world population does $P(t)$ predict for 10 years from the present?

    c. What world population does *P(t)* predict for 50 years from now?

    d. According to the rule, how long will it take the world population to double?

    e. According to the rule, when was the world population only 1 billion?

4. The rule for *P(t)*, giving world population as a function of time, predicts a population growing without upper limit. For instance, it predicts that the population will have reached 100 billion in about 175 years.

    a. Do you believe that the world population will ever grow as large as 100 billion?

    b. If so, do you believe 175 years to be a good estimate of the time it will take? Explain your reasoning.

    c. What do your answers say about the significance and limitations of models for population growth such as this one?

5. S I T U A T I O N   Suppose that 10 kilograms of a radioactive substance are stored. An exponential function *A* giving the number of kilograms remaining *t* years later has decay constant *a* = 0.8. Since 80% is left after a year, this means that 20% of the substance decays each year.

    a. Write the rule *A(t)* for the exponential function as described.

    b. Use the rule in part a to evaluate or to solve each of the following. Then explain briefly what each solution tells about the situation.

        i. $A(0)$

        ii. $A(1)$

        iii. $A(4.5)$

        iv. $A(t) = 2$

        v. $A(t) \leq 0.1$

    c. Sketch a graph of *A(t)*. You need not plot specific points—only show the basic trend in the relation.

4 Summary

In this chapter you have studied properties of functions with rules in the form

$$f(x) = Ca^x,$$

where a and C are positive numbers. You have seen many situations in which those functions are good models for relations among variables. The basic properties of exponential expressions include:

1. For any number a and any positive integer n,

   $$a^n = a \times a \times a \times \ldots \times a \times a \ (n \text{ factors});$$

2. For any number a except 0, $a^0 = 1$;
3. For any number a except 0 and for any integer n, $a^{-n} = \dfrac{1}{a^n}$; and
4. For any number a except 0 and for any integers m and n,

   $$a^m \times a^n = a^{m+n}.$$

Although we have given no formal description of the meaning of the power a^x where a is positive and x is not an integer, you have used technology to evaluate such powers and to display graphs. You have seen that for $C > 0$ and $a > 0$:

1. the graph of an exponential function Ca^x is a **smooth curve**;
2. the curve lies entirely **above the horizontal axis**; and
3. if $a > 1$, the function outputs **increase** as inputs increase, but if $0 < a < 1$, the outputs **decrease** as inputs increase.

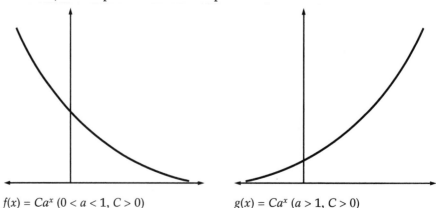

$f(x) = Ca^x \ (0 < a < 1, C > 0)$ $\qquad\qquad$ $g(x) = Ca^x \ (a > 1, C > 0)$

Notice that in each of the above points, C and a are both positive. The base, a, must always be positive, but the constant, C, can be negative. Explore on your own to see what affect the sign of this constant has on tables and graphs of this type of function. In many situations where exponential functions are good models of relations among variables, the input variable measures time. The value of the output variable at time $t + 1$ is calculated from the value at time t by multiplication with the constant factor a. In function notation this property of exponential functions is written

$$f(x + 1) = a \times f(x).$$

This type of relation gives a table of pairs, $(x, f(x))$. If the input changes with a step of 1, then the output in one row of the table can be calculated by multiplying the preceding output by the constant a. That pattern occurs in situations such as growth of population, growth of a savings account balance, or decay of a radioactive substance.

Review Exercises

1. Evaluate each of the following exponential expressions. Give your answer in whole-number, fractional, or decimal form.
   a. 2^5 b. 5^3 c. $(2.5)^2$ d. 10^6
   e. $(0.3)^2$ f. $(0.8)^2$ g. 8^2 h. 3.2×10^5
   i. 3^{-1} j. 3^{-2} k. 2^{-3} l. $12\,345^0$

2. Write each of the following numbers in exponential form. (Use exponents other than 1.)
   a. 8   b. 16   c. 27   d. 125   e. 1/9

3. Write each of the following exponential expressions in the simpler form a^x.
   a. $(t^5)(t^3)$ b. $(r^7)(r^5)$ c. $(a^{-3})(a^2)$ d. t^5 / t^3

4. Write a^5 as a product of the form $a^m \times a^n$ in 8 different ways.

5. Sketch graphs of the following exponential functions. On each graph, mark and give the coordinates of the point of intersection with the vertical axis.
   a. $f(x) = 15(2.5^x)$ b. $g(x) = 4(0.8^x)$

6. **SITUATION** The spread of information (frequently misinformation or rumors) often follows an exponential pattern. Suppose that at 9 a.m. the five officers of the student council are the first to learn about secret plans to close the school for remodeling at the end of the year. They each tell a friend one hour later. Then each student who now knows the story tells another who had not yet known. In this way the number of people who know about the plans doubles every hour throughout the school day.

   a. If $N(t)$ gives the number of people who know the story t hours from its start, what is the rule for $N(t)$?

   b. For each of the following calculations, equations, or inequalities, write a question that can be answered using the expression, then answer your question.

      i. $N(0)$ **ii.** $N(4)$

     iii. $N(t) = 320$ **iv.** $N(t) \leq 100$

7. Given below are rules and graphs for three functions. Match each graph with the rule it fits best. Then, describe a real situation and two quantitative variables whose relation might be modeled by such a function rule and graph. A sample answer is given at the bottom of the page.

   Rules: $f(x) = 2.50x$; $g(x) = -4.9x^2 + 20x + 10$; $h(x) = 45\,(0.5^x)$

   Sketches:

   a.   **b.**   **c.**

   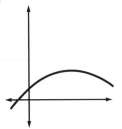 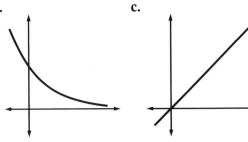

   Sample Answer: The sketch at the right is of a linear function with rule of the form $f(x) = mx + b$. It could represent something like the profit of a concert as a function of ticket sales.

   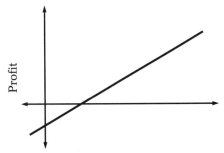

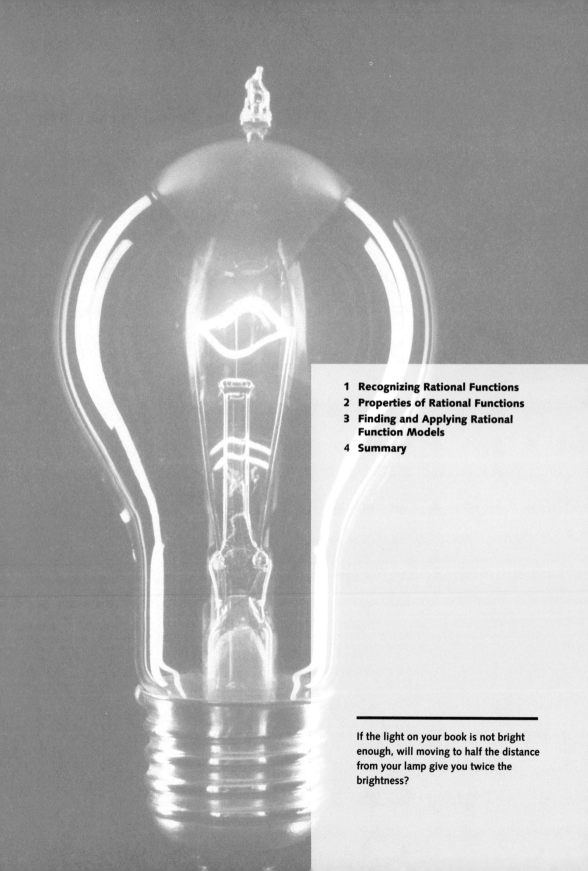

If the light on your book is not bright
enough, will moving to half the distance
from your lamp give you twice the
brightness?

6 Rational Functions

In the preceding chapters you have studied relations among quantitative variables that can be modeled well by linear, quadratic, or exponential functions.

You used rules of the form $f(x) = mx + b$ to describe and predict relations between ticket price and ticket demand for shows and sporting events, the income and profit from simple businesses, and the use of fuel by airliners or speed boats.

You used rules of the form $g(x) = ax^2 + bx + c$ to analyze the flight of a baseball, the stopping distance of a speeding car, and the profit prospects of a business for various price and sales levels.

You used rules of the form $h(x) = Ca^x$ to describe and predict the growth of populations, the decline in value of a used car, and the speed of a parachutist in free fall.

The linear, quadratic, and higher degree polynomials that you have studied earlier are all members of a larger family called the **rational functions**. In this chapter you will study properties and applications of several other important functions in this family.

1 Recognizing Rational Functions

From your previous studies, you know many questions that must be answered to understand what makes any new family of functions interesting or important, such as:

1. What are the typical patterns in *graphs,* *tables of values,* and *rules* of the functions?

2. What kinds of relations among variables are best *modeled* by the functions being studied?

As you study the situations in this section, try to find properties of these new kinds of rational functions that make them similar to and different from those you have studied in earlier chapters.

Exploration I

One of the simplest, yet practically important, principles of science is the relation between the primary forces when you use a **lever**. For example, levers are involved whenever you use a screwdriver to open a can of paint, or a wrench to tighten or loosen a nut on a bolt. The basic components in each of these physical systems are shown in the diagram at the right.

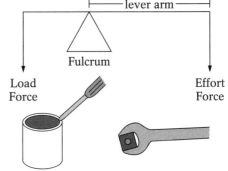

In each case, the **effort force** works against the resisting **load force** by pressing the **lever arm** to rotate about the **fulcrum**.

The effort required to balance or overcome a resisting load force depends on the length of the lever arm (the distance between the fulcrum and the effort force). You might already know that the required force decreases as the length of the lever arm increases. That is, by using a longer screwdriver or wrench, the twisting and prying tasks for which they are used become easier.

What you probably do not know is the numerical relation between the two variables *length of lever arm* and *effort force*. A simple experiment will help to reveal that relation.

To perform this experiment you will need a balance beam apparatus similar to the one pictured on the right. In simplest form it consists of a meter stick, supported at its center, and a collection of weights that can be hung from the meter stick at various points, perhaps in a bucket or tray.

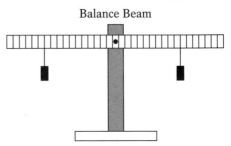

Balance Beam

Experiment 1. Begin by choosing a weight of 2 units. Hang the weight at a point 6 units to the left of center, or fulcrum, on the beam. This sets the load force. Choose a distance to the right of the fulcrum, and by experimentation, find the weights required to balance the beam. Record your results, lever arm length and weight, in a table, then choose another distance. Repeat this process until you have completed at least five rows in your table.

If you are using the kind of balance beam and weights that are common in schools, you might be able to find balance at only a few distances. Look for a pattern in those results that suggests others which cannot be directly tested with your apparatus.

After you have completed at least five rows in your table, plot the data pairs, (lever arm length, weight), on a coordinate grid.

1. In the table and graph of data, (lever arm length, weight), what is the general pattern relating the two variables L (lever arm length) and W (weight)?

2. What algebraic rule describes weight required to balance the beam as a function W of lever arm length L?

Experiment 2. Next place a weight of 3 units at a point 8 units to the left of the fulcrum. Then, by experimentation, determine the weights needed to balance the beam at various distances on the right of the fulcrum. Record at least five sets of results in a table, then plot the data pairs, (lever arm length, weight), on a coordinate grid.

1. In the table and graph of data, (lever arm length, weight), what is the general pattern relating the two variables L (lever arm length) and W (weight)?

2. What algebraic rule describes weight required to balance the beam as a function W of lever arm length L?

Next, experiment with some different combinations of load weights and positions, such as a weight of 2 units at a point 9 units to the left of the fulcrum. As you experiment, look for a general rule that describes the relations involved in all of these balance beam experiments. Write a short report summarizing your ideas. In linear, quadratic, and exponential function rules, the basic operations performed to calculate outputs for given inputs are addition, subtraction, and multiplication. However, you may have found in the previous experiments that the balancing weight (force) required is related to the length of the lever arm by a rule of the form

$$W(L) = \frac{a}{L} \text{ or } W = \frac{a}{L}.$$

There are many other important relations between variables in which the function rule involves division by the input variable, or a fraction with the input variable in its denominator. A function of this type is called a **rational function** because of the ratio comparison implied when one number is divided by another.

In this chapter you will study relations between variables that can be modeled by rational functions with rules of the forms

$$f(x) = \frac{a}{x}, \ g(x) = \frac{a}{x^2}, \ h(x) = \frac{a}{x^3}, \ ...,$$

and several basic variations on these forms. The common feature is that the input variable appears in the denominator of a fraction. The patterns that arise in tables and graphs of these functions have interesting differences from those of linear, quadratic, and exponential functions. The next two explorations give examples that illustrate the patterns and possibilities.

Exploration II

The quantitative variables and relations in this exploration are among the most important in all of algebra and its applications—they provide the algebraic models for problems involving distance, rate of motion, and time.

SITUATION 1.1

What is the proper speed limit for interstate highways in the United States? What about state and local roads and city streets? These are questions that affect each of us as we drive to and from school or work or on long distance trips.

Highway planners and police argue for lower limits to reduce gasoline consumption and to improve safety. Drivers of private cars and trucks often argue for higher limits to save time on their various trips.

How does a change in average speed for a trip affect the time required for that trip? Consider the case of a 360-kilometer trip from New York City to Washington, D.C.

You have already studied the key relation among distance, rate, and time:

$$d = r \times t$$

From this equation you can see that $t = d/r$ and that $r = d/t$. Thus the rule giving trip time as a function of average speed or driving rate should be

$$t(r) = \frac{360}{r} \text{ or } t = \frac{360}{r},$$

where r is the rate or speed of the car in kilometers per hour and $t(r)$ or t is the time in hours to complete the 360-kilometer trip.

1. The speed limit on highways between New York, NY and Washington, D.C. is about 90 kilometers per hour (kph). However, heavy traffic or bad weather could reduce the average speed to a much slower (and safer!) rate. Some daring drivers travel at much faster average speeds.

   Produce a table of data pairs, (speed, time), for a sample of possible average speeds. Describe any interesting patterns in the table. What do you think might be the overall relation between average speed and time for the trip?

2. Produce a graph of the function relating speed and time for a reasonable set of average speeds. Sketch the shape of that graph. Explain what the shape of the graph shows about the overall relation between average speed and time for the trip from New York to Washington.

3. For each of the following questions:
   –use function notation to write an algebraic calculation, equation, or inequality that can be used to answer the given question.
   –solve the problem, using the function rule, table, or graph, then write a sentence stating your answer to the original question.
   a. An express bus makes the trip from New York to Washington at an average speed of 70 kilometers per hour. How long does it take for the 360 kilometer trip on that bus?
   b. A train makes the 360 kilometer run from New York to Washington in 2.75 hours. What is its average speed?

    c. What is the slowest average speed you could travel to complete the trip in 6 hours or less?

    d. What average speed is necessary to complete the trip in 3.5 hours?

4. From earlier chapters you know that for a linear function, outputs change at a constant rate as inputs change, and for a quadratic function, outputs change at varying rates as inputs change. As you complete the following, think about the pattern of change for the rational function $t(r) = 360/r$.

    a. If the New York-to-Washington express bus increases its average speed by 10 kph, from 70 to 80, how many minutes sooner does it complete the trip?

    b. If the train described in Exercise 3 increases its average speed by 10 kph, from 130 kph to 140 kph, how many minutes sooner does it complete the trip?

    c. Copy and complete the following table and use it to explore, systematically, the way outputs change as inputs change for the rational function $t(r) = 360/r$.

Speed (kph)	Time(hours)	Difference
20		
40		$t(40) - t(20) =$
60		$t(60) - t(40) =$
80		
100		
120		

    d. As the average speed for the trip increases, what happens to the time required for the trip?

    e. As the average speed for the trip increases, does trip time change at a constant rate? Record any patterns that you have observed.

Making Connections

Consider how the function $t(r) = 360/r$ is similar to, and how it is different from, other types of functions you have studied in algebra. Then answer the following.

1. As the input values r increase, how do the output values $t(r)$ change, and how does that pattern compare to linear, quadratic, and exponential functions?

2. How does the shape of the graph of $t(r)$ compare to the shape of the graphs of linear, quadratic, and exponential functions?

Exploration III

The loudness or intensity of **sound** is a quantitative variable of interest to people of all ages in many different settings. Many young people like loud music on their radios and unmuffled roars from the tailpipes of their cars. Older people generally prefer quieter music, well-muffled cars, and control of environmental noises from jet planes and refuse trucks early in the morning.

Most arguments about the loudness of sound are not based on actual measurement of sound intensity, but physicists have devised several scales and instruments for measuring loudness. One common scale uses units called *watts per square meter.* Another compares a given sound to the softest sound that humans can detect. This ratio is reported in units called *decibels.* The following table shows typical measures of sound intensity for a number of common sounds.

Sound	Decibels	Watts/m²
Jet takeoff	120	1.0
Loud thunder	110	0.10
Niagara Falls	95	0.0032
Busy city street	75	0.000032
Normal conversation	60	0.0000010
Purring cat	20	0.0000000001

SITUATION 1.3

You may have noticed that the loudness of any sound is a function of the listener's distance from the source of that sound. As a listener moves farther from a radio, a noisy car, or a thunderstorm, the loudness of the sound decreases.

In general, sound intensity in watts per square meter is a function I of the distance d in meters with rule of the form

$$I(d) = \frac{c}{d^2} \text{ or } I = \frac{c}{d^2},$$

where c is a constant determined by the sound source and the environment.

If a radio has the volume turned up high, the intensity of its sound might be given by a rule like

$$I(d) = \frac{0.01}{d^2}.$$

Use this rule to answer the following.

1. What is the pattern relating distance and sound intensity values in this situation?

2. Sketch a graph showing the relation between distance and sound intensity in this situation.

3. What can you say about the sound intensity for very small distances from the radio source? For larger distances from the source?

4. Write the calculation, equation, or inequality that can be used to answer each of the following questions. Then use an appropriate tool (or tools) to find the answer.
   a. At what distance is the radio sound intensity 0.01 watts per square meter?
   b. At what distance is the radio sound intensity 0.0025 watts per square meter?
   c. What is the sound intensity at a distance of 0.5 meters from the radio?
   d. What is the sound intensity at a location 0.25 meters away from the radio?

Making Connections

1. How is the sound intensity function similar to and different from the lever function and the travel time function?

2. How are all three of these rational functions alike?

3. How are they different from other types of functions?

Exercises

1. SITUATION   Distance running is a very popular American sport. In many areas there are frequent races for all kinds of runners at distances from 5K (km) to 30K (km). (Runners often use K instead of km when referring to the length of a race.) In such races, each finisher is told his or her time. From that time and the race distance, one can calculate the runner's average running speed.

   Consider the case of a 20 kilometer (20K) cross-country race.
   a. What is the rule giving average speed in kilometers per hour as a function r of time t in hours?
   b. Calculate the average speed for racers whose time in hours varies from 1 to 3 in steps of 0.25.
   c. Plot the data points, (time, speed), from part b on a coordinate grid.
   d. If the winning runner completes the 20K race at an average speed of 18 kilometers per hour, how long is the winning running time?
   e. What average speed is required if the running time is reduced to:
       i.   1 hour?
       ii. 0.5 hour?
       iii. 0.1 hour?
   f. Which of the following changes in race time causes the greater change in average speed:
       i.   an increase from 1.5 hours to 2 hours, or
       ii. an increase from 4 hours to 4.5 hours?
   g. How is the rule in part a different from the rules for a 10K race and for a 30K race?

2. SITUATION   In shorter races the winning times are often less than one hour, and the rate or average speed is often reported in a unit that might seem unusual in other contexts: *minutes per kilometer*. Consider the case of a 10K race.
   a. In this race what is the average speed in minutes per kilometer for a runner whose time is 50 minutes? 60 minutes? 70 minutes?
   b. Write a rule giving the rate in minutes per kilometer as a function of time for a 10K race.
   c. Make a table of values and a graph of this function. How is this function similar to or different from others you have studied?

3. **SITUATION** California has experienced many earthquakes. In 1994 many roads and buildings collapsed, leaving cars and people pinned in the wreckage. When heavy-duty cranes are not available to help with lifting, strong beams can be used as simple levers. The diagram shown here illustrates how such a lever is often set up.

   Suppose that the force required to free one trapped person is a function of lever length in meters, with rule $F(L) = 300/L$.

   a. What effort force is required if one pushes on the lever at a point 1 meter from the fulcrum?
   b. What force is required 2 meters from the fulcrum?
   c. What force is required 3 meters from the fulcrum?

4. **SITUATION** One of the noisiest jobs in road construction work is using an air hammer to break up concrete. Suppose that the intensity of sound in watts per square meter is a function I of distance d in meters from the air hammer with rule

$$I(d) = \frac{0.0324}{d^2}.$$

   a. Calculate and graph the intensity of this sound as heard from distances of 0.5, 1, 2, 3, 4, 5, and 10 meters.
   b. At what distance does this sound have the same intensity as a normal conversation, 0.000001 watts per square meter?
   c. What is the general effect on intensity of doubling the distance from this sound? Check several pairs of distances to see if you can find a pattern.

5. **SITUATION** A travel agent has offered the senior class a chance to charter a plane for a class trip to Los Angeles. The cost of chartering the plane will be $60,000, for any number of students n traveling (up to a limit of 150).

   a. What is the cost per student for each of the following numbers of students who sign up to go?
      i.   20   ii.   40   iii. 60   iv. 100
   b. What is the rule that expresses cost per student as a function of number of students?
   c. Which has the greater effect on cost per student: an increase from 10 to 11 students or an increase from 100 to 101 students?

    d. Sketch the general shape of the graph of the cost per student function. Explain how it shows the correctness of your answer to part c.

6. **S I T U A T I O N**   On the trip to school, a bus travels a total of 25 kilometers on a highway with speed limit 90 kilometers per hour (kph) and a total of 8 kilometers on a city street with speed limit 50 kph.

    a. How long does the trip take if the bus travels at the speed limit on each segment?

    b. List at least five factors that might make actual average speed less than what the limit would allow.

    c. For each segment of the trip, write a rule giving the time as a function of the average speed for that segment.

    d. Produce two tables of values, (average speed, time)—one table for each segment of the trip. Use average speeds from 60 to 90 kph in steps of 10 kph on the highway segment and average speeds from 20 to 50 kph in steps of 10 kph on the city street segment.

    e. For each segment, which of the changes in average speed of 10 kph cause the greatest change in time ? What are the greatest changes in time?

2   Properties of Rational Functions

In Section 1, you began to identify patterns in tables and graphs for functions with rules in the form

$$f(x) = \frac{a}{x} \text{ and } g(x) = \frac{a}{x^2}.$$

However, your study of the behavior of these rational functions was limited by constraints from the real-world situations modeled by those rules.

    For example, in analyzing the relation between average speed and time for the 360 kilometer trip by car or train from New York City to Washington, D.C., we focused on the function rule $t(r) = 360/r$. The numerator is fixed at 360, representing the fixed length of the trip in kilometers. Further, it only made sense to consider speeds between 0 and 200 kilometers per hour because the top speed of the fastest American car or train is under 200 kph.

In this section you will explore a variety of rational functions, each of which has a rule with one of the forms

$$f(x) = \frac{a}{x}, \; f(x) = \frac{a}{x^2}, \; \ldots, \; f(x) = \frac{a}{x^n}, \; \text{and } f(x) = \frac{a}{x+b}.$$

As you explore, search for numerical and graphical properties that are true for all functions of each type, regardless of the specific situation being modeled.

Exploration I

In this exploration, we will study a function with a rule like $f(x) = a/x$. However, we do not want to be limited in our choice of inputs, for example, to only those inputs whose values are reasonable average speeds for a car or train. Nor do we want the value for a to be restricted to a value such as the distance between two specific cities like New York and Washington. Instead, we will search for answers to questions like the following.

1. What happens in a table and a graph of $f(x) = a/x$ when x takes on negative values?
2. What happens when we try to let the input x be zero?
3. What happens as x assumes large positive values?
4. What happens when we change the function by choosing a different numerator a?

   Begin with the table that follows. This table displays some values of the input and output variables for the rational function $f(x) = a/x$ when $a = 75$.

Input	Output	Input	Output
x	$f(x) = \frac{75}{x}$	x	$f(x) = \frac{75}{x}$
−100	−0.75	1	75.00
−50	−1.50	5	15.00
−10	−7.50	10	7.50
−5	−15.00	50	1.50
−1	−75.00	100	0.75
0	undefined		

1. **For Class Discussion:** What numerical patterns do you see in the preceding table? As inputs increase from −100 to 100, what happens to the corresponding outputs?

2. Relying on the patterns in the table, sketch a graph that you believe fits the function with rule $f(x) = 75/x$.

Experiment 1. To explore the effect of the value of a on tables and graphs for rational functions of the form $f(x) = a/x$, your tasks are as follows:

−examine tables of data pairs, (input, output), and identify patterns that can be predicted from the value of a;

−describe how patterns observed in tables of data pairs, (input, output), match the trends and symmetries of graphs for these rational functions.

To help with these tasks, you might review the connection between time, speed, and distance for the 360 kilometer trip from New York to Washington. The time required for that trip is a function of the average speed r of the car; the rule is $t(r) = 360/r$. What condition in that situation changes when the numerator is a number different from 360?

The following is a list of four rational functions with rules of the form $f(x) = a/x$. Values of a range from 1 to 100. Use technology as needed to construct a table of data pairs, (input,output), for each function. Inspect function tables displaying inputs ranging from −100 to 100, but be sure to look more closely at those values of x that are close to the number zero—an input for which no output is defined.

The test functions:

$$f(x) = \frac{1}{x}, \; g(x) = \frac{5}{x}, \; h(x) = \frac{50}{x}, \text{ and } j(x) = \frac{100}{x}.$$

1. Describe what happens to the outputs in each of the four function tables as inputs increase from −100 to 100.

2. What do the patterns in the tables of values, (input, output), suggest about possible maximum (or minimum) outputs for these functions?

3. Are there any input values for which each of these functions has the same output? If so, what are they?

4. Describe what happens as inputs get very close to zero. What happens when 0 itself is used as an input?

5. How do the *differences* in values of the numerator a lead to *differences* in the output values in the various function tables?

Using a graphing utility, generate a graph for each of the four test functions. Look for ways in which patterns and properties in the tables of values for rational functions of the form $f(x) = a/x$ are reflected in the graphs of those functions.

As you have discovered from looking at the tables of values for these test functions, ($f(x) = 1/x$, $g(x) = 5/x$, $h(x) = 50/x$, and $j(x) = 100/x$), the outputs are very different for these functions. Finding a suitable scale for the output axis that displays all functions at once can be tricky. We suggest that you choose a graphing window with input range $x = -10$ to $x = 10$, then experiment with several different scales on the vertical axis.

On one set of axes draw a rough sketch of each function. Show the general pattern and relative positions of the graphs of the four functions. Be sure to label each function graph.

6. What do the patterns in the function graphs suggest about possible maximum or minimum output values for these functions? Describe what happens as x gets large. Tell what happens as x nears zero.

7. Do any of these graphs suggest that the outputs for the corresponding functions change at a constant rate as the inputs increase? Explain how the graphs support your conclusion.

8. How does the change in value of the numerator a from one function to another lead to *differences* in the function graphs?

Exercises

1. The rules for the two functions $f(x) = 4x$ and $g(x) = 4/x$ may *look* rather similar, but the tables and graphs of the functions are very different.

   a. For each of the two functions, use your calculator to make a table of data pairs, (input, output), and then use the table to sketch a graph of the function. Use inputs from –10 to 10 in steps of 2.

   b. Describe at least two ways that the tables and graphs of these two functions are different.

2. The rules for the two functions $g(x) = 4/x$ and $h(x) = x/4$ may *look* fairly similar, but the tables and graphs are very different.

   a. For each function, build a table of data pairs, (input, output), and then sketch the graph. Use your calculator to make the table, with inputs from –10 to 10 in steps of 2.

   b. Describe at least two ways that the tables and graphs of these two functions are different.

3. Use the fact that $x/4 = (1/4)x$ to explain why $h(x) = x/4$ is more like $f(x) = 4x$ than it is like $g(x) = 4/x$.

4. Investigate what the sign (+ or –) of a tells about a table and the graph of a rational function of the form $f(x) = a/x$.

   a. Complete a table for the pair of test functions $f(x) = 1/x$ and $g(x) = -1/x$, then graph the functions on the same coordinate grid.

   b. Describe ways in which the tables and graphs of these functions are alike.

   c. Describe ways in which the tables and graphs are different.

   d. Make a similar table and graph for the functions $f(x) = 5/x$ and $g(x) = -5/x$.

   e. Describe how the tables and graphs of these functions are alike, and how they are different. With functions of your own design, test your hunches about the effect of the sign of the numerator in a rational function $f(x) = a/x$. Be prepared to share your findings with the class.

Making Connections

In Exploration I you found that the tables and graphs for all rational functions of the form $f(x) = a/x$ have similar patterns. The function $f(x) = 2/x$ is a typical example of such a function. Its table and graph are displayed below.

x	$f(x) = 2/x$
−20.0	−0.1
−10.0	−0.2
−5.0	−0.4
−1.0	−2.0
−0.5	−4.0
−0.1	−20.0
0	undefined
0.1	20.0
0.5	10.0
1.0	2.0
5.0	0.4
10.0	0.2
20.0	0.1

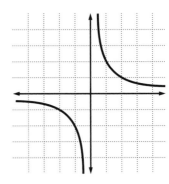

$f(x) = 2/x$

The horizontal (input) scale unit and the vertical (output) scale unit are both 1.

Summarize your observations about the behavior of *all* functions of the form $f(x) = a/x$, where a is a constant different from zero, by completing the following statements.

1. Our study of the tables of input-output pairs and function graphs for rational functions of the form $f(x) = a/x$ suggests that as inputs increase, outputs tend to....

2. When the input is zero for $f(x) = a/x$, the output is....

3. As positive inputs for $f(x) = a/x$ get smaller and smaller and approach zero, outputs tend to....

4. As negative inputs for $f(x) = a/x$ approach zero, outputs tend to....

5. The graph of the rational function $f(x) = a/x$ has the following symmetry properties....

6. Consider $f(x) = a/x$ and $g(x) = b/x$, where a and b are both positive numbers and $a > b$. The tables and graphs of these two functions *differ* in the following ways....

Exploration II

You have seen that interesting patterns in function tables and graphs occur when a variable appears in the denominator of a fraction in a function rule. Now, you will explore the properties of function graphs with rules of the form

$$f(x) = \frac{a}{x^2}, \; f(x) = \frac{a}{x^3}, \text{ and } f(x) = \frac{a}{x^4}.$$

In an example of Section 1, the relation between sound intensity and the listener's distance d from a radio was described by the rule $I(d) = 0.01/d^2$. This rule is one member in the family of rational functions of the form

$$f(x) = \frac{a}{x^2},$$

where a is some fixed number and x is the input variable.

To get started on a study of other functions in that family, copy and complete the following table that displays some inputs and outputs for the function $f(x) = a/x^2$ when $a = 100$.

Input x	Output $f(x) = \frac{100}{x^2}$	Input x	Output $f(x) = \frac{100}{x^2}$
−20		5	
−15		10	
−10		15	
−5		20	
0	undefined		

1. **For Class Discussion:** How do you think tables and graphs for functions with rules of the form $f(x) = a/x^2$ are similar to and different from functions with rules of the form $f(x) = a/x$?

2. What must always be true about the *sign* (+ or −) of the denominator when outputs are evaluated for a function with rule of the form $f(x) = a/x^2$? How does this property affect the function table and graph?

3. As x gets large, what happens to a/x^2?

4. How does a graph of $f(x) = 100/x^2$ compare with one for $g(x) = 100/x$? Try sketching the graphs of $f(x)$ and $g(x)$ on the same set of axes.

Experiment 2. To test your hunches about the patterns in tables and graphs of functions with rules like $f(x) = a/x^2$, complete the following explorations comparing two pairs of rational functions. First, compare

$$f(x) = \frac{25}{x^2} \text{ and } g(x) = \frac{25}{x},$$

then compare

$$h(x) = \frac{5}{x^2} \text{ and } j(x) = \frac{5}{x}.$$

In each comparison note any similarities and any differences that you observe.

1. Generate a graph for the function $f(x) = 25/x^2$. Display the graph using a window that shows $x = -10$ to $x = 10$ and output values from -50 to 50. Then plot $g(x) = 25/x$ on the same set of axes.

2. Make a rough sketch of both graphs on paper. Be sure to label each graph clearly.

   Describe any apparent similarities and differences between the two functions as revealed in the graphic display.

3. Next list any similarities and differences in tables of data pairs, (input, output), for the functions $f(x)$ and $g(x)$. Explore function tables of your own choosing, but be sure to consider a variety of negative and positive inputs.

4. Now repeat your graph and table comparisons for the pair of functions $h(x) = 5/x^2$ and $j(x) = 5/x$. How are the tables and graphs for these two functions similar and how are they different?

   What do your results suggest about the general pattern of a table of values, (input, output), for any function with rule of the form $f(x) = a/x^2$? How is that pattern represented in the graph for the function?

Experiment 3. Consider the rational function with rule

$$h(x) = \frac{75}{x^3}.$$

Do you suppose that it and other functions with rules of the form a/x^3 behave like either $f(x) = 75/x^2$ or $g(x) = 75/x$?

1. Test your prediction by studying a table of data pairs, (input, output), and the graph for each of the following test functions. Use technological help as needed.

   Test functions:

   $$h(x) = \frac{75}{x^3}, \; j(x) = \frac{25}{x^3}, \text{ and } k(x) = \frac{5}{x^3}.$$

   A table with the following headings may help you compare outputs of these functions.

x	$h(x) = \frac{75}{x^3}$	$j(x) = \frac{25}{x^3}$	$k(x) = \frac{5}{x^3}$

2. Based upon ideas from function tables and graphs, make any necessary adjustments to your predictions about properties of functions with rules of the form $f(x) = a/x^3$. Does $f(x) = a/x^3$ behave more like a/x^2 or a/x? Explain.

Experiment 4. For a mathematician who has just studied three functions that are related by a pattern of similar features, the next natural step is to continue the pattern and look for general properties of all such functions.

Do you suppose that a function of the form $f(x) = a/x^4$ behaves more like a function of the form $f(x) = a/x$, of the form $f(x) = a/x^2$, or of the form $f(x) = a/x^3$?

Plan and carry out an investigation of rational functions of the form $f(x) = a/x^4$. Write a report describing your findings. Be sure to specify the test functions you selected, and to include any function tables and sketches of graphs that you found to be strong supporting evidence for your conclusions. You may wish to state your conclusions by completing the statement:

$f(x) = a/x^4$ behaves most like . . . since . . .

Exploration III

You have now observed a variety of similarities and differences in the tables and graphs of rational functions with rules in the form $f(x) = a/x^n$ for various values of the constant numerator a and the constant, integral exponent n. Other than polynomials, these are the simplest and most common types of rational functions, but they are not the only ones.

In the exercises of Section 1 you studied the following business situation.

> A travel agent has offered the senior class a chance to charter a plane for a trip to Los Angeles. The total cost of chartering the plane is $60,000, for any number of students traveling (to a limit of 150).

In this situation, the cost per student is a function of the number of students n, expressed by the rule $C(n) = 60\,000/n$.

Suppose that 15 class officers and faculty sponsors have already promised to go on the trip and that the variable x represents the *number of additional students* to be recruited. Then the cost per student depends on x and can be expressed using the rule

$$f(x) = 60\,000/(15 + x) \text{ or } f(x) = \frac{60\,000}{15 + x}.$$

To evaluate the output for this function corresponding to an input x, the first step is to add x to 15. Then 60,000 is divided by that sum. This order of operations is indicated clearly by the parentheses $(15 + x)$ in the first form of the rule for $f(x)$. The accepted grammar rules of algebra imply that the fraction bar in the second form of the rule for $f(x)$ should produce the same result: evaluate the numerator, evaluate the denominator, and then carry out the division.

Do you have ideas about how the table and graph for this new type of rational function are similar to and different from those for $C(n) = 60\,000/n$? What different patterns in the tables and graphs would you expect from other such functions in which the rule a/x is modified by *adding (or subtracting) a constant term in the denominator*? That is, can you predict the behavior of a rational function in the form

$$f(x) = \frac{a}{x + b},$$

where both a and b are constant for a given situation?

Experiment 5. Design and carry out your own exploration of the properties of rational functions of the form $f(x) = a/(x + b)$. Summarize your observations in a written report; include the following points of interest. Be prepared to support your findings with numerical or graphical data that you generate.

1. Analyze and describe patterns in the tables and graphs of functions with rules like $f(x) = a/(x + b)$. Describe the shapes of the graphs of these functions.

2. How are functions with rules like $f(x) = a/(x + b)$ *similar* to those with rules of the form $g(x) = a/x^n$?

3. How do functions like $f(x) = a/(x + b)$ *differ* from those like $g(x) = a/x^n$?

Exercises

1. In a 500 kilometer auto trip, the time required to complete the trip can be considered a function t of average speed s with rule $t(s) = 500/s$. What change in this situation is represented by each of the following variations on the rule for $t(s)$?
   a. $t(s) = 300/s$ **b.** $t(s) = 1000/s$ **c.** $t(s) = 150/s$

2. Suppose the intensity of a particular sound is a function I of distance from its source d, with rule $I(d) = 0.3/d^2$. What change in this situation is represented by each of the following variations on the rule for $I(d)$?
   a. $I(d) = 0.6/d^2$ **b.** $I(d) = 0.03/d^2$ **c.** $I(d) = 3/d^2$

3. If 15 people have already agreed to go on a charter plane trip that costs \$60,000, the cost per person is a function C of the number of additional passengers recruited x. This function can be expressed using the rule $C(x) = 60\ 000/(15 + x)$. What change in this situation is represented by each of the following variations on the rule?

   a. $C(x) = \dfrac{40\ 000}{15 + x}$ **b.** $C(x) = \dfrac{60\ 000}{30 + x}$

   c. $C(x) = \dfrac{50\ 000}{25 + x}$

4. The following are rules and graphs of four rational functions. No tic marks or scales are given on the graphs. However, the same scales are used on each and you should be able to use your knowledge about graphs for such functions to match rules and graphs. Be prepared to explain your reasons for each match.

Rules: $f(x) = \dfrac{8}{x}$, $g(x) = \dfrac{2.5}{x^2}$, $h(x) = \dfrac{2.5}{x}$, and $j(x) = \dfrac{2.5}{x+2}$.

a.

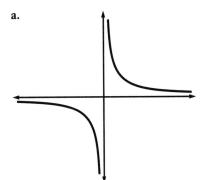

b.

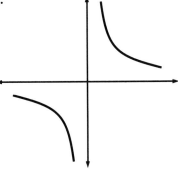

c.

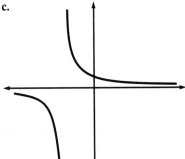

d.

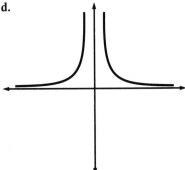

5. The following are tables of values, (input, output), for five different functions. The rule for each function has one of the following forms:

$$f(x) = \frac{a}{x}, \quad g(x) = \frac{a}{x^2}, \quad h(x) = ax, \quad j(x) = ax^2, \quad \text{and } k(x) = \frac{a}{x+b}.$$

Match each rule type with the table it could generate and, if possible, determine the number a and (if applicable) the number b. Be prepared to explain your reasoning for each match.

a.

Input	−3	−2	−1	0	1	2	3
Output	54	24	6	0	6	24	54

b.

Input	−3	−2	−1	0	1	2	3
Output	−2	−3	−6	undefined	6	3	2

c.

Input	−3	−2	−1	0	1	2	3
Output	0.66...	1.5	6.0	undefined	6.0	1.5	0.66...

d.

Input	−3	−2	−1	0	1	2	3
Output	−3	−6	undefined	6	3	2	1.5

e.

Input	−3	−2	−1	0	1	2	3
Output	−18	−12	−6	0	6	12	18

Making Connections

Your explorations of this section should have given you a clear picture of the patterns that can be expected in the tables and graphs of rational functions with rules of the form $f(x) = a/x^n$. For $a > 0$ the graphs have one of two basic symmetric shapes.

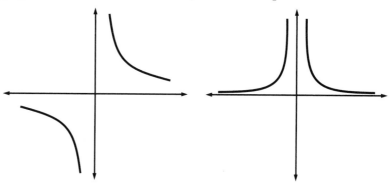

The graphs above show the patterns to be expected in any table of inputs and outputs. Copy and complete the following statements describing those patterns.

1. For the input 0, the output is ....

2. For values of x near 0, the absolute value of the outputs are...

3. For inputs x with large absolute value, the outputs are....

4. The functions have (a maximum, a minimum, both a maximum and a minimum, or neither a maximum nor a minimum) output value(s).

5. The outputs (do or do not) change at a constant rate as the inputs change.

 ## Exploration IV *Making Connections*

As you have seen in Section 1 of this chapter, simple rational functions provide useful algebraic models for many different quantitative relations. Complete the following exploration to see how the new kinds of rational function models are different from those that you have studied in earlier chapters.

Use the following set of test functions to compare and contrast properties of linear, quadratic, exponential, and rational functions. Record your observations in a chart similar to the one below. Enlist technological aid as needed when constructing tables of values and function graphs.

Test functions:

Linear	Quadratic	Exponential	Rational
$f(x) = 2x + 3$	$g(x) = x^2 + 10x + 4$	$h(x) = 3^x$	Choose examples
$f(x) = -2x + 3$	$g(x) = -x^2 + 10x + 4$	$h(x) = (0.3)^x$	of the forms a/x, a/x^2, a/x^3, a/x^4

Summary of observations:

Function form	Rate of change	Symmetry feature	Max/min value	Special features
$f(x) = mx + b$ (Linear)				
$f(x) = ax^2 + bx + c$ (Quadratic)				

Summary of observations (*continued*):

Function form	Rate of change	Symmetry feature	Max/min value	Special features
$f(x) = a^x$ (Exponential)				
$f(x) = a/x$ (Rational)				
$f(x) = a/x^2$ (Rational)				
$f(x) = a/x^3$ (Rational)				
$f(x) = a/x^4$ (Rational)				

3 Finding and Applying Rational Function Models

In the first two sections of this chapter you studied several functions with rules of the form $f(x) = a/x^n$ and $f(x) = a/(x + b)$. Different values of a and b and of the exponent n lead to different graphs and tables of function values—which model different relations among variables in scientific or business situations. For any particular example of a rule like those, you should now be able to predict the shape of the graph and the pattern of entries in any table of values.

Exploration I

One of the basic properties of functions with rules like $f(x) = a/x^n$ $(n \geq 1)$ or $g(x) = a/(x + b)$ is that as values of x increase (with $x > 0$), function output values decrease toward 0. This property makes those rules good models for situations in which the intensity of radiation from an energy source (like sound waves) decreases as distance from the source increases.

That general relation applies well to the illumination from a light source like a camera flash or streetlight. The greater your distance from the light source, the lower the illumination provided by the light. Of course, it is very important in photography or in the design of street lighting to be able to measure that decrease in illumination.

To understand the best model for illumination as a function of distance, it helps to think of a small light bulb in the middle of a very large dark space. From the light source, the light energy travels in all directions. The rate of flow, or *luminous flux*, is commonly measured in a unit called a **lumen**. A higher measure of lumens means a brighter light source.

Imagine a light source of L lumens. At a distance of r meters from the source, you can think of the light as shining evenly on the inside of a spherical balloon with radius r meters. Thus the original energy is spread over a surface of area $4\pi r^2$ square meters, according to the geometry formula for the surface area of a sphere.

The *illumination* provided by the bulb is measured in **lumens per square meter**. Thus the illumination can be measured by calculating

$$I = \frac{L}{4\pi r^2}.$$

For any particular light source, the illumination at a distance of x meters can be expressed by a rational function f with rule in the form $f(x) = a/x^2$.

SITUATION 3.1

The illumination provided by a single 100-watt light bulb is a function I of the distance d in meters from that light bulb. A typical rule for such a function is

$$I(d) = \frac{90}{d^2}.$$

Use this rule to answer the following.

1. Calculate a table of values for this function as d (in meters) varies from 1 to 10 in steps of 1.

2. Sketch a graph of the function I.

3. For this light bulb consider the illumination on a screen whose initial position is 1 meter from the light source. By answering the following questions, describe how the illumination changes when the screen is moved farther away from the bulb. Record both your calculations and your results.

    a. Approximately how far from the source is the screen if the illumination is only one-half as much as when the screen is in its original position?

    b. Approximately how many meters away from the bulb is the screen when the illumination is only one-fourth as much as when the screen is in its initial location?

    c. What is the approximate distance between the source and the screen if the illumination is only one-eighth as much as at the initial position?

4. Write and solve an equation or inequality that can be used to answer each of the following questions; then find the answer.

    a. At what distance is the illumination due to this light bulb 20 lumens per square meter?

    b. At what distance is the illumination due to this light bulb at least 60 lumens per square meter?

5. What happens to the illumination of a light source as your eye moves very close to it (0.1 meter, 0.01 meter, *etc.*)?

6. Consider the illumination questions of Situation 3.1 again, this time for a 60-watt light source. A typical function rule is $I(d) = 70/d^2$. How are the table, graph, and answers to various questions for this function similar to and different from those of the function in Situation 3.1?

7. *Optional*: As a final activity in this exploration, get a flashlight and a light meter. Test the illumination of the flashlight at various distances in a dark room.

 ## Exploration II

Many physical principles relate variables with rational function rules that are somewhat more complex than the simpler forms you have studied so far. One of the most important sources of such rules is the force of gravity that holds our moon in orbit and each of us safely on the surface of the earth.

SITUATION 3.2

When astronauts ride the NASA space shuttles into orbit around the earth, they experience weightlessness that allows them to float around inside the shuttle like balloons bouncing around a room. This weightlessness is caused, in part, by the shuttle's distance from the earth. In fact, the weight of any object decreases as it rises above the earth's surface and out of the planet's gravitational pull, according to the rational function with rule of the form

$$w(h) = \frac{w_e}{\left(1 + \dfrac{h}{6400}\right)^2}.$$

In this rule, w_e is the object's weight in kilograms of force at sea level on the earth's surface and h is the height in kilometers of the object above sea level. The number 6400 represents the average radius in kilometers of the earth at sea level (rounded to the nearest hundred for simplicity). Use this rule in the following questions that explore the relation between distance above the earth's surface and weightlessness.

1. Suppose that an astronaut is carried into space by a space shuttle. Assume that the astronaut's weight on earth at sea level is about 65 kilograms of force (a unit which we abbreviate by "kgf").
   a. Write the rule giving the astronaut's weight as a function of shuttle altitude.
   b. Using your rule, calculate $w(h)$ for $h = 0$ to $h = 40\,000$ in steps of 4000, and record your results in a table. Sketch the graph of $w(h)$.
   c. Communication satellites are often put in orbits that make them appear to stay fixed in the same position (relative to locations on earth). The altitude of such an orbit is approximately 35,000 km. What is the astronaut's weight when working on a satellite in such an orbit?
   d. At what altitude does the astronaut's weight equal:
      i. one-half of what it was at sea level?
      ii. one-quarter of what it was at sea level?
      iii. one-eighth of what it was at sea level?

2. Calculate the difference between the astronaut's weight while at sea level and while flying in a commercial jetliner 10 km above sea level.

3. If a satellite weighs 1000 kgf at sea level, how does its weight change as it is raised to higher orbits?

   a. What is the rule relating altitude and weight in this case?

   b. At what altitude is the satellite weight:

      i. one-half its weight on earth at sea level?

      ii. one-quarter its weight on earth at sea level?

      iii. one-eighth its weight on earth at sea level?

   c. How do your answers to the three questions in part b compare with your answers to the three questions in Exercise 1d?

4. Recall from Exercise 1b the following outputs $w(0)$ and $w(40\,000)$. If weight decreased *linearly* as a function of altitude, what function rule would fit these two data points? Use the form Weight = ___ altitude + ___ .

   How do the numerical values and the graph of this linear function compare to those of the rational function studied in Exercise 1?

The questions you have answered in this exploration give some sense of the way that gravity acts on objects in orbit around the earth. For such orbiting objects there is another important factor that helps create the weightless sensation of space travel—the centripetal force experienced by anything spinning in a circular or elliptical path. Full consideration of weightlessness requires accounting for that force also, but our simplifcation still shows much about the effects of gravity.

Exploration III

In many situations, you will have reasons to believe that two variables are related by a rational function, but no clear sign of the rule that fits the relation. Some technological tools have an option that gives a "best fitting" rule of the form $y = a/x^n$, but others do not. Few allow you to choose from all the possible rules. Depending on the technology you are using, you may have to do some guess-and-test searching when trying to find a reasonable rule.

The exploration that follows gives you another experience in finding relations among quantitative variables. You will start by collecting some experimental data. Display those data in a table and a graph, look for patterns in the data, and finally use the patterns to make predictions.

Experiment. Whether you ride on a bicycle, in-line skates, or a skateboard, the time required to coast down a hill depends on the steepness of the hill and a variety of other factors. To discover the effect of hill steepness alone, scientists might devise an experiment in which all other factors are held constant. They can then look for patterns in the data, (steepness, time).

The directions that follow outline such an experiment. The vehicle is a skateboard; the recommended "hill" is a 3-meter ramp; and only the ramp height will be varied. Look for a pattern that helps you predict run time as a function of ramp height.

To run the experiment you will need (1) a flat board, 3 meters long and wide enough to serve as a skateboard ramp, (2) an adjustable prop that can raise one end of the board to heights from 0 to 100 centimeters, (3) a skateboard, (4) at least two stopwatches, (5) a meter stick, and (6) a group of 3–5 students to conduct the experiment for the class.

Skateboard Experiment Layout

3 meters

Before you start recording data, conduct two or three practice runs to check procedures. At the start of each run the skateboard should not be pushed at all—simply release it from the top of the ramp. Timers should start their watches when the skateboard is released and stop them when it reaches the end of the course. To make an accurate estimate of the time for each ramp height, it may be helpful to average the results of several runs and have more than one timer for each run.

Make several runs for ramp heights of 15 cm, 30 cm, 45 cm, and 60 cm. Record the average of the runs at each height in a table. Then each student should plot the data pairs (ramp height, run time) on a coordinate grid and sketch a curve that fits the pattern in the points.

1. Describe the overall pattern in the situation relating ramp height and run time.

2. The pattern in your table and graph can be used to make estimates for run times with other ramp heights. Make estimates for run times if the ramp height is 25 cm, 40 cm, and 55 cm. Record your estimates in a table and check their accuracy by conducting actual skateboard runs at those heights. Include a column in your table for the actual run time. Your table should look similar to the following.

Ramp height (in cm)	Estimated run time (in sec)	Actual run time (in sec)
25		
40		
55		

3. Now use your results to answer the following questions about run time as a function of ramp height.

a. What factors might explain differences between your time estimates and the actual run times?

b. How well could you estimate run time for a 90 cm ramp height?

c. How well could you estimate run time for a 1 cm ramp height?

There are usually several reasons to look beyond the pattern in a table or graph and find an algebraic rule for a relation between the variables. First, an algebraic rule is usually simpler to record and communicate than a table or graph. Second, the rule often can be used to calculate accurate estimates for other interesting outputs. Third, the rule can be used to compare a particular experiment to other similar experiments and to reveal basic scientific principles that explain the observed results. The problem is deciding what sort of function rule provides the best model of the observed data.

4. For each of the following function types, find a rule that you believe best fits the skateboard data pairs, (height, time). While deciding which rule you believe to be best, experiment with rules that you guess-and-test, as well as with any rule obtained from technology. Each of the first three function types—linear, quadratic, and exponential—has some features that do not fit the probable relation between ramp height h and time t.

a. Linear function of best fit: $(T(h) = \underline{\quad} h + \underline{\quad})$.
Describe the limitations of this function as a model.

b. Quadratic function of best fit: $(T(h) = \underline{\quad} h^2 + \underline{\quad} h + \underline{\quad})$.
Describe the limitations of this function as a model.

c. Exponential function of best fit: $(T(h) = \underline{\quad})$.
Describe the limitations of this function as a model.

d. Rational function model of best fit: $(T(h) = \underline{\quad} / \underline{\quad})$.
Remember, many function-fitting tools are not able to supply this type of function. You may need to use a guess-and-test strategy.
Describe the limitations of this function as a model.

5. In the skateboard experiment the principal force causing motion is gravity. The same force that pulls a batted ball or a parachutist to earth pulls the skateboard down its ramp. The physical laws governing gravity can be used to derive a relation between ramp height and run time. It turns out that, under reasonable assumptions, run time is related to ramp height by a rule of the form

$$T(h) = \frac{a}{\sqrt{h}}.$$

Find a model of this form that fits your skateboard data well, and describe any limitations of this function as a model:

Caution: A function with rule of the form $a/\sqrt{h}$ is *not* a rational function. The appearance of the input variable h under a square root symbol disqualifies it. The four "rational operations" are addition, subtraction, multiplication, and division. A rule for a rational function of an input variable x may involve any combination of rational operations applied to x and any real numbers, but may not involve $\sqrt{h}$ or any other root of the input variable.

Exercises

1. SITUATION The moon is much smaller than the earth, so its gravitational attraction is not as great. An object with earth weight w_e has moon weight only $0.16\ w_e$. Further, since the radius of the moon is only about 1740 kilometers, one rule for the function that gives moon weight at an altitude of h kilometers above the moon's surface is

$$W(h) = \frac{w_m}{\left(1 + \dfrac{h}{1740}\right)^2}.$$

Use this information about gravity to answer the following questions.

a. What does a person who weighs 100 on the surface of the earth weigh on the surface of the moon?

b. What does that person weigh at an altitude 1740 km above the moon?

    c. What does that person weigh at an altitude 1740 km above the earth?

    d. At what altitude above the moon does the person weigh only half as much as on the surface of the moon?

2. SITUATION The following table gives data obtained in two skateboard experiments. In one, the skateboard had a weight attached to it; in the other, no weight was attached.

Height of ramp (cm)	Unweighted skateboard time (seconds)	Weighted skateboard time (seconds)
15	4.76	4.33
30	2.93	2.78
45	2.27	2.12
60	1.95	1.84

Using a ramp height of h cm, let $U(h)$ be the time for the unweighted skateboard's run and $W(h)$ be the time for the weighted skateboard's run.

    a. On separate grids, plot the four data points given for $U(h)$ and $W(h)$, and sketch curves that fit the patterns in those data.

    b. What type of function seems likely to provide the best fitting model of these data? Explain your choice.

    c. How are the two graphs similar and how are they different?

3. In the experiments of Exercise 2, the length of the skateboard ramp was 3 meters. Use that fact to calculate the average *speed* (in m/s) for each skateboard at each ramp height. Record your calculations in a table similar to the following.

Height of ramp (cm)	Unweighted skateboard average speed (m/s)	Weighted skateboard average speed (m/s)
15	3 / 4.76 = 0.63	
30		
45		
60		

    a. Use your table to graph average speed as a function of ramp height for both weighted and unweighted skateboards.

    b. What function type is likely to provide the best model of the relation between ramp height and average speed? Explain your answer.

4. In the exploration of illumination as a function of distance from a light source, you were given rules in the form $I(d) = a/d^2$ and asked relevant questions about the relations. In order to make calculations and answer the questions, you needed to know the specific number a. Sometimes in the future, however, the constant a may not be provided for you.

   You can determine the constant a in the illumination function rule if you have some experimental data relating d and $I(d)$. Each of the following sets of data is related to a rule of the form

   $$I(d) = a/d^2$$

   for different values of a. In each of the following, use the given data to find a value for a that gives a rational function rule that fits the data well.

a.	d	$I(d)$	b.	d	$I(d)$	c.	d	$I(d)$
	1	300		2	125		5	60.0
	2	75		3	56		8	23.0
	4	19		5	20		15	6.7

5. Suppose you are using a flashlight to help you read an outdoor sign at night. As you move the light closer to the sign, which of the following changes causes the greater increase in illumination of the sign:
   a. moving the light from 8 meters to 6 meters from the sign, or
   b. moving the light from 6 meters to 4 meters from the sign?

   Explain your answer using properties of the table, graph, or rule for the illumination function.

4 Summary

Your explorations in this chapter should have given you a clear picture of the patterns to be expected in the tables and graphs of rational functions with rules of the form $f(x) = a/x^n$, where the number a is positive and the exponent n is a positive integer. Each graph has one of the following basic symmetric shapes.

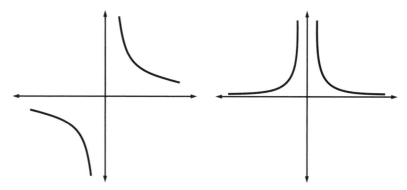

The graphs above show the patterns that can be expected in any table of inputs and outputs. Several important properties of these functions are summarized in the following statements.

1. There is no output corresponding to the input 0.

2. For positive values of the variable x, as the input increases, the output decreases. Moreover, inputs close to 0 give large outputs, while huge inputs give tiny outputs.

3. Each of these functions has neither a maximum value nor a minimum value.

4. For any of these functions the outputs do not change at a constant rate as the inputs vary.

5. If the exponent n is an even integer, the graph of the function is symmetric with respect to the vertical axis; but if n is an odd integer, then the graph is symmetric about the point (0, 0).

Furthermore, you have seen that these properties make rational functions very useful as models of relations among quantitative variables—especially in situations where some type of energy or force, like sound or light or gravity, decreases as a function of distance from its source.

Review Exercises

1. SITUATION   In various localities the size of a billboard is restricted in order to control "sign pollution." Suppose that an advertising agency has been asked to design a billboard that has area 40 square meters.

The area A of any rectangle can be calculated by multiplying its base b and height h. The familiar formula is $A = bh$. That formula can be used in two other equivalent forms when the questions focus on finding the base or the height given a fixed area:

$$b = \frac{A}{h} \text{ or } h = \frac{A}{b}.$$

What base and height dimensions are possible with area 40 m²? You can quickly calculate some pairs that work, for example, 5 by 8, 10 by 4, or 20 by 2.

 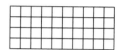

To get a systematic picture of all possibilities, it helps to proceed as follows:

a. With $A = 40$, write the height h as a function of the base b.

b. Copy and complete a table of data pairs, (base, height), from 1 to 45 in steps of 2.

c. Sketch a graph of the relation between base b and height h.

d. Use the data in your table to decide whether all rectangles of area 40 square meters have the same perimeter $(2b + 2h)$. If so, what is that perimeter? If not, estimate the smallest perimeter.

e. How does the required height change if the base is made very large? How is your answer shown by the graph?

f. How does the required height change if the base is made very small? How is your answer shown by the graph?

g. If the base is increased by 1 meter, how does the height change? Is that change 1 meter also? What if the base is decreased by 1 meter?

h. If $A = 40$, what rule expresses the base as a function of the height?

i. Would any of your answers to parts b through g change if height and base were interchanged in the function rule and in the questions? Explain your reasoning.

2. SITUATION In a bicycle race across America the total distance is about 5000 kilometers. The average speed for each racer depends on the time taken to complete the race.

a. What is the average speed for a racer who takes the following time to complete the race:

i. 200 hours? **ii.** 150 hours? **iii.** 100 hours?

b. How long does it take a racer to finish if the average speed is:
  i. 40 kilometers per hour?
  ii. 20 kilometers per hour?
c. What is the overall relation between time to complete the race and average speed?
d. Sketch a graph of the function S giving average speed as a function of time.

3. SITUATION   Suppose one of the bicycle riders in Exercise 2 finishes the 5000 km race in a total of 100 hours of riding time, but then is given penalties for getting off the course. The cyclist's *effective average speed* can be considered the final score (time with penalties added) divided by the length of the race.
  a. If p is the number of hours of penalty time assigned to that rider, write a rule giving effective average speed for the race as a function of p.
  b. What affect does each of the following penalties have on the effective average speed:
    i. the first penalty hour?
    ii. the second penalty hour?
    iii. the third penalty hour?

4. SITUATION   The local hockey club is planning to reserve time each week on a skating rink. The cost for the season is $4000. The club now has 25 members and wants to sign up new members to help reduce the cost per member.
  a. What is the cost per member if the club signs up 5 new members?
  b. What is the cost per member if the club gets 10 new members?
  c. Write a rule giving cost per member as a function of the *number of new members* who are signed up.
  d. What number of new members makes the cost $50 per member?
  e. Which of the following membership changes causes the greater change in cost per member:
    i. an increase from 15 to 16 new members, or
    ii. an increase from 45 to 48 new members?
  f. Sketch a graph of the function giving *cost per member* as a function of the *number of new members*.
  g. Does it make sense to consider a fraction or mixed number, such as $32\frac{1}{3}$, as an input? What limitations to the input values might you need to consider when using your rules for this situation?

5. Given below are six function rules and sketches of their graphs—but not in the same order. The graphs are not labeled, and no scales or tic marks are given on the diagrams. Match each rule to the sketch that best represents it by using your knowledge of graph shapes for function types.

$$f(x) = \frac{10}{x} \qquad g(x) = \frac{10}{x^2} \qquad h(x) = \frac{10}{x+2}$$

$$j(x) = 2x \qquad k(x) = 1.5x^2 \qquad m(x) = 3(1.5^x)$$

a.

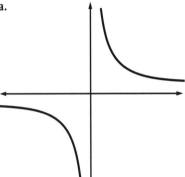

b.

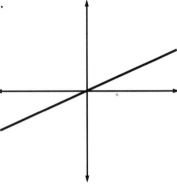

c.

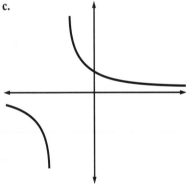

d.

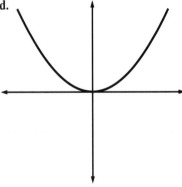

e.

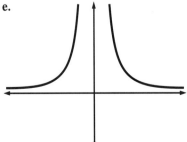

f.

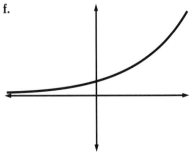

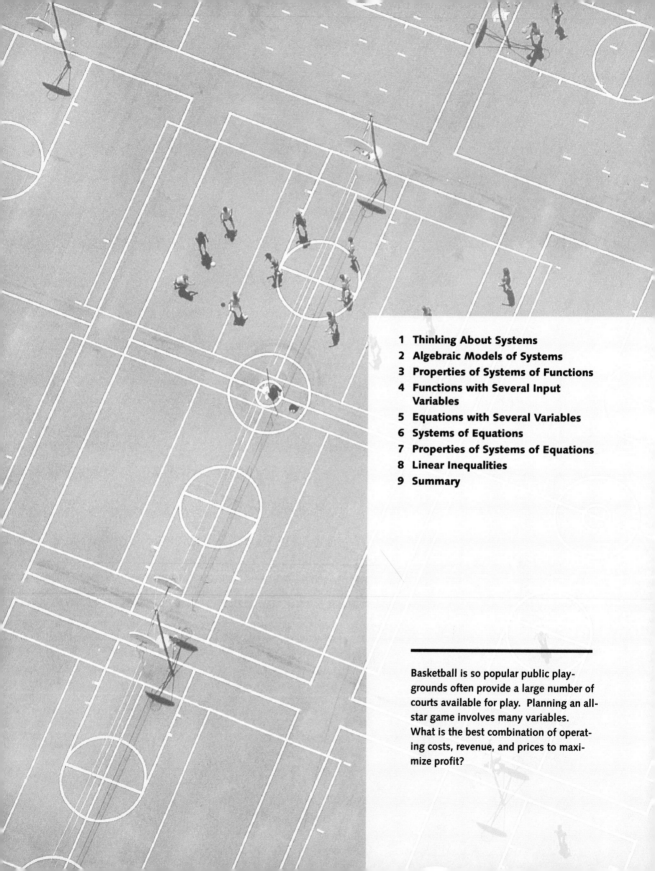

Basketball is so popular public playgrounds often provide a large number of courts available for play. Planning an all-star game involves many variables. What is the best combination of operating costs, revenue, and prices to maximize profit?

7 Systems of Functions and Equations

In the earlier parts of this algebra course, you studied relations between variables in quantitative problems. You were asked to identify situations in which one variable y *is a function of* another variable x or in which the value of y *depends on* the value of x. Then you used tables of data pairs, (x, y), graphs of those data, and symbolic rules relating x and y to find overall patterns and specific information about the relations. You studied how linear, quadratic, exponential, and rational functions model quantitative relationships.

Most problem situations in the previous chapters involved only one function relating two variables. Simple two-variable models like those are very useful. However, many important problem situations involve more than two variables or more than one relation among the variables. The mathematical models for these multi-variable, multi-relation situations are called systems.

In this chapter, you will extend your knowledge of variables, functions, equations, and inequalities, and learn how to construct and analyze models that are algebraic systems.

1 Thinking About Systems

The first step in modeling a system is identifying the variables and relations that describe and predict conditions in the system. That task usually requires some special knowledge about the situation, which you may or may not have, but it is often possible to make an intelligent start with very little specific data or knowledge.

Exploration

Earth science, biology, chemistry, and physics are important fields of science. Recent concerns about the health hazards caused by pollution have made us all aware of how these separate sciences are related when used to study our *environment*. One aspect of environmental science is the study of *ecosystems*—communities of animals, plants, and bacteria and their physical and chemical surroundings.

Environmental scientists study a variety of important ecosystems ranging from small lakes, rivers, and parks to very large land and water regions like the Chesapeake Bay and the Amazon River rainforest. The mathematical ideas that you learned earlier in this course provide tools to help you start on the task of describing and predicting life in an ecosystem's community of organisms and the physical and chemical features of the system's environment.

SITUATION 1.1

Sleepy Creek State Forest is a 22,000 acre wildlife region in the mountains of West Virginia. Contained within the forest, a small stream running between two mountain ridges has been dammed to form a lake. The forest is home to many kinds of animals and plants. A few roads and campsites bring people there to hunt, fish, and hike. The forest contains a beautiful and interesting wildlife ecosystem to study and enjoy.

1. Add your own ideas to the following list of animals one might expect to find in the Sleepy Creek ecosystem.

   Animals: snakes, mice, song birds

2. Add your own ideas to the following list of plants one might expect to find in the Sleepy Creek ecosystem.

   Plants: oak trees, poison ivy

3. Add your own ideas to the following list of physical and chemical properties of the forest ecosystem that would affect animal and plant life.

   Environmental conditions: rainfall, temperature

4. When biologists study ecosystems, they often focus on describing and predicting the size of animal and plant populations. They want to know how populations are related and how environmental conditions affect them.

   For example, in the Sleepy Creek State Forest, the population of mice is affected by the population of snakes, because snakes eat mice. The population of mice is also affected by the food supply of seeds from various plants and by weather conditions like temperature and snow depth in winter.

   a. Select one of the types of animals from your list and explain how you expect that animal's population to be related to other animal and plant populations and environment conditions. Write your answer in the form, "The population of ... depends on ... because...."

   b. Select one of the types of plants from your list and explain how you expect that plant's population to be related to other animal and plant populations and environment conditions. Write your answer in the form, "The population of ... depends on ... because...."

5. When biologists know what kinds of animals inhabit an ecosystem, they often study the relation between several animal populations by analyzing census data—counts of the number of each animal in the system at various times. Counting the number of mice, snakes, squirrels, or deer in a forest is not so easy as counting the number of people in an apartment building, but there are methods for making good estimates.

The data in the following table show how populations of several animals in Sleepy Creek State Forest might vary over a ten-year period. All numbers are given in thousands; that is, 700 in the "mice" column stands for 700,000, 0.15 in the "fox" column stands for 0.15 × 1000 = 150 fox, *etc.* Time is measured in years from the start of the population study, year 0.

Populations in thousands

Year	Mice	Snakes	Rabbits	Fox	Squirrels
0	700	35	50	0.15	75
1	860	23	30	0.19	95
2	890	21	26	0.20	99
3	780	29	40	0.17	85
4	620	41	60	0.13	65
5	500	50	75	0.10	50
6	540	47	70	0.11	55
7	700	34	51	0.15	73
8	855	22	29	0.18	94
9	890	21	25	0.19	100
10	770	28	39	0.17	83

From these data, you can see how each population varies over time. Study the data and record any interesting patterns you observe. Also, try to find patterns that suggest how populations are related to one another.

6. Naturalists know that mice are a source of food for snakes, so the populations of mice and snakes might be related to each other.

You learned in earlier work that numerical patterns are often easier to see when displayed in graphic form. Plot the population data for Sleepy Creek mice and snakes on a grid such as the one shown here.

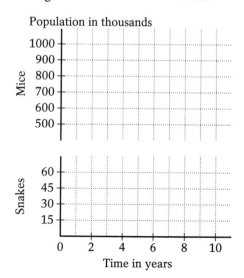

Population in thousands

Because there are so many more mice than snakes, it is hard to choose a population scale that displays both functions well. In cases like this it is common to put a break in the axis, so that the visual patterns of the two functions can be compared. Note the scales given on this split diagram.

You will have eleven data points for each population. Although population surveys might occur only once each year, the populations actually change gradually throughout the year. Thus it is reasonable to connect the eleven points to estimate the population at times between surveys and to help reveal patterns of change.

a. If $M(t)$ and $S(t)$ are the respective populations of mice and snakes in year t, then in what years is the difference $M(t) - S(t)$
   i. the greatest?
   ii. the smallest?
b. When does the snake population increase and when does it decrease? What are the changes in the population of mice during those years?
c. Is the pattern in part b what you would expect? Explain.

7. Rabbits and squirrels do not seem to compete with each other for survival, so you might expect that their populations are not related in any simple way. Plot the population data for rabbits and squirrels on a grid. Reviewing the data in the table in Exercise 5 may help you choose a scale for both axes..

Answer the following questions about the rabbit population $R(t)$ and the squirrel population $Q(t)$, both in year t. Explain what each answer tells about the populations of rabbits and squirrels.
a. For which t is $R(t) > Q(t)$?
b. For which t is $R(t) < Q(t)$?
c. For which t is $R(t) \approx Q(t)$?

Exercises

The key idea in thinking about systems is that a change in one variable is usually related to changes in several other variables as well. Each of the following exercises asks you to explain how changes in one variable for an ecosystem affect changes in several other variables.

1. What relations does the following diagram show between the population of Sleepy Creek hawks and the populations of mice, snakes, and rabbits?

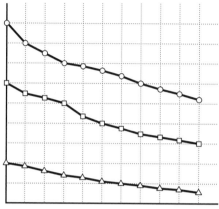

o Population of mice
□ Population of rabbits
△ Population of snakes

Population of hawks

2. What relations among rainfall, average stream flow, mosquito population density, and number of campers in the summer months at Sleepy Creek Forest are suggested by data in the following table?

Rainfall (cm)	Stream flow (l/sec)	Mosquitoes (per m²)	Campers
5	250	4	450
15	560	6	420
25	800	12	330
35	1200	35	150
45	1500	40	85

3. During a recent dry summer, biologists noticed that oak trees produced above-average numbers of acorns, which are prime food for squirrels. What conclusion might you make about the effect of dry weather on squirrel populations?

4. What population effects would you expect from an increase of polluting chemicals in Sleepy Creek?

5. Study the population data for rabbits and fox in Exercise 5 of the exploration.

   a. Make a graph showing the patterns of those populations for a ten-year period. You may wish to use the split-axis technique because the numbers in these two populations are so different.

   b. Describe the patterns you notice in the table and graph.

   c. Do you believe that fox depend on rabbits as a source of food? How do the data in the table or graph support your conclusion?

6. Owls are known to feed on snakes, mice, rabbits, and squirrels. Sketch a graph that shows how the populations of owls (the predators) and one of these other animals (the prey) might vary over a ten-year period.

7. **Simulation** In the preceding questions you have looked at the relations between several animal populations as time passes. This is a first step in analyzing complex systems like wildlife preserves. However, accurate models require consideration of more variables, and more relations among those variables, than we have considered here.

Scientists have built simulation models of many ecosystems. Such models allow you to experiment by changing animal populations, plant life, and physical or chemical conditions of a system and observing the effects of those changes.

If possible, get a copy of the software for such a simulation and explore it to get further understanding of quantitative systems and their properties.

2 Algebraic Models of Systems

The relations among ecosystem variables are often difficult to model with algebraic rules. However, in other systems it is easier to find rules that relate the important variables. The following exploration asks you to study one such system involving the economics of a high school basketball all-star game.

Exploration I

Every year at the end of the high school basketball season there are many all-star games in major cities around the country. One of the most famous is the Dapper Dan game in Pittsburgh, PA. It is a nationwide attraction for players, coaches, news media, and college recruiters.

In planning for this game the organizers must consider many variables and make many decisions. It looks like a complex system to model, but most variables and decisions can be sorted into two major categories—those that affect *operating expenses* or *costs* of the game, and those that affect *income* or *revenue* from the game.

Operating expenses or costs include items such as rent for the stadium, travel expenses for players, pay for officials, advertising of the game, and printing of tickets and programs. Sources of income or revenue include sale of tickets, programs, and concessions, as well as possible television broadcast rights.

The critical variable in both cost and revenue functions is *attendance* at the game, and a critical factor influencing attendance is *ticket price* charged for admission. You have seen in earlier chapters how these variables can be used as inputs for cost and revenue functions. The following examples present possible cost and revenue functions for this game.

1. The fixed and variable *operating costs* can be combined to give a function that can be expressed using the rule $C(x) = 3x + 30\ 000$, where x is the number of people who attend the game.

2. The sources of *revenue* can be combined to give a function that can be expressed using the rule $R(x) = -0.005x^2 + 40x$, where x is the number of people attending the game.

3. The expected attendance is related to the *ticket price* for the game. The function rule might be expressed $T(x) = -0.005x + 40$. Again, the input x is the attendance.

4. Finally, *profit* for the game is the difference between revenue and operating costs, $P(x) = R(x) - C(x)$.

Organizers of the game are obviously interested in how various ticket prices and attendance figures might affect operating costs, revenue, and profits. There are several ways to study this system of variables and functions.

Method I. Tables of Values for Systems of Functions In studying individual functions, you have found useful information in tables of data pairs. Copy the following table. Use your technological tools to display the relations among the key variables for the all-star game.

Attendance	Operating costs	Revenue	Profit	Ticket price
x	$3x + 30\ 000$	$-0.005x^2 + 40x$	$R(x) - C(x)$	$40 - 0.005x$
0	30 000	0	$-30\ 000$	40
1 000	33 000	35 000	2 000	35
2 000				
3 000				
4 000				
5 000				
6 000				
7 000				
8 000				
9 000				
10 000				

1. There is a great deal of information in the completed table. Study the data relating attendance and operating costs, revenue, profit, and ticket price. Then use your observations to answer the following questions.

   How does each of the following variables change as attendance increases from 0 to 10,000?
   a. Operating costs
   b. Revenue
   c. Profit
   d. Ticket price

2. Use the information in your table of values to estimate answers for the following questions.
   a. What attendance is related to a ticket price of $20?
   b. For what attendance values are revenues greater than operating costs?
   c. For what attendance is revenue equal to operating costs?
   d. What are the attendance levels when the profit is a positive number?
   e. What attendance gives maximum profit?
   f. What are the attendance levels when the operating costs are greater than the revenue, and how does that relate to profit?

For some of these questions, the numerical answers can only be estimated on the basis of the information in the completed table. However, in much the same way you "zoomed in" on more accurate answers for questions about single functions, you could produce more detailed tables for this system of functions and obtain better approximations to the desired answers.

Method II. Graphing Systems of Functions In Section 1, you studied interactions of animal populations in Sleepy Creek State Forest by graphing several population functions on the same diagram. When you know algebraic rules for the functions of a system, you can produce those graphs easily with technological help.

1. Generate a graph that displays the following three functions of attendance at the all-star game. Then sketch the functions on a coordinate grid. Remember, for attendance x, use the following rules.

   Operating costs: $C(x) = 3x + 30\ 000$.
   Revenue: $R(x) = -0.005x^2 + 40x$.
   Profit: $P(x) = (-0.005x^2 + 40x) - (3x + 30\ 000)$.

   The rule for the profit function may appear rather complicated. However, you can get a simpler form from a computer-algebra program by entering $R(x) - C(x)$ and instructing the tool to produce a simplified form. This can be done using a command such as SIMPLIFY. The response probably will be the following simpler, but equivalent, rule.

   $$\text{Profit: } P(x) = -0.005x^2 + 37x - 30\ 000.$$

   Hint: For help choosing a good graphing window and scales for the axes, look at the table of values for these variables and find the smallest and the largest values in the input and output columns.

2. On your sketch, mark points that correspond to the answers to the following questions.
   a. Mark with the letter B (for *break even*) any points where cost is equal to revenue. Write the approximate coordinates of those points and a sentence explaining what you can tell about the situation from these coordinates.

b. Mark with the letter pairs *MR* (for *maximum revenue*) and *MP* (*for maximum profit*) the points where revenue and profit are largest. Write the approximate coordinates of those points and a sentence explaining what you can tell about the situation from these coordinates.

c. Mark with the letter pair *BE* (for *break even*) any points where profit is equal to 0. Write approximate coordinates for those points and a sentence explaining what you can tell about the situation from these coordinates.

d. Using their numerical coordinates, label the three points for the cost, the revenue, and the profit when the attendance x is equal to 2000. Explain how $C(2000)$, $R(2000)$, and $P(2000)$ are related, and how that relation is shown on the graph.

Method III. Symbol Manipulation in Systems of Functions   In earlier chapters, you have seen that many questions about functions require solving equations or inequalities and that symbol manipulation technology is useful in finding the solutions. The same idea applies to systems of functions.

For example, the question *For what attendance does revenue equal operating costs?* corresponds to the equation

$$R(x) = C(x) \text{ or } -0.005x^2 + 40x = 3x + 30\ 000.$$

You can solve this equation using a computer-algebra program to get the roots

$$x = 3700 + 100\sqrt{769} \text{ and } x = 3700 - 100\sqrt{769},$$

which are approximately $x = 6473$ and $x = 927$, respectively.

1. For each of the following questions about the all-star game, write an equation or inequality that can be used to answer the question. Then use your computer-algebra program to find a solution (where possible).

   a. What attendance x gives revenue equal to operating costs plus $5000?

   b. For what attendance x are operating costs less than $50,000?

   c. For what attendance does profit equal $5000?

2. Sometimes you may find an algebraic equation or inequality that

cannot be solved by your computer-algebra program. Try the following example. Record both the equation or inequality you enter on your system and the response you receive.

For what attendance levels are the revenues greater than the costs?

Exploration II

In Exploration I you studied one model of the profit prospects for an all-star basketball game. You used tables, graphs, and symbol manipulations to answer questions about that system model.

It is quite likely that the all-star game organizers might make different decisions about how to operate—for instance, different allowances for advertising, for players' expenses, or for expected television revenue. Then they would have a different model and different predicted results.

SITUATION 2.2

Suppose that the organizers of the all-star game focus on ticket price t as the key input variable and that their assumptions lead to the following system of functions.

1. Ticket sales for the game are a function of ticket price t with rule $S(t) = 8000 - 200t$;

2. Operating costs for the game depend on ticket sales, which depend on ticket price t, with rule $C(t) = 54\,000 - 600t$;

3. Revenue for the game depends on ticket price t with rule $R(t) = -200t^2 + 8000t$; and

4. Profit for the game depends on revenue and costs, and can be described by the rule $P(t) = (-200t^2 + 8000t) - (54\,000 - 600t)$ or, in equivalent form, $P(t) = -200t^2 + 8600t - 54\,000$.

Use these function rules to answer the following questions about the system of functions that describe the new model for the all-star game.

1. Consider the demand function $S(t) = 8000 - 200t$ relating ticket price and ticket sales to answer the next questions. You might find a table, a graph, or symbolic manipulation technology helpful.
   a. How does attendance change as ticket price is increased?

    b. Does attendance change at a constant rate as ticket price increases?

    c. At what price is no one willing to purchase a ticket?

2. For ticket prices between $0 and $40, create tables and graphs of the cost, revenue, and profit functions. Make sketches of the three graphs on the same diagram, labeling each graph with its function name.

Use the table and graph from Exercise 2 to answer the following.

3. How do each of the following variables change as ticket price increases from 0 to 40?

    a. Operating costs

    b. Revenue

    c. Profit

4. On your sketch, mark points that correspond to the answers to the following questions.

    a. Mark with the letter *B* (for *break even*) any points where cost is equal to revenue. Write the approximate coordinates of those points and a sentence explaining what you can tell about the situation from these coordinates.

    b. Mark with the letter pairs *MR* (for *maximum revenue*) and *MP* (for *maximum profit*) the points with the greatest revenue and the greatest profit, respectively. Write the approximate coordinates of those points and a sentence explaining what you can tell about the situation from these coordinates.

    c. Mark with the letter pair *BE* (for *break even*) any points where profit is equal to 0. Write approximate coordinates for those points and a sentence explaining what you can tell about the situation from these coordinates.

    d. Label with their coordinates the three points that show the costs, the revenue, and the profit when the price of a ticket is $15. Explain how $C(15)$, $R(15)$, and $P(15)$ are related and how that relation is shown on the graph.

Exercises

As you have seen in other chapters, one of the key tasks in building a mathematical model of a situation is writing symbolic rules that relate the variables. These exercises give you further practice in translating business choices and decisions into algebraic function rules. All exercises relate to parts of the all-star game of Exploration I. Notice that the function names all have two letters. Your technological tools may not accept a function name using two letters, but such names are often used in models to make it more clear what they represent.

1. Suppose the game organizers expect that the people attending the game will spend an average of $2.50 on food and drinks from the concession stands. Concession income can be considered a function *CI* of attendance at the game *x*. Write a rule to show this relation.

2. Suppose the game organizers expect that one-quarter of the people who attend the game will buy a game program for $2. Income from programs can be considered a function *PI* of attendance *x*. Write a rule to show this relation.

3. To hold the all-star game, the organizers must hire workers such as officials, scorekeepers, ticket takers, ushers, and security guards. Some of these workers must be hired regardless of how many tickets are sold for the game, but others are hired in numbers intended to handle the crowds. The number of additional personnel depends on expected ticket sales.
   a. List some jobs that must be done regardless of the number of tickets sold.
   b. List some jobs for which additional people might be hired if a large crowd is expected.
   c. Suppose that fixed personnel costs are $1500 and that additional workers will be hired at an average rate of $0.25 per ticket sold. Then total workers costs will be a function *WC* of tickets sold *x*. Write a rule to show this relation.

4. Suppose the all-star game organizers have a stadium-rental contract that requires payment of $4000 plus $0.50 per ticket sold. Then the total rental cost for the stadium is a function *RC* of ticket sales x. Write a rule to show this relation.

5. Suppose that the organizers plan to sell a special t-shirt only at the game. They intend to order 500 shirts at $4 each in one-size-fits-all. They plan to charge $10 for each t-shirt and to pay the t-shirt sellers $1.50 per shirt sold. The organizers' profit from t-shirts is a function of *TP* of *n*, the number sold. Write a rule to show this relation.

(**Note:** The organizers must order 500 in advance, but some might not be sold!)

3 Properties of Systems of Functions

The methods you used in answering questions about the Sleepy Creek ecosystem and the all-star game can be used to study systems of functions in very different situations. The simplest kind of system involves two functions, $f(x)$ and $g(x)$. There are three basic questions that are often asked about those functions:

1. For which x is $f(x) = g(x)$?

   For example, for which attendance figures does revenue for the all-star game equal operating costs?

2. For which x is $f(x) < g(x)$?

   For example, when is the Sleepy Creek rabbit population smaller than the squirrel population?

3. For which x is $f(x) > g(x)$?

   For example, which ticket prices for the all-star game give revenue greater than operating costs?

When you encounter a system of two functions, it helps to look at the rules of the given functions and make intelligent guesses about the answers to these three questions. It may be helpful to think about the questions in terms of graphs.

For example, suppose a problem situation leads to the system of functions $f(x) = 2^x$ and $g(x) = -x + 11$. From previous work you should recall that $f(x)$ is an exponential function for which outputs increase as x increases. On the other hand, $g(x)$ is a linear function for which outputs decrease as x increases. The graph of this exponential function intersects the output axis at the point $(0, 1)$, and the graph of this linear function crosses the vertical axis at $(0, 11)$.

The diagram on the right shows the graphs of the given exponential and linear functions. The diagram suggests that $f(x) < g(x)$ for $x < 3$, $f(3) = g(3)$, and $f(x) > g(x)$ for $x > 3$.

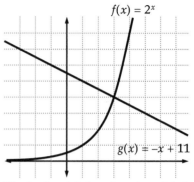

$f(x) = 2^x$

$g(x) = -x + 11$

The horizontal (input) scale unit is 1 and the vertical (output) scale unit is 2.

 ## Exploration

This exploration is designed to help you develop "symbol sense" about systems—the ability to look at rules for functions in a system and then to estimate the way that tables and graphs for those functions relate to each other. In each case you will be given two function rules. Use computer tables, graphs, and symbol manipulations to answer the given questions about those functions.

In any problem situation, the interesting relations between two functions might occur over a wide range of input and output values. You have clues to those interesting points from the numbers given in the situation. Since the exercises of this exploration have no situation clues, we have planned them so that the interesting patterns generally occur for inputs between –10 and 10.

1. Consider the functions $f(x) = x - 4$ and $g(x) = -0.5x + 5$.
    a. What type of functions are $f(x)$ and $g(x)$?
    b. Sketch the graphs of these two functions on the same coordinate system.
    c. For which x is $f(x) = g(x)$?
    d. For which x is $f(x) < g(x)$?
    e. For which x is $f(x) > g(x)$?

2. Consider the functions $f(x) = \frac{1}{4}x - 5$ and $g(x) = \frac{1}{4}x + 3$.
    a. What type of functions are $f(x)$ and $g(x)$?
    b. Sketch the graphs of these two functions on the same coordinate system.
    c. For which x is $f(x) = g(x)$?
    d. For which x is $f(x) < g(x)$?
    e. For which x is $f(x) > g(x)$?

3. Consider the functions $f(x) = x^2 - 3$ and $g(x) = -x^2 + 5$.
   a. Sketch the graphs of these two functions on the same coordinate system.
   b. What type of functions are $f(x)$ and $g(x)$?
   c. What is the graph of each of these functions called?
   d. For which x is $f(x) = g(x)$?
   e. For which x is $f(x) < g(x)$?
   f. For which x is $f(x) > g(x)$?

4. Consider the functions $f(x) = x^2 - 4$ and $g(x) = x + 2$.
   a. What type of functions are $f(x)$ and $g(x)$?
   b. Sketch the graphs of these two functions on the same coordinate system.
   c. For which x is $f(x) = g(x)$?
   d. For which x is $f(x) < g(x)$?
   e. For which x is $f(x) > g(x)$?

5. Consider the functions $f(x) = 1/x$ and $g(x) = x$.
   a. Sketch the graphs of these two functions on the same coordinate system.
   b. What type of functions are $f(x)$ and $g(x)$?
   c. For which x is $f(x) = g(x)$?
   d. Solve the inequality $f(x) < g(x)$.
   e. Solve the inequality $f(x) > g(x)$.

6. Consider the functions $f(x) = x^2$ and $g(x) = x - 3$.
   a. What types of functions are $f(x)$ and $g(x)$?
   b. Sketch the graphs of these two functions on the same coordinate system.
   c. What are the roots of the equation $f(x) = g(x)$?
   d. Solve the inequality $f(x) < g(x)$.
   e. Solve $f(x) > g(x)$.

7. Consider the functions $f(x) = 5\,(0.7)^x$ and $g(x) = -x + 5$.
   a. Sketch the graphs of these two functions.
   b. What type of functions are $f(x)$ and $g(x)$?
   c. Find each root of $f(x) = g(x)$.
   d. For which x is $f(x) < g(x)$?
   e. For which x is $f(x) > g(x)$?

In each of the seven exercises of this exploration, you have determined how two given functions are related. You have recorded the inputs for which the two outputs are equal, the first output is the greater, and the first output is the smaller. You probably have some ideas about what relations can be expected for various systems of functions—two linear functions, a linear function and a quadratic function, two quadratic functions, a linear function and an exponential function, *etc.*

Write some systems of functions, then use technology-generated tables and graphs to explore these systems. See whether you can discover patterns for any pair of function types. Try to find systems whose graphs never cross and other systems whose graphs intersect once, twice, three times, or even four times.

Exercises

For each of Exercises 1 through 4 you are given the rules for two functions of a system. Match each system with its graph in the following diagrams a through d. Then label each individual function graph with its rule.

a.

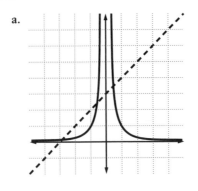

b.

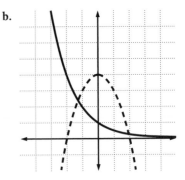

c.

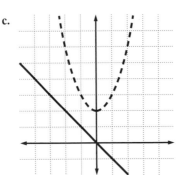

d.

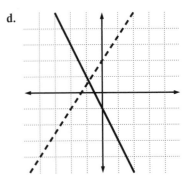

1. $f(x) = -2x - 1,\ g(x) = 1.5x + 2$

2. $f(x) = \dfrac{1}{x^2},\ g(x) = x + 3$

3. $f(x) = (0.5)^x,\ g(x) = -x^2 + 4$

4. $f(x) = -x,\ g(x) = x^2 + 2$

5. For each system in Exercises 1 through 4, use the graphs to estimate answers for the following questions.
   −For which x is $f(x) = g(x)$?
   −For which x is $f(x) < g(x)$?
   −For which x is $f(x) > g(x)$?

6. If $f(x)$ and $g(x)$ are both linear functions, for how many values of x can $f(x) = g(x)$? Draw sketches to illustrate each possibility.

7. If $f(x)$ is a linear function and $g(x)$ is a quadratic function, for how many values of x can $f(x) = g(x)$? Draw sketches illustrating each possibility.

8. If $f(x)$ and $g(x)$ are both quadratic functions, for how many inputs x can $f(x) = g(x)$? Draw sketches illustrating each possibility.

9. Suppose $f(x)$ is a linear function and $g(x)$ is an exponential function. Recall that exponential functions have rules of the form $g(x) = Ca^x$, where a is positive. How many roots can the equation $f(x) = g(x)$ have? Draw sketches to illustrate each possibility.

10. If $f(x)$ is a linear function and $g(x)$ is the rational function $g(x) = 1/x$, how many roots can the equation $f(x) = g(x)$ have? Draw sketches illustrating each possibility.

11. If $f(x)$ and $g(x)$ are both exponential functions, each having a rule in the form Ca^x with positive base a, for how many values of x can $f(x) = g(x)$? Draw sketches to illustrate each possibility.

Making Connections

Answer the following questions based on your observations in the exploration of systems of two functions with a common input variable.

1. How can you use technology-generated tables to solve equations of the form $f(x) = g(x)$ or inequalities of one of the forms $f(x) < g(x)$ or $f(x) > g(x)$?

2. How can you use a graphing utility to solve an equation of the form $f(x) = g(x)$ or an inequality of one of the forms $f(x) < g(x)$ or $f(x) > g(x)$?

3. How can you use symbol manipulation technology to solve equations of the form $f(x) = g(x)$ or inequalities of the form $f(x) < g(x)$ or $f(x) > g(x)$?

4. What are the possible forms of the solutions for
   a. an equation, $f(x) = g(x)$?
   b. an inequality, $f(x) < g(x)$ or $f(x) > g(x)$?

4 Functions with Several Input Variables

Each of the systems you have studied in this chapter involve two or more functions with a common input variable. Many systems focus on a single output variable that depends on two or three different input variables.

Exploration I

The number of tickets sold for an all-star basketball game depends on the price of tickets, but it also depends on the advertising for the game.

SITUATION 4.1

Suppose that research on the advertising and ticket prices from past games suggests the following function rule:

Ticket sales = 5000 + 0.2(Advertising expenses) – 100(Ticket price).
Using a for advertising expenses and p for ticket price, both in dollars, we can write ticket sales as a function S of two variables as follows:
$$S(a, p) = 5000 + 0.2a - 100p.$$

Use this rule to answer the questions that follow.

1. Calculate and explain the meaning of each of the following:
   a. $S(5000, 10)$
   b. $S(5000, 15)$
   c. $S(15\,000, 10)$

2. Write the following questions in function notation, then find the answers.
   a. How many tickets are expected to be sold if the advertising expense is $7500 and the ticket price is $20?
   b. How many tickets are expected to be sold if the ticket price is $8 and advertising expense is $12,500?

3. Use a guess-and-test strategy to search for answers to the following questions about the relation among advertising, ticket price, and ticket sales for the all-star basketball game. Many symbolic manipulation tools permit definition of a function with several input variables. It may be helpful to enter the rule for $S(a, p) = 5000 + 0.2a - 100p$ in the symbolic manipulation tool you use.
   a. Find three combinations of advertising expense and ticket price that give ticket sales of 8000. A suggested table heading to help your guess-and-test strategy is shown here.

Guess		Test	Decision
a	p	$S(a, p)$	Is $S(a, p)$ = 8000?

   b. Find three combinations of advertising expense and ticket price that give ticket sales greater than 8000.

Guess		Test	Decision
a	p	$S(a, p)$	Is $S(a, p) > 8000$?

4. Use a symbolic manipulation tool to explore output of $S(a, p)$ for many different inputs, then copy and complete the following statements about the relation of ticket price and advertising expenditure to ticket sales.
   a. If the ticket price is fixed at $10 and if advertising expenses increase from $0 to $20,000, then ticket sales....
   b. If advertising expenses are fixed at $10,000 and if the ticket price is increased from $0 to $30, then ticket sales....

Exploration II

Two of the most common functions with multiple inputs are the formulas for the perimeter and the area of a rectangle. If a rectangle has length L and width W, its area A and perimeter P are given by the following function rules.

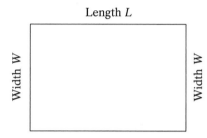

Area: $A(L, W) = L \times W$

Perimeter: $P(L, W) = (2 \times L) + (2 \times W)$
$= 2 \times (L + W)$

SITUATION 4.2

The Pennsylvania State Department of Environmental Resources has a tree nursery where evergreens are grown from seeds to a size that can be planted in parks around the state.

Suppose that to protect the seedlings from hungry deer, the nursery manager wants to enclose a rectangular field with a special high fence. The fencing costs \$15 per meter of length. The manager would like to choose dimensions for a field that has a large growing area and can be fenced at low cost. The functions that help in the study of these conditions are as follows, for a field of length L meters and width W meters.

Area: $A(L, W) = L \times W$

Cost: $C(L, W) = 15 \times \text{Perimeter}$
$= 15 \times 2 \times (L + W)$
$= 30 \times (L + W)$

Use these function rules to answer the questions that follow.

1. Calculate the following function outputs and explain what the results tell about the nursery fencing situation.
   a. $A(30, 20)$ and $C(30, 20)$
   b. $A(20, 15)$ and $C(20, 15)$
   c. $A(40, 25)$ and $C(40, 25)$

2. Find the dimensions L and W of four different fields with area 400 m², and find the fencing cost for each.

    Note: In printed books such as this one, italics are often used for variables (for example, x and y). In typed or written materials, this cannot be done. You should be careful to use information in each situation to decide whether a letter represents a name for a variable or an abbreviation for a unit of measurement.

3. Find the dimensions of a field that has area 400 m² and for which you believe the fencing cost is minimum. Use your answers from Exercise 2 to begin, then explore with some other dimensions L and W.

    What dimensions do you believe give minimum cost for fencing that surrounds an area of 600 m²?

Exploration III

Many American adults are concerned about the effects of food on health. When choosing what to eat, some people think about cholesterol, fat, and fiber as much as, or more than, taste. The study of nutrition is another situation in which several inputs (foods) combine to produce several related outputs like protein, carbohydrates, fats, and various vitamins. Your body is a complex input-output system.

SITUATION 4.3

One important part of a balanced diet is protein. Nutritionists recommend daily protein intake from 40 to 80 grams, depending on a person's age, size, and level of activity. The main sources of protein are meat, cheese, and whole grain cereals or bread.
The following values are the amounts of protein in a typical food of each given type. For example, each gram of whole grain cereal or bread contains 0.10 grams of protein, each gram of cheese contains 0.5 grams of protein, and each gram of lean meat contains about 0.25 grams of protein.
Using these values, a function rule can be written that expresses the total number of grams of protein in a meal with b grams of whole grain cereal or bread, c grams of this cheese, and m grams of this meat.

Write a rule $P(b, c, m)$ that you believe correctly models the given information about protein content of various foods. Use your rule to answer the questions that follow.

1. Calculate the following function outputs and explain what each tells about the nutritional value of various possible meals.
   a. $P(30, 60, 90)$
   b. $P(20, 60, 120)$

2. Write the following in function notation, then calculate the answers.
   a. How much protein is in a meal with 40 grams of cheese, 80 grams of lean meat, but no cereal or bread?
   b. How much protein is in a meal with 60 grams of whole grain cereal or bread, 120 grams of cheese, but no meat?

3. Calculate each of the following protein function outputs.
   a. $P(100, 60, 80)$
   b. $P(0, 80, 80)$
   c. $P(100, 40, 120)$
   d. $P(50, 80, 60)$
   e. $P(150, 0, 180)$
   f. Explain what you can tell from the pattern in your calculations.

4. Use your rule $P(b, c, m)$, giving protein content of a meal with b grams of whole grain cereal or bread, c grams of cheese, and m grams of lean meat, to do the following questions.
   a. Find two input triples, (b, c, m), for which the protein content is 50 g.
   b. Find two input triples, (b, c, m), each of which yields 40 g of protein.

5. Exercises 1 through 4 have focused on calculating the protein content of various possible meals. List several other factors that should be considered in deciding what foods to serve and eat.

Exercises

The explorations of this section illustrate situations in which a system, or part of a system, can be modeled by functions with several input variables. The following exercises give more practice with this idea, its notation, and the typical questions asked about such functions.

1. **SITUATION** Suppose that planners of a rock concert charge $15 for tickets and $10 for t-shirts sold at the concert.
   a. Concert revenue is a function of the number t of tickets sold and the number *s* of t-shirts sold. Write a rule for the revenue function.
   b. Calculate and explain the meaning of each of the following.
      i. *R(500, 80)*   **ii.** *R(1000, 200)*   **iii.** *R(4000, 750)*
   c. At what rate does total revenue change
      i. if ticket sales increase but t-shirt sales remain constant?
      ii. if t-shirt sales increase but ticket sales remain constant?
   d. Find two different combinations of ticket and t-shirt sales that give total revenue of $20,000.

2. **SITUATION** The volume of a rectangular box is a function V of its length L, its width W, and its height H with rule $V(L, W, H) = L \times W \times H$ or simply $V(L, W, H) = LWH$.

   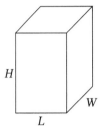

   a. Calculate and explain the meaning of each of the following:
      i. $V(2, 3, 5)$ **ii.** $V(3, 2, 5)$
      iii. $V(4, 6, 0)$
   b. Find dimensions of three boxes, each of which has a volume of 60 cm³.
   c. Calculate the following volumes, then find a pattern in your results and write a conclusion that describes the pattern.
      i. $V(1, 7, 3)$ **ii.** $V(2, 7, 3)$
      iii. $V(4, 7, 3)$ **iv.** $V(8, 7, 3)$
   d. Calculate the following volumes, then find a pattern in your results and write a conclusion that describes the pattern.
      i. $V(1, 2, 3)$ **ii.** $V(2,4 ,6)$
      iii. $V(4, 8, 12)$ **iv.** $V(8, 16, 24)$

3. **SITUATION** If you go on a long hike or bike trip, it helps to take along foods that are healthy sources of energy, such as peanuts and raisins. Peanuts have about 6 calories per gram, and raisins have about 4 calories per gram.
   a. Write a rule for the calorie content $C(p, r)$ of a mixture with p grams of peanuts and r grams of raisins.
   b. Calculate and explain the meaning of each of the following.
      i. $C(50, 80)$ **ii.** $C(80, 50)$ **iii.** $C(75, 0)$

    c. Find three combinations of peanuts and raisins that give an output of 500 calories.

    d. How does the calorie output of a snack change if the amount of both peanuts and raisins is doubled?

    e. How does the calorie output of a snack change if only one of the two inputs, peanuts or raisins, is doubled?

4. SITUATION The actual speed of an airplane is a function of two variables, the speed due to the plane's engines and the speed of the winds in which the plane is flying. Suppose that the engines of a plane are set so that the plane would fly p kilometers per hour if there were no wind. Then suppose the wind is actually blowing w kilometers per hour at the altitude where the plane is flying.

    a. Write the rule for a function $ST(p, w)$ giving the plane's speed across the country if the wind is a tailwind (that is, pushing the plane from the rear).

    b. Write the rule for a function $SH(p, w)$ giving the plane's speed across the country if the wind is a headwind (that is, blowing against the plane from the front).

5 Equations with Several Variables

Functions with two or more input variables are often more interesting than functions with a single input variable, but they can also be more difficult to deal with.

Having several input variables makes it harder to build a table that shows outputs for all the combinations of inputs you wish to use, and the familiar method of graphing data pairs, (input, output), does not apply when the input itself is an ordered pair or triple of numbers. These problems make it especially difficult to answer questions of the form, "Find the inputs that give specified output."

Consider a rectangle: its perimeter P is a function of its length L and its width W with rule $P(L, W) = 2L + 2W$. How can we display and study the outputs of this function for all inputs (L, W)? What about for values of L and W between 0 and 10?

Since the inputs are ordered pairs, (L, W), there is a way to make a table of values for $P(L, W)$ based on the method of graphing that uses ordered pairs as coordinates. We begin by displaying values of L along a horizontal axis and values of W along a vertical axis. Then at each point determined by these coordinates, (L, W), we write the output of the function $P(L, W)$. For example, the red 34, in the diagram shown here, indicates that $P(10, 7) = 34$. The red 14 shows that $P(2, 5) = 14$.

W	10	22	24	26	28	30	32	34	36	38	40
	9	20	22	24	26	28	30	32	34	36	38
	8	18	20	22	24	26	28	30	32	34	36
	7	16	18	20	22	24	26	28	30	32	34
	6	14	16	18	20	22	24	26	28	30	32
	5	12	14	16	18	20	22	24	26	28	30
	4	10	12	14	16	18	20	22	24	26	28
	3	8	10	12	14	16	18	20	22	24	26
	2	6	8	10	12	14	16	18	20	22	24
	1	4	6	8	10	12	14	16	18	20	22
		1	2	3	4	5	6	7	8	9	10	L

This new kind of table for a function of two input variables contains a great deal of information arranged in a way that reveals interesting patterns. Discuss the following questions with your classmates.

- –If we move across any row of the table, the perimeter value increases by 2 for each increase of 1 in the length L. Similarly, if we move up any column, the perimeter value increases by 2 for each increase of 1 in the width W. Can you explain these patterns?

- –If we move diagonally—that is, 1 step to the right and 1 step up— the perimeter entries increase in steps of 4. Can you explain this pattern?

- –There are 9 different input pairs, (L, W), that give a perimeter of 24. In fact, most of the listed perimeters are produced by several different input pairs. Can you explain this pattern?

- –When a single perimeter value appears at several places in the table, those entries appear to lie on a diagonal from upper left to lower right. Can you explain this pattern? Can you predict some pairs, (L, W), besides those already in the table, that give perimeter equal to 36? Equal to 28?

–The table gives only outputs for whole number inputs from 1 to 10, but you might imagine how a bigger table could include outputs for fractional inputs or for inputs larger than 10. In other situations, you may also need negative inputs. What is $P(4.5, 7.5)$? Where would that value appear in the table?

–Can you tell, from the pattern of entries shown, what $P(11, 2)$ is?

The ordered pairs, (L, W), that produce perimeter equal to 24 answer a familiar kind of algebra question—they are roots of the equation

$$P(L, W) = 24 \text{ or } 24 = 2L + 2W.$$

That is, they are the input pairs that produce output 24. The patterns in the table for $P(L, W)$ suggest two interesting possibilities:

1. an equation of this type might have many roots, and

2. it might be possible to picture the roots of such an equation by points on a coordinate grid.

Exploration I

The simplest, yet important, family of functions with one input variable includes those with rules of the form $f(x) = ax + b$. For functions with two input variables, the simplest important family includes those with rules of the form $f(x, y) = ax + by$. The following exploration focuses on properties of this family of functions, equations that can be used to answer important questions about the functions, and several methods of illustrating or displaying the solutions of those equations.

The aim of this exploration is to discover properties of all functions with rules of the form $f(x, y) = ax + by$. However, in each example there are specific situations in which the particular function can be useful.

Experiment 1. First, consider the function $f(x, y) = y - x$. For any two input numbers, the output is their difference (with the first subtracted from the second).

This function might be used to calculate the profit of a business where x and y represent the expenses and the revenue, respectively. The units can be anywhere from hundreds to millions of dollars, depending on the size of the business and the amount of time in question.

1. Copy and complete the following table showing outputs for inputs from 1 to 10 in steps of 1.

y	1	2	3	4	5	6	7	8	9	10
10	9	—	—	—	—	—	—	—	—	—
9	8	—	—	—	—	—	—	—	—	—
8	—	—	—	—	—	—	—	—	—	—
7	—	—	—	—	—	—	—	—	—	—
6	—	—	—	—	—	—	—	—	—	—
5	—	—	—	—	—	—	—	—	—	—
4	—	2	—	—	—	—	—	—	—	—
3	—	—	—	−1	—	—	—	—	−6	—
2	—	—	—	—	—	—	—	—	—	—
1	0	—	—	—	—	—	—	−7	—	—

x: 1 2 3 4 5 6 7 8 9 10

a. Circle each entry of "2" in the table and describe the pattern of those entries.

List all of the ordered pairs, (x, y), included in the table for which $f(x, y) = 2$. Each of these ordered pairs is a **root** of the equation $f(x, y) = 2$ or $y - x = 2$.

b. Now circle each entry of "–5" in the table and describe the pattern of those entries.

List all of the ordered pairs, (x, y), included in the table for which $f(x, y) = -5$. These ordered pairs are all roots of the equation $f(x, y) = -5$ or $y - x = -5$.

2. The table displays all outputs of the function $f(x, y) = y - x$ for integer inputs from 1 to 10. If you focus attention on a specific equation related to this function, it is possible to produce a graph of the equation on the familiar coordinate grid you have used so often in this course. Let us now explore this new application of the coordinate grid.

a. On a coordinate diagram such as the one shown here, put a dot (•) on each grid point for which $f(x, y) = 2$ and put a square (■) on each grid point for which $f(x, y) = -5$.

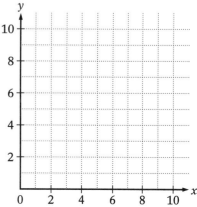

b. Describe the pattern formed by the dots (•) and the pattern formed by the squares (■).

c. Write a paragraph describing any connections you can see between your coordinate diagram and the earlier table of the function $f(x, y) = y - x$. Be sure to include the following: what information from the table is shown in the graph; what information from the table is not shown in the graph; what information is shown in the graph, but not in the table.

Experiment 2. Next, consider the function $g(x, y) = x + 2y$. For any input pair, the output is equal to the first input plus twice the second. For inputs where $x < 2y$, you can think of $g(x, y)$ as the perimeter of an isosceles triangle with base of length x and congruent sides of length y. Although this does not illustrate the function when $x \geq 2y$, we still can include these inputs in our study.

1. Copy and complete the following table showing outputs for x and y from 1 to 10 with step size 1.

y \ x	1	2	3	4	5	6	7	8	9	10
10	21	—	—	—	—	—	—	—	—	—
9	19	—	—	—	—	—	—	—	—	—
8	—	—	—	—	—	—	—	—	—	—
7	—	—	—	—	—	—	—	—	—	—
6	—	—	—	—	—	—	—	—	—	—
5	—	—	—	—	—	—	—	—	—	—
4	—	10	—	—	—	—	—	—	—	—
3	—	—	—	10	—	—	—	—	15	—
2	—	—	—	—	—	—	—	—	—	—
1	—	—	—	—	—	—	—	10	—	—

a. Circle each entry of "11" in the table and describe the pattern of those entries.

List all of the ordered pairs, (x, y), included in the table for which $g(x, y) = 11$. These ordered pairs are all roots of the equation $g(x, y) = 11$ or $x + 2y = 11$.

b. Now circle each entry of "20" in the table and describe the pattern of those entries.

List all of the ordered pairs, (x, y), included in the table for which $g(x, y) = 20$. Each of these ordered pairs is a root of the equation $g(x, y) = 20$ or $x + 2y = 20$.

2. **a.** On a coordinate grid such as the one shown here, put a dot (•) on each grid point for which $g(x, y) = 11$ and a square (■) on each grid point for which $g(x, y) = 20$.

    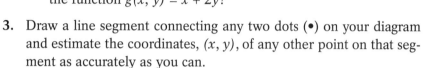

    b. Describe the pattern formed by the dots (•) and the pattern formed by the squares (■).

    c. How is the information in this graph similar to, and how is it different from, the information in the table of values for the function $g(x, y) = x + 2y$?

3. Draw a line segment connecting any two dots (•) on your diagram and estimate the coordinates, (x, y), of any other point on that segment as accurately as you can.

    Calculate $g(x, y)$ for the ordered pair you have just chosen and explain what that result may indicate about roots of the equation $x + 2y = 11$.

4. Draw a line segment connecting two of the squares (■) on your diagram and estimate the coordinates, (x, y), of any other point on that segment as accurately as you can.

    Calculate $g(x, y)$ for the ordered pair you have just chosen and explain what that result may indicate about roots for the equation $x + 2y = 20$.

5. Repeat the test of Exercises 3 and 4 by connecting other pairs of dots and other pairs of squares and calculating $g(x, y)$ for points on the connecting segments. Then answer the following questions based on your findings.
    a. How many roots do equations of the form $g(x, y) = c$ or $x + 2y = c$, where c is a constant, appear to have?
    b. What kind of pattern appears to occur in equations of the form $g(x, y) = c$ or $x + 2y = c$?

Experiment 3. Now consider the function $h(x, y) = x - y$ for x and y from -6 to 6. This function takes two inputs and finds their difference, with the second input subtracted from the first.

This function might be used to calculate the change in temperature at a weather station during a 24-hour period, where x is the temperature (in degrees Celsius) at noon one day and y is the temperature (in degrees Celsius) at noon the day before.

Suppose that in this case you are asked only the single question, "For what pairs of noontime temperatures is there a change of 3°?" This question can be answered by examining the equation $h(x, y) = 3$ or $x - y = 3$.

You may have seen in Experiments 1 and 2 that to find roots of a single equation with two variables, you do not need a picture of all input and output combinations for the function. You can find the pattern in the roots by using the familiar kind of coordinate graph.

1. Copy and complete each of the following ordered pairs so that it is a root of the equation $h(x, y) = 3$ or $x - y = 3$.

     a. $(-3, \underline{\hspace{1em}})$ b. $(0, \underline{\hspace{1em}})$ c. $(\underline{\hspace{1em}}, 2)$

2. Plot solid dots ($\bullet$) on a coordinate grid to represent the three roots for $x - y = 3$ that you found in Exercise 1.

3. Connect the three solid dots plotted in Exercise 2 with straight line segments. Then estimate the coordinates, (x, y), of two other points, one on each connecting segment.

     Calculate $h(x, y)$ for each estimated coordinate pair. Explain what you can infer about roots of the equation $x - y = 3$ from your results.

4. Complete each of the following ordered pairs to give a root of the equation $h(x, y) = -1$ or $x - y = -1$.

     a. $(-4, \underline{\hspace{1em}})$ b. $(0, \underline{\hspace{1em}})$ c. $(\underline{\hspace{1em}}, 5)$

5. Plot squares ($\blacksquare$) on the diagram you created in Exercise 2 to represent the three roots for $x - y = -1$ that you found in Exercise 4.

6. Connect the three squares plotted in Exercise 5 with straight line segments. Then estimate the coordinates of two points, one on each connecting segment.

     Calculate $h(x, y)$ for each estimated coordinate pair. Explain what you can infer about roots of the equation $x - y = -1$ from your results.

Making Connections

In Exploration I you studied functions with two input variables and rules of the form $f(x, y) = ax + by$. You used a new kind of table to display outputs of such a function for many different input pairs, (x, y).

The most common questions about functions with two input variables require finding roots of equations like $f(x, y) = c$ or $ax + by = c$. You have seen that such an equation can be pictured by graphing the points whose coordinates, (x, y), are roots for the equation.

The set of these points is also called the *graph of the equation*. For the equations in Exploration I you may have discovered that each graph is a straight line. This was no accident! A function with a rule in the form $f(x, y) = ax + by$ is called a *linear function* and an equation in the form $ax + by = c$ is called a *linear equation*.

Copy and complete the following table for the linear function $f(x, y) = 2x - y$, then graph the equation $f(x, y) = 5$ or $2x - y = 5$.

```
y   5  |  __     __    __    __     __
    4  | -2      __    __    __     __
    3  | -1      __    __    __     __
    2  |  0      __    __    __     __
    1  |  1    3     5    __     __
       |_____
          1     2     3     4     5    x
```

1. What information is given better by the table than by the graph? What questions are easier to answer when using the table?

2. What information is given better by the graph than by the table? What questions are easier to answer when using the graph?

Exploration II

In Exploration I, you looked very closely at equations and graphs for only three linear functions. You probably formed some important ideas about how these functions and equations behave, but it would certainly help to be able to look at many more examples.

Fortunately, there are programs for graphing linear equations with two variables. For many, all you have to do is enter the equation (or perhaps just the coefficients a, b, and c) and the program produces a graph.

This exploration is designed for use with a technological tool that graphs linear equations in two variables like $ax + by = c$.

1. **a.–h.** Graph at least 8 different linear equations for various combinations of the coefficients a, b, and c. For each example you select, record the following:

   i. the equation;

   ii. at least three roots of the equation; and

   iii. any other properties of the graph that you feel are interesting.

2. Look back over the examples you have studied. Can you find some connections between the equation and its graph which allow you to predict graphs of other equations. Make a list of your ideas.

Exploration III

From your exploration of linear functions and equations, you should now have good ideas about the solutions and graphs that can occur with two input variables. However, more than two input variables were involved in some of the examples of functions and systems that you studied in earlier sections.

The graphs of functions and equations with three or more variables are considerably more complex than those with two variables. Technological tools are helpful with the more complex situations, but you should be able to estimate what outcomes are reasonable when you use such tools. The next situation and the questions that follow should give you some useful ideas about how to make these estimates.

SITUATION 5.1

The concession stand at a basketball game sells three things—the price for soft drinks is $1.00, for hot dogs is $2.00, and for popcorn is $1.50. The revenue for the stand is a function R of the numbers of the three items sold, x, y, and z, where x represents the number of drinks sold, y represents the number of hot dogs sold, and z represents the number of popcorn orders.

1. Write a rule for $R(x, y, z)$ and use it to answer the questions that follow.

2. Calculate the revenue received from selling 200 soft drinks, 75 hot dogs, and 100 boxes of popcorn.

3. How does revenue change from the result in Exercise 2 in each of the following cases?
   a. Soft drink sales increase by one, and hot dog and popcorn sales remain fixed at 75 and 100 respectively.
   b. Hot dog sales increase by one, and soft drink and popcorn sales remain fixed at 200 and 100 respectively.
   c. Popcorn sales increase by one, and soft drink and hot dog sales remain fixed at 200 and 75 respectively.

4. The ordered triple (200, 75, 100) is one root of the linear equation $R(x, y, z) = 500$. Find five other roots and check them by calculating $R(x, y, z)$ for each. Your answers to the questions in Exercise 3 may be helpful in finding different roots.

5. Make up a different set of concession prices. List the prices and then write the new revenue function for those prices.

   Write an equation that involves this new revenue function, and then find and check four roots of that equation.

6. Describe a change in this situation that would lead to a revenue function with four input variables.

7. Write your ideas about the number of roots for a linear equation with three or more variables. How are these roots related?

Exercises

1. This diagram shows the graphs of two linear equations. For each equation, use its graph to find three ordered pairs that satisfy the equation, including one for which the value of at least one variable, x or y, is not a whole number. Then check each root by substituting the numbers x and y into the equation.

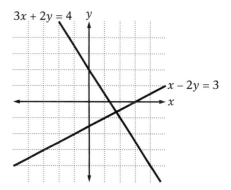

2. For each of the following linear equations, find two roots by a guess-and-test method. Then graph those two roots and draw the line they determine, that is, the line that goes through both points. Using that line, estimate and check two other roots for the equation, including at least one that involves fractional values of the variables.

   a. $2x - 3y = 6$

   b. $-2x + 4y = 6$

3. For each of the following linear functions, determine how the outputs change when one of the input variables increases or decreases, but the others are held constant. Use several different input combinations for your tests. Describe your observations and explain the relations between the function rule and any patterns that you find.

   a. $f(x, y) = 3x + 6y$ **Hint:** If x increases by 1, how does $f(x, y)$ change?

   b. $g(x, y, z) = 2x + 3y + 4z$.

4. SITUATION A bus company has two kinds of buses that it charters for tour groups. The smaller seats 30 passengers; the larger seats 45.

   a. Write the rule for a function $C(S, L)$ that gives the maximum number of passengers that can be carried if S small buses and L large buses are used.

   b. Write an equation that can be used to answer the following question: How many buses of each type can be used to carry a tour group of 360 passengers?

   c. Find three different roots for the equation in part b.

   d. Check the fact that $C(1.5, 7) = 360$, and explain what this result says about the question posed in part b.

5. SITUATION The Ace Car Rental Company buys new cars each year. For the year ahead they plan to purchase compact cars for $7500 each, mid-size cars for $15,000 each, and luxury cars for $30,000 each.

   a. Write a rule giving the company's expense for new cars as a function of the numbers of compact, mid-size, and luxury cars they buy.

   b. Write a calculation or equation required to answer the question: What combinations of new cars can be purchased for $750,000?

   c. Show that (40, 20, 5) is one combination that can be purchased for $750,000.

> **d.** How does the company's total order for cars change if it keeps the total expense at $750,000 but
> **i.** increases the number of compact cars by 4?
> **ii.** increases the number of mid-size cars by 2?
> **iii.** buys one more luxury car?

6. The equations below are not linear. For each of the equations find five roots and plot those roots on a coordinate graph as you have done for the linear examples in this section.
  a. $xy = 12$
  b. $x^2 - y = 4$

6 Systems of Equations

In the situations of Section 5, you saw how questions about functions with several input variables were answered by solving equations involving those variables. In models of many important situations, the key questions require analyzing several equations, each of which relates several variables. These conditions are called **systems of equations**. The explorations of this section show how such systems arise and what it means to solve them.

Exploration I

SITUATION 6.1

Everyone should eat many kinds of food during a given day. Each food provides some part of the nutrients a person needs to live a healthy, active life.

Consider only the possibilities for breakfast, a meal at which many people eat the same food each day. The following table gives some information about typical breakfast foods and their nutritional values.

Nutrition from one gram of each food

	Cereal	Fruit juice	Milk	Pastry
Calories	2.5	0.5	0.7	4.0
Protein (grams)	0.15	0.01	0.04	0.05
Carbohydrate (grams)	0.75	0.15	0.05	0.7
Fat (grams)	0.05	0.02	0.04	0.1

The table entries show the number of calories and the numbers of grams of protein, carbohydrate, and fat available in one gram of each listed food. For example, 1 gram of fruit juice can produce 0.5 calories and contains 0.01 grams of protein, 0.15 grams of carbohydrate, and 0.02 grams of fat. Information like this appears on most food packages, bottles, and cans, though not always in this form.

1. Suppose you drink j grams of fruit juice. Write the expressions that give the following nutritional values of that juice.
   a. Number of calories
   b. Number of grams of protein
   c. Number of grams of carbohydrate
   d. Number of grams of fat

2. Suppose you eat a breakfast consisting of c grams of cereal, j grams of fruit juice, m grams of milk, and p grams of pastry. Write function rules giving the following food values from this meal.
   a. Calories, $CAL(c, j, m, p) =$
   b. Protein, $PRO(c, j, m, p) =$
   c. Carbohydrate, $CARB(c, j, m, p) =$
   d. Fat, $FAT(c, j, m, p) =$

3. Calculate and explain the meaning of the following function values. You might find it handy to define each of the four functions in a technological tool with a computer-algebra program, to make these and the following computations easier.
   a. $CAL(20, 100, 200, 40)$
   b. $PRO(10, 150, 200, 60)$
   c. $CARB(20, 100, 250, 0)$
   d. $FAT(15, 120, 150, 50)$

The questions in Exercise 3 require quite a bit of arithmetic, but once the values of input variables are substituted in the function rules, the outputs come in a straightforward manner. You can find the calorie, protein, carbohydrate, and fat outputs of any given breakfast easily.

The reverse problem involves planning amounts of each food type to give specified outputs of calories, protein, *etc.* Finding inputs to produce specified outputs is not so easy. It requires solving a system of four equations with four variables.

4. Suppose you are to plan a breakfast that can produce 540 calories and contains 17.5 grams of protein, 92 grams of carbohydrate, and 20 grams of fat. Write a system of four equations that models this problem, and number your equations *(1)* to *(4)* for later reference.

Solving this system of equations means finding values of the input variables, *(c, j, m, p)*, that make all four equations true statements. The values must be the same for all four equations—we are planning *one* breakfast that can produce 540 calories *and* contains 17.5 grams of protein *and* contains 92 grams of carbohydrate *and* contains 20 grams of fat.

5. In other cases where you have met new kinds of equations to solve, one simple strategy often has been useful: guess-and-test. This situation might seem too complicated for that strategy; but before going on, try a few guesses.

   a. Test the guess (20, 100, 200, 50) for the four equations you found in Exercise 4. Record your results in a table with headings similar to the following.

Equation	Test	Decision

   Does the input (20, 100, 200, 50) satisfy the system of four equations?

   b. Make a guess of your own. Record the inputs for your guess and use a table such as the one in part a.
   Does the input that you guessed satisfy the system of four equations?

   c. Make another guess of your own. Record the inputs for your guess and use a table such as the one in part a.
   Is this guess a root of the system of equations?

6. Because it is hard to guess the solution for a system with many equations and many variables, mathematicians have developed various procedures for finding the solution systematically. Many of those procedures have been incorporated into computer-algebra programs.

Use a computer-algebra program to solve the system of equations you have written. Then report the result and test it as you have tested your guesses in Exercises 4 and 5.

Does your result check as a root of the system of equations?

7. Now write and solve a system of equations in the four variables, c, j, m, p, to answer the following question. Number your equations, then use a table to test your solution.

Plan a breakfast that can produce 300 calories and contains 10 grams of protein, 60 grams of carbohydrate, and 10 grams of fat.

Do the values that you tested actually satisfy the system of equations?

8. Now write your own question about the amounts of foods and beverages that yield specified nutritional output.
   a. Record your question.
   b. Write a system of four equations that can be used to answer your question. Label your equations *(1)* through *(4)*.
   c. Use a technological tool with a computer-algebra program to solve the system. Record the root your tool provides.
   d. Test the solution to insure that it is correct. Show your test.
   e. Summarize your answer in a complete sentence.

Exploration II

In Exploration I, you studied a situation in which important questions are modeled by systems of equations. Many other situations with multiple input and output variables require the same type of mathematical models and solution procedures, namely:

1. identifying the input and output variables;
2. writing function rules that relate the variables;
3. writing equations that match specific questions about the input and output variables; and
4. finding values of the input variables that produce each of the desired system outputs.

The breakfast nutrition situation happened to involve four input variables (the quantities of cereal, fruit juice, milk, and pastry to be consumed) and four output variables (the amount of energy, protein, carbohydrate, and fat given by those foods and beverages). There are other important systems with two, three, five, ten, or even several thousand input variables and equations.

In many of these systems it is possible to find a single combination of input variables that meet all of the desired output conditions. In other cases there may be many suitable combinations of inputs and, in some cases, none at all. The situations of this exploration give practice in setting up system-of-equation models, using technological tools to find solutions, and interpreting those solutions in the original problem situations.

SITUATION 6.2

The promoters of a summer concert hold their event at a pavilion. Tickets are available for the lawn or for covered seats close to the stage. The promoters set ticket prices at $15 for covered seating and $8 for general admission, with seating space only on the lawn. Research from past concerts indicate that each ticket sold for the covered seating yields an average of $2.50 in concession sales; each ticket sold for general admission yields an average of $4 in concession sales.

1. Using c and n to represent the number of tickets sold for covered seating and non-covered seating respectively at a concert, write expressions for each of the following output variables.
   a. Revenue from ticket sales, in dollars
   b. Revenue from concession sales, in dollars
   c. Total number of tickets sold

Use these relations to answer the questions that follow.

2. Suppose 500 covered seating and 300 general admission tickets are sold. Calculate the following.
   a. Revenue from ticket sales, in dollars
   b. Revenue from concession sales, in dollars

For Exercises 3 through 6, write a system of equations that model the given questions about this situation. Use technology to solve the system, then check and interpret the result.

3. What combination of ticket sales for covered and for non-covered seating gives ticket revenue of $13,700 and concession revenue of $3,350?

4. What combination of ticket sales for covered and for non-covered seating gives ticket revenue of $16,000 and total number of tickets sold 1,200?

5. How many tickets for covered seating and for non-covered seating are sold if a total of 1350 tickets are sold and if the total revenue is $21,700 from ticket sales and concessions combined?

6. What combination of ticket sales for covered and for non-covered seating gives $10,600 in ticket revenue, $2,300 in concession revenue, and a total of 900 tickets sold?

SITUATION 6.3

A baseball team is planning a special promotion at one of its games. All fans who arrive early will get a team athletic bag or a jacket, as long as the supply lasts. The promotion manager knows that each athletic bag costs $5.50 and each jacket costs $9.75.

1. Using b to represent the number of athletic bags and j to represent the number of jackets, write an expression for each of the following
   a. Total cost of bags and jackets
   b. Number of fans who get a bag or jacket

Use these expressions to answer the following questions.

2. The team promotion manager claims that 1000 team bags and 500 team jackets can be bought for a cost of only $10,000. Check this claim, showing any calculations that you make. Is the promotion manager right?

For Exercises 3 through 5, write a system of equations that model the given questions about this situation. Use technology to solve the system, then check and interpret the result.

3. What combination of team bags and jackets can provide for a total of 3500 fans at a cost of $23,500?

4. What combination of team bags and jackets can provide for a total of 800 fans at a cost of $3,550?

   (**Note:** Be careful when you try to interpret this mathematical solution to the real-world problem of planning for the baseball game promotion.)

5. Write your own question about the economics of this baseball game promotion that requires solving a system of equations. Write and solve the system that models your question, then check and interpret your results.

6. Write another question that would be of interest in this situation, but requires a different kind of system, such as inequalities.

SITUATION 6.4

A hiker plans to carry only high-energy, light-weight food—a mixture of peanuts and raisins with some cocoa powder to make drinks from water. The nutritional value of these foods is shown in the following table.

Nutrition available from each gram of food

	Cocoa	Peanuts	Raisins
Calories	3.5	6.0	3.0
Protein (grams)	0.03	0.25	0.04
Carbohydrate (grams)	0.7	0.2	0.8

1. Choose variable names for the input variables in this situation and record your choices.

2. Name each output variable and give expressions for calculating each of those output variables from the inputs you have identified.

Use these expressions to answer the following questions.

3. On the packages for cocoa, peanuts, and raisins, the recommended single serving is 28 grams of cocoa (before water is added), 28 grams of peanuts, and 85 grams of raisins. Calculate the calorie, protein, and carbohydrate yield from a snack that consists of a single serving of each food.

For Exercises 4 through 7, write a system of equations that model the given questions about this situation. Use technology to solve the system, then check and interpret the result.

4. What combination of cocoa, peanuts, and raisins can provide a snack with 850 calories, 20 grams of protein, and 120 grams of carbohydrate?

5. What combination of cocoa, peanuts, and raisins gives a snack with 1200 calories, 30 grams of protein, and 150 grams of carbohydrate?

6. What combinations of cocoa, peanuts, and raisins can yield a snack with 1000 calories and 125 grams of carbohydrate?

7. What combination of raisins and peanuts only *(no cocoa!)* gives a snack with 640 calories, 12 grams of protein, and 68 grams of carbohydrate?

The examples of systems of linear equations in this exploration give some hints about the possible results when you search for solutions of such systems. You should have found several examples where exactly one combination of values for the input variables led to the desired output values. In at least one case, you may have found that there is no combination of the inputs giving the desired combination of output values. In at least one other case, the results may have suggested that many possible input combinations give the desired outputs.

The study of this variety of results in systems makes up a very important branch of mathematics called *linear algebra*. In Section 7 you will get a first look at that theory in some simple cases.

Exercises

1. **SITUATION** The Land Ahoy Boat Company makes three types of inflatable boats—a single-person model, a two-person model, and a four-person model. Each model goes through three departments in the Land Ahoy plant—materials cutting, assembly, and packing. The time required in each department for one boat of each model is given in the following table.

Manufacturing time in hours

	Cutting	Assembly	Packing
Single-person model	0.5	0.6	0.2
Two-person model	1.0	0.9	0.3
Four-person model	1.5	1.2	0.5

a. While planning a production schedule, the plant manager decides how many boats of each model to make. Using s, t, and f for the numbers of single-person, two-person, and four-person boats to be made, write expressions for the total times required for cutting, assembly, and packing, respectively.

b. If $s = 30$, $t = 20$, and $f = 15$, what time is needed in each department?

c. Suppose that each of the three operations in the plant require skilled labor. The plant has workers who can provide a total of 380 hours a month in the cutting department, 330 hours a month in assembly, and 120 hours a month in packing. Write a system of equations that models the question,

   "How many boats of each type can be manufactured in a month?"

d. Make a guess at the solution of your system of equations in part c and test that guess.

e. Try to improve the guess you made in part d and test the new guess.

f. Suppose that some workers go on vacation, reducing the time available in cutting to 340 hours a month and in assembly to 290 hours a month. How does this change the system of equations in part c?

2. SITUATION The principal of a school wants to improve student attendance and achievement, so she plans to give prizes for school leaders in each category. The students with the best attendance will win $20 gift certificates; the students with the best grades will win $25 gift certificates. The principal has a budget of $1500 for this prize competition. She plans to give a total of 65 prizes.

   a. Using x for the number of attendance prizes and y for the number of top-grades prizes, write equations that model the conditions.

   b. Using the guess-and-test method, find a root of this system of equations.

    c. How do your equations in part a change if the total prize budget is $1000 and the number of prizes is 45?

    d. How do your equations in part a change if the prizes are reduced in value to $15 and $20 respectively?

In traditional algebra books, systems of linear equations often model the following kinds of puzzle problems. For each of Exercises 3 through 10, identify the variables whose values are to be found. Using single-letter names for those variables, write equations that model the problem conditions. Then make two guess-and-test tries to find a solution to the system of equations.

3. SITUATION Anita is 20 years older than Betty. If you double Betty's age and add it to Anita's age, the total is 41 years. How old are Anita and Betty?

4. SITUATION The total of Chuck's and Darren's ages is 37 years. Chuck's age is one year more than twice Darren's age. How old are Chuck and Darren?

5. SITUATION At a basketball game the box office sold a total of 350 tickets. The tickets cost $5 for adults and $3 for students, and box office revenue was $1550. How many adult tickets were sold? How many student tickets were sold?

6. SITUATION Eddie has a collection of nickels, dimes, and quarters worth a total of $3.75. There are 40 coins in the collection. How many are nickels? How many dimes? How many quarters?

7. SITUATION Felice has a job where she is paid $5 per hour regularly and $7.50 per hour for overtime (more than 8 hours on any single day). One week she worked a total of 26 hours and was paid $145. How many of her hours that week were regular time and how many were overtime?

8. SITUATION A certain two-digit number has the property that the sum of its digits is 11 and the difference of its digits is 5. What are the tens and ones digits in that number?

9. SITUATION An airplane flies 850 kilometers per hour with a tailwind and 550 kilometers per hour against that same wind. What is the speed of the wind and the speed of the plane without any wind?

10. S I T U A T I O N A boat travels upstream, against the current, for 2 hours, and covers 80 kilometers. The same boat travels downstream, with the current, and covers the same 80 kilometers in 1.5 hours. What is the speed of the current in the river? How fast would the boat travel in water with no current?

    Hint: Let x kph be the speed of the boat with no current and y kph be the speed of the current, and remember that distance traveled is the product of speed and time.

11. The previous ten exercises have focused on your ability to write systems of equations that model problem conditions. Use the systems you have written and a technological tool with a computer-algebra program to find the required solutions.

7   Properties of Systems of Equations

To use computing technology wisely when solving algebra problems, it helps to be able to look at the symbolic form of the problem and make some estimate of the solution. If you know what sort of solution, graph, or table of values to expect, you can choose the best tool to solve the problem. Also, you can often catch mistakes due to errors in typing instructions into your technological tool.

The examples of Section 6 have shown that a system of linear equations might be satisfied by one combination of input values, many such combinations, or, in some cases, by no combination at all. To see why this can happen it is helpful to look first at the simplest kind of system—two linear equations with two input variables.

The next exploration shows how the various solution possibilities are related to the properties of the equations and their graphs.

Exploration I

In the following series of examples you will be asked to solve various systems of linear equations—by a technological tool with a computer-algebra program, tables of values, or graphs—and to sketch the system graphs for each.

Example 1. Consider first the following system of two linear equations.

$$x - 2y = -4$$
$$2x + y = 2$$

These two equations are graphed on the diagram shown here. The two lines cross at the point whose coordinates, $(0, 2)$, are the only root for both equations and so for the system.

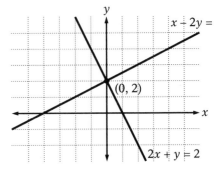

Now use your computer-algebra and graphing tools to study the examples that follow. Look for the answers to two main questions.

1. What are the patterns in graphs of a system whose solution contains one root? many roots? no root?

2. What are the patterns in coefficients and constant terms of equations in a system for which the solution contains one root? many roots? no root?

Example 2. Solve the following system of two linear equations with two input variables x and y, and record the solution.

$$x + y = 4$$
$$2x - y = 5$$

Sketch graphs of the two equations.

Example 3. Find the roots of the following system and record the solution.

$$x + 2y = 6$$
$$5x + 2y = -6$$

Sketch graphs of the two equations.

Example 4. Find the roots of the following system and record the solution.

$$2x - y = -1$$
$$-2x + 5y = -15$$

Sketch graphs of the two equations.

Example 5. Find the roots of the following system and record the solution.

$$x + 2y = 4$$
$$1.5x + 3y = 9$$

Sketch graphs of the two equations.

Example 6. Find the roots of the following system and record the solution.

$$2x - y = -1$$
$$6x - 3y = 5$$

Sketch graphs of the two equations.

Example 7. Find the roots of the following system and record the solution.

$$x + 3y = 4$$
$$2x + 6y = 8$$

Sketch graphs of the two equations.

Example 8. Find the roots of the following system and record the solution.

$$6x - 3y = 12$$
$$2x - y = 4$$

Sketch graphs of the two equations.

Among the seven systems of linear equations you solved and graphed in Examples 2 through 8, you should have found three for which the solution consists of a single ordered pair, (x, y). Two others had no solutions; that is, there are no satisfactory ordered pairs. The remaining two had solutions with many satisfactory ordered pairs.

1. **System Solutions and Graphs** Recall the first question of this exploration: What are the patterns in graphs of a system whose solution contains one, many, or no roots? Look back over your responses for Examples 2–8 and then complete the following sentences describing the connection between system solutions and the graphs of the equations.

a. If a system of two equations in two variables has a solution consisting of exactly one ordered pair, (x, y), then the graphs of the separate equations....

b. If the solution of a system of two equations in two variables consists of many ordered pairs, (x, y), then the graphs of the separate equations....

c. If a system of two equations in two variables has no root, then the graphs of the separate equations....

2. **System Solutions and Equations** Recall the second question: What are the patterns in coefficients and constant terms of equations in a system for which the solution contains one root, many roots, or no roots? Sort the systems you have studied into three lists. Use the list heading shown below, and write the equations for each system under the appropriate heading.

   One root   Many roots   No root

3. Now use these examples to help you answer the following questions about systems with equations in the following form.

$$ax + by = c$$
$$dx + ey = f$$

Try to discover how the values of the coefficients a, b, and d, e, together with constants c and f, determine the type of solution the system has. In your answers to the following questions, describe any patterns in those values that you observe.

a. What pattern seems to fit each system with many roots?

b. What pattern seems to fit each system that has no root?

c. What pattern seems to fit each system with exactly one root?

Now you might have some insight about how the coefficients and constant terms in the equations of a system can be used to predict the type of solution and its graph. To test and perhaps clarify your ideas, look at the examples in the exploration that follows.

Exploration II

In this exploration, you are given ten systems, each consisting of two linear equations with two variables. For each system:

    a. Make a prediction about whether the system solution consists of many, one, or no ordered pairs, (x, y), and make a prediction about the appearance of the system graph.

    b. Use technological tools to graph and solve the system; describe the pattern of the graph, and report the solution.

1. $x + y = 0$ and $x - y = 2$ **2.** $x + y = 1$ and $3x + 3y = 3$

3. $x - y = 2$ and $4x - 4y = 10$ **4.** $3x + 2y = 5$ and $x - 3y = 6$

5. $x - 3y = 6$ and $2x - 6y = 12$ **6.** $3x + 2y = 5$ and $4x + 3y = 6$

7. $1.5x + 3y = 6$ and $x - 2y = 0$ **8.** $x - 2y = 0$ and $3x - 6y = 2$

9. $x - 2y = 0$ and $1.5x - 3y = 0$

10. $x + 0y = 4$ and $0x + y = 2$ **(Note:** This system is usually written in the simpler form $x = 4$, $y = 2$.)

Making Connections

Look back over the ten systems you have just studied to check your ideas about how the equations of a system determine the graph and solution type. Record your revised conclusions by copying and completing the following statements.

For a system of two linear equations in the form

$$ax + by = c$$
$$dx + ey = f$$

where a, b, c, d, e, and f are numbers:

1. the graph consists   **2.** the solution consists

    a. of two intersecting lines if....   **a.** of a single root if....

    b. of two parallel lines if....   **b.** of many roots if....

    c. of a single line if....   **c.** of no root if....

Exercises

1. Consider the linear equation $5x + 8y = 3$. Write equations that combine with this equation to give systems with graphs consisting
   a. of two intersecting lines;
   b. of two parallel lines;
   c. of a single line.

2. Consider the linear equation $7x - 3y = 8$. Write equations that combine with this equation to give systems whose solutions consist
   a. of a single root;
   b. of many roots;
   c. of no root.

3. Consider the linear equation $0.5x + 8.2y = 3.14$. Write equations that combine with this equation to give systems with graphs consisting
   a. of a pair of intersecting lines;
   b. of a pair of parallel lines;
   c. of a single line.

4. Consider the linear equation $3.2x + 1.8y = 2.7$. Write equations that combine with this equation to give systems whose solutions consist
   a. of a single root;
   b. of many roots;
   c. of no root.

5. The system with two linear equations $3x + 4y = 5$ and $6x + 8y = 10$ has a solution with many roots. What connection between the equations makes every root of one equation also a root of the other?

6. The system with two linear equations $3x + 4y = 5$ and $6x + 8y = 11$ has a solution with no roots. What connection between the equations makes it impossible for any root of one equation to be also a root of the other?

7. Consider the equation with three variables $x + 2y + 3z = 4$.
   a. Write another equation with three variables x, y, and z that you are certain has many roots in common with the given equation.
   b. Write another equation with three variables x, y, and z that you are certain has no roots in common with the given equation.

8 Linear Inequalities

In Sections 6 and 7, you were asked to focus on solving systems of linear equations. That meant finding inputs that produce a particular output target. In many situations the goal is to produce outputs less than or greater than some output targets. These problems are modeled well by algebraic inequalities.

 ## Exploration I

The exercises of this exploration illustrate the kinds of questions that require using inequalities to answer and the kinds of solutions that such inequalities have.

SITUATION 8.1

You may recall from Section 6 that if you have a breakfast with c grams of cereal, j grams of fruit juice, m grams of milk, and p grams of pastry, the number of calories in the meal can be calculated by the function rule

$$\mathrm{CAL}(c, j, m, p) = 2.5c + 0.5j + 0.7m + 4.0p.$$

Use this function rule to answer the questions that follow.

1. Many people count calories in order to maintain a limit set by a diet. Suppose that a person aims to keep the calorie count of breakfast below 500. This goal is met by inputs that satisfy the inequality

    $$CAL(c, j, m, p) < 500 \text{ or } 2.5c + 0.5j + 0.7m + 4.0p < 500$$

    Test each of the following inputs to see which (if any) satisfy the inequality, and record your results in a similar table. You might find it helpful to enter the calorie function in a technological tool with a program that uses computer algebra or that will generate tables.

Guess input (c, j, m, p)	Test output $2.5c + 0.5j + 0.7m + 4.0p$	Decision Is output < 500?
(50, 200, 100, 25)	$125 + 100 + 70 + 100$	yes
(100, 100, 200, 20)		
(50, 0, 250, 100)		
(0, 150, 100, 75)		
(0, 250, 0, 200)		
(0, 0, 0, 150)		
(0, 200, 0, 0)		

2. Suppose the diet is changed to a limit of not more than 450 calories for breakfast. Rewrite the inequality from Exericse 1, taking this condition into account, and then use your inequality to answer the questions that follow.

   a. Does a breakfast with 50 grams of cereal, 200 grams of juice, 150 grams of milk, and 50 grams of pastry meet this condition? Show your calculations to check.

   b. The input (50, 100, 100, 50) does yield fewer than 450 calories. Suppose that these quantities of cereal, juice, and milk are held constant, but the amount of pastry is varied. What is the largest number of grams of pastry that can be eaten? Show your calculations in the search.

3. Look back at the rule for the calorie function and at your search for solutions to inequalities involving that function. In trying to limit the calorie content of a breakfast, which food is most important? Explain your answer.

SITUATION 8.2

A theater sells tickets in three categories. Adults under the age of 65 pay $6, children under 12 years of age pay $3, and senior citizens pay $4. The number of tickets of each type sold varies from day to day. If the numbers of each type are labeled with a, c, and s respectively, the ticket revenue is a function R of the three variables.

1. Write a rule for this function, and use your rule to answer the questions that follow.

2. Suppose a theater manager wants to have at least $1500 in ticket revenue every day. Write an inequality that matches this condition.

    a. Does the input combination (200, 150, 50) satisfy your inequality? Show your calculations.

    b. Does the input (200, 50, 50) satisfy your inequality? Show your calculations.

    c. Find three other combinations of ticket sales—two that meet the revenue goal and one that does not. Record your results in a table with the following headings.

    Guess input *(a, c, s)*   Test output   Decision

    d. What combination of adult, child, and senior citizen tickets meets the revenue target with smallest total number of tickets sold? Explain why you believe your answer is correct.

3. Suppose the president of a company wants to treat some employees and their families to a special theater party. What combinations of adult, child, and senior citizen tickets can keep the ticket cost not more than $500?

    a. Write an inequality that can be used to answer this question.

    b. Find three combinations of input variables that are roots of this inequality. Record your answers and the test in a table with the following headings.

    Guess input *(a, c, s)* Test output

    c. What combination of tickets meets the cost restriction with the largest number of people admitted? Explain your answer.

    d. i. Write a question of your own about the theater's ticket revenue function that can be modeled by an inequality.
      ii. Write an inequality that models your question
      iii. Find at least two inputs that satisfy your inequality, and explain what each indicates about the situation.

Exploration II

Your answers to the exercises in Exploration I probably suggest that the solution to any inequality includes many combinations of input variables. The important algebra question is whether there is a pattern that helps locate all these combinations. To begin answering this question, it helps to look again at the results in some simple cases with two input variables.

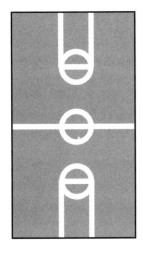

SITUATION 8.3

The basketball coaches at a university run summer camps for athletes. They charge $200 for a camper who lives at home and commutes each day; they charge $300 for a camper who lives on campus in a dormitory. Revenue for the camp is a function R of the number of commuter campers x and the number of dormitory campers y.

1. Write a rule for the revenue function.

Use your function rule to answer the questions that follow.

2. Suppose the coaches need revenue of at least $60,000 to cover their fixed costs. For each of the following, write an expression that matches the given conditions. Then find four pairs, (x, y), that meet the condition.
   a. The total revenue is exactly $60,000.
   b. The total of all revenue is at least $60,000.

3. One very helpful clue to the patterns in solutions of inequalities is revealed by the graphs of those solutions. A coordinate diagram displaying part of the graph of the equation $200x + 300y = 60\,000$ is shown here.
   a. Copy the diagram, then put a solid dot (•) at 15 or more grid points whose coordinates satisfy the inequality $200x + 300y > 60\,000$.

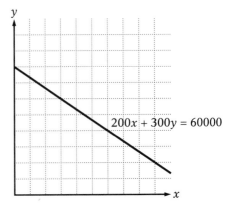

The horizontal (input) scale unit and the vertical (output) scale unit are both 25.

b. On your diagram, put a square (■) at 12 or more grid points whose coordinates satisfy the inequality $200x + 300y < 60000$.

c. What does this example suggest about the likely patterns in graphs of solutions for linear inequalities?

d. You have marked only grid points. Choose at least three points that are not grid points but have positive coordinates, and test them to see whether they behave like the nearby grid points.

Exploration III

As was the case in exploring patterns of graphs for systems of linear equations, it is helpful to find patterns in solutions of linear inequalities by some graphing experiments. In each of the following examples you will study a linear equation of the form $ax + by = c$ and the two related inequalities $ax + by < c$ and $ax + by > c$. In each case perform the following steps.

a. Sketch the graph of the equation $ax + by = c$.

b. Shade the part of the diagram containing points that have coordinates that satisfy the inequality $ax + by < c$.

c. In a different color, shade the part of the diagram containing points with coordinates that satisfy the inequality $ax + by > c$.

d. Find coordinates for three points, one on the graph of the equation and one each on the graphs of the two inequalities. Test those coordinates to see that they meet roots of the appropriate conditions.

Technology that helps with graphing inequalities may be useful. If such technology is not available, you may do a careful guess-and-test search to locate the solution regions for the inequalities.

Example 1.   Consider the equation $x + y = 2$ and the related inequalities $x + y < 2$ and $x + y > 2$. Graph the solutions, then test three points as described above. A table such as the following may be helpful for recording your results

Equation/inequality	Selected point	Test
$x + y = 2$		
$x + y < 2$		
$x + y > 2$		

Example 2.   Graph the equation $x - 2y = 4$ and related inequalities $x - 2y < 4$ and $x - 2y > 4$. Test three points as described above.

Example 3.   Graph the equation $3x + 2y = 5$ and related inequalities $3x + 2y < 5$ and $3x + 2y > 5$. Test three points as described above.

Example 4.   Graph the equation $-2x + y = 1$ and related inequalities $-2x + y < 1$ and $-2x + y > 1$. Test three points as described above.

Example 5.   Next, consider the *system* of inequalities $x + y > -1$ and $-2x + y < 0$.

1. Graph the two related equations $x + y = -1$ and $-2x + y = 0$.

2. Shade any points of the diagram where coordinates satisfy *both* of the inequalities.

3. Locate two points whose coordinates satisfy both inequalities in the system and test those points in both inequalities.

Making Connections

Review the answers you have given to the questions in Explorations I and II before completing the following statements about linear inequalities.

1. If $f(x, y) = ax + by$ is a function with two input variables, then solving the inequality $f(x, y) < c$ or $ax + by < c$ means finding....

2. A linear inequality with two input variables has (many, one, or no) solutions.

3. The graph of a linear inequality with two variables is....

4. A linear inequality with three or more variables usually has (many, one, or no) solutions, each consisting of....

Exercises

1. SITUATION On the Metroliner train from Washington to New York, coach tickets cost $75 and first-class tickets cost $90.
    a. Write a rule giving Metroliner ticket revenue for a trip as a function of the numbers of coach tickets and first-class tickets sold. For each of the following, write an equation or an inequality that can be used to answer the question. Then use a guess-and-test strategy to find three different pairs that satisfy your equation or inequality.
    b. What combinations of coach and first-class tickets give exactly $19,500 revenue?
    c. What combinations of coach and first-class tickets give at least $10,000 revenue?
    d. What combinations of coach and first-class tickets give less than $5,000 in revenue?

2. SITUATION A photographer who takes pictures of students at the school homecoming dance offers the pictures for sale. The photographer charges $2.50 for each wallet-size picture and $8.75 for each portrait-size picture. Using w for the number of wallet-size pictures and p for the number of portrait-size pictures, write equations or inequalities that can be used to answer each of the following questions.
    a. What combinations of picture orders bring the photographer income of at least $250?
    b. What combinations of pictures can a student order and keep the bill under $35?

3. SITUATION In basketball, a player can score a free throw (one point), a field goal (two points), or a three-point field goal.
    a. Using x for the number of free throws, y for the number of (regular) field goals, and z for the number of three-point field goals, write a rule giving the total number of points scored in a season by a player.
    For each of the following, write an equation or inequality that can be used to answer the question, then find three solutions.
    b. What combinations of free throws, field goals, and three-point field goals give the player at least 250 points for a season?
    c. What combinations give fewer than 100 points for a season?

4. **SITUATION** A used music store sells all tapes at $4.50, all compact disks at $9, and all LP records at $7.25.

    a. Using t for the number of tapes, c for the number of compact disks, and r for the number of records sold in a day, write a rule giving the store's daily revenue as a function of t, c, and r.

    For each of the following, write an equation or inequality that can be used to answer the question. Then use a guess-and-test strategy to find three different triples that satisfy your equation or inequality.

    b. What combinations of tape, disk, and record sales give at least $500 daily revenue?

    c. What combinations of tape, disk, and record sales give daily revenue less than $250?

    d. What combinations of tape, disk, and record sales yield daily revenue between $300 and $400?

5. **SITUATION** A jar of coins contains n nickels, d dimes, and q quarters. For each of the following, write an equation or an inequality that can be used to answer the question. Then use a guess-and-test strategy to find three different triples that satisfy your equation or inequality.

    a. For what combinations of coins is the total value of the money in the jar at least $20?

    b. What combinations of coins make the total value of money in the jar at most $5?

    c. For what combinations of coins is the money in the jar worth between $5 and $10?

In each of Exercises 6 through 8, graph the solution of the given inequality. On each graph, shade the solution region, place a heavy dot at one point which has coordinates that satisfy the inequality, and then show by calculation that the coordinates are a solution.

6. $x + y < 1$

7. $2x + 3y > 6$

8. $x - y < 4$

9 Summary

The situations in this chapter have introduced several new algebraic structures: systems involving several related functions, functions with several input variables, and equations or inequalities with several variables. You have seen ways to use tables of function values, graphs of several functions, guess-and-test searches, and technology with computer-algebra programs to answer important questions in those structures.

For a system model with two output variables that depend on a single input variable, the important questions fall into a few categories that require you:

1. to find all inputs x for which $f(x) = g(x)$, $f(x) < g(x)$, and $f(x) > g(x)$.

2. to find inputs x for which the difference $f(x) - g(x)$ is largest or smallest.

You saw how technology-generated tables, graphs, and symbol manipulations can be used to answer these types of questions. You also studied a variety of examples to learn just what sorts of answers should be expected for various kinds of situations.

For systems modeled by functions with several input variables, the important questions also take familiar forms. For example, if inputs and outputs are related by a function f with inputs (x, y, z), the key questions might require you:

1. to find outputs for given inputs—that is, calculate $f(x, y, z)$.

2. to search for inputs that give specified output—solve equations of the form $f(x, y, z) = d$ and solve inequalities of the form $f(x, y, z) < d$ or $f(x, y, z) > d$.

3. to determine how outputs change as inputs change.

In cases with two input variables and a single function of those variables, it is possible to study system questions by a new kind of coordinate graph. In this kind of graph, you mark all points whose coordinates satisfy the given equation or inequality conditions. The graphs of such equations and inequalities usually include many points lying in predictable patterns.

Review Exercises

1. **SITUATION** A fast food company sells hot dogs, coffee, soft drinks, and ice cream from vending carts located at many street corners in a large city.
   a. List at least five variables that are likely to affect the sales from one of these carts on a given day. For each variable, explain what you believe the cause-and-effect relation is.
   b. Changes in the weather probably affect sales of the different foods available from a vending cart. Consider the sales of a single cart at a single location, with varying daily temperature.

   Copy and complete the following table showing how you believe the sales for a cart might depend on the average temperature during working hours of a day. Use your own judgment about numbers. There are no "right answers", but you should be able to explain your reasoning.

   Vending cart sales

Temperature (° Celsius)	Hot dogs	Coffee	Soda	Ice cream
0	40	95	15	8
5				
10				
15				
20				
25				
30				
35				

   c. On a single coordinate diagram sketch graphs of four product sales functions: $h(t)$ for hot dogs, $C(t)$ for coffee, $S(t)$ for soft drinks, and $i(t)$ for ice cream, showing how each varies as temperature changes.
   d. Based on the data and graphs you have given in parts b and c, find specific examples of the patterns $f(x) = g(x)$ and $f(x) < g(x)$ and $f(x) > g(x)$. Explain what each example tells about sales at the vending cart.

2. Consider the functions $f(x) = 0.5x^2 - 3$ and $g(x) = 0.75x + 2$. Values for these functions are given in the following table and graph.

Use the table and graph to estimate answers for each of the following.

x	$f(x) = 0.5x^2 - 3$	$g(x) = 0.75x + 2$
−10	47	−5.5
−8	29	−4.0
−6	15	−2.5
−4	5	−1.0
−2	−1	0.5
0	−3	2.0
2	−1	3.5
4	5	5.0
6	15	6.5
8	29	8.0
10	47	9.5

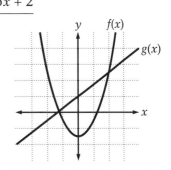

The horizontal (input) scale unit and the vertical (output) scale unit are both 2.

a. For what inputs x are $f(x) = g(x)$?
b. For what inputs x are $f(x) < g(x)$?
c. For what inputs x are $f(x) > g(x)$?
d. Is there a maximum difference $f(x) - g(x)$? If so, what is it? If not, explain why.
e. Is there a minimum difference $f(x) - g(x)$? If so, what is it? If not, explain why.

3. SITUATION Suppose the street vendor described in Exercise 1 charges $1.25 for hot dogs, $0.50 for a cup of coffee, $0.75 for a soft drink, and $1.00 for ice cream. The daily revenue of a cart *(in dollars)* is a function of the number of each product sold.
a. Write a rule for the revenue $R(h, s, c, i)$.
b. Calculate $R(20, 30, 40, 15)$, and explain what it indicates about the business.
c. Find three roots of the equation $R(h, s, c, i) = 85$, and explain what all such roots tell about the business.
d. Find three input combinations that satisfy the following inequalities, and explain what all such combinations indicate about the business.
   i. $R(h, s, c, i) < 50$
   ii. $R(h, s, c, i) > 100$

4. **SITUATION** When weather is reported in the winter time, we are often given something called the *wind chill* temperature. One formula for wind chill combines the actual temperature t (in degrees Fahrenheit) and the wind speed w (in miles per hour) to give wind chill: $WC(t,w) = t - 1.5w$.

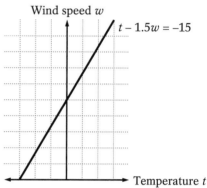

Wind speed w

$t - 1.5w = -15$

Temperature t

The horizontal (input) scale unit is 5 and the vertical (output) scale unit is 2.

a. A portion of the graph of the equation $t - 1.5w = -15$ is shown here. Find coordinates of three points on that graph, and explain what each tells about wind chill.

b. Copy the diagram on your paper (you may want to use graph paper). On your copy, graph the equation $t - 1.5w = 0$, and explain what the points of that graph indicate about wind chill. Give coordinates of three points on the graph.

c. On the same diagram, shade the region in which points have coordinates satisfying the inequality $WC(t, w) < 0$. Explain what those points indicate about wind chill.

5. Graph each of the following equations and inequalities.
   a. $3x + 2y = 6$
   b. $3x + 2y < 6$
   c. $-3x + 2y = 6$
   d. $-3x + 2y > 6$
   e. $x - y = 3$
   f. $x - y < 3$

6. For each of the following, write a system with two input variables and two equations that have the stated quantity of roots. Sketch the graph of your systems.
   a. Exactly one root
   b. No roots
   c. Many roots

7. **SITUATION** A taxicab driver can work a total of 22 shifts in each month. The driver can choose how many of those shifts are to be day shifts and how many are to be night shifts. Night shifts pay an average of $120 each and day shifts pay an average of $90 each. The driver prefers to work days, if possible, but needs a monthly income of $2220 to pay for rent, food, and other bills.

   a. Using d for the number of day shifts and n for the number of night shifts, write a rule giving the driver's monthly income as a function of d and n.

   b. Write equations or inequalities that express the conditions describing the maximum number of shifts worked and income required.

   c. Use technology with a computer-algebra program or some other method to solve the system.

   d. How would the objective, the algebraic model, and the solution change if you did not know that the driver preferred working day shifts?

Operating a pet hotel involves much careful planning. How many walls are needed to construct holding wards?

8 Symbolic Reasoning: Equivalent Expressions

Interesting relations among quantitative variables arise in many different situations and in many different forms. In some cases the relation is given by a table of data pairs, (input, output); in other cases the relation is pictured by a coordinate graph produced by plotting data pairs. In still other cases the relation is described in words that express scientific or economic principles.

Regardless of how you discover a function relating variables, it is usually helpful to write the function in a symbolic form like

$$f(x) = x^2 + 3x, \ g(t) = \frac{5t + 2}{t^2}, \ h(s) = -45s$$

and so on. With a function rule expressed in symbolic form, you can use computing technology to build a table, draw a graph, or solve an equation or inequality. You can also use properties of numbers and operations to answer questions about the function through reasoning based only on the form of the rule.

This chapter introduces the basic procedures for reasoning about algebraic expressions for function rules. These procedures offer you another approach to answering important questions about the functions that you encounter.

1 Equivalent Expressions

SITUATION 1.1

A national pet-hotel chain is planning to build units for use in a series of franchises. Each of the units for small pets is a row of two-meter by two-meter square wards. The wards are connected as shown in the partial floor plan below.

Walls for these units come only in two-meter panels, and the number of two-meter panels needed depends on the number of wards to be included in the unit. Because the management plans to build many units of different sizes, the manager wants to have a rule relating the number of wards and the number of panels.

Two of the staff members, Suzanne and Roberto, have the task of finding a rule that gives the number of panels needed for a unit. Being independent thinkers, the two staff members come up with two different formulas.

Roberto suggests that the rule

$$p_1(w) = 2w + 2 + (w - 1)$$

represents the relation p, between the number of wards w and the number of panels. (We use the subscript here because we will be looking at many different rules). He explains that he started by counting the panels shown horizontally in the diagram.

Each ward has two panels shown horizontally in the diagram above. For four wards there are two times four, or eight, two-meter panels. In general, for w wards there should be $2w$ of these panels. Next, Roberto counted the end walls, another two panels.

His count is now $2w + 2$. Finally, Roberto counted the inside walls. In the example with four wards, there are three inside walls.

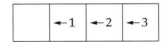

In general, the number of inside walls is one fewer than the number of wards. So, for w wards there are $w - 1$ inside walls. Roberto's rule states that the total number of panels required for w wards is

$$p_1(w) = 2w + 2 + (w - 1).$$

Suzanne, taking a "construction" point of view, finds a different rule. After thinking about how to construct the wards with prefabricated chunks of wall, she concludes that the number of panels depends on the number of wards w. She then writes the rule

$$p_2(w) = 1 + 3w$$

where $p_2(w)$ is the number of panels needed for w wards.

Suzanne reasons that she can build the unit by starting with the leftmost wall and adding modular pieces like those shown below. Each module has 3 panels. For four wards, she needs to attach four of these pieces, so the total number of panels to be attached is $3 \times 4 = 12$.

In order to build w wards, she needs to attach w of these pieces. Since each piece consists of 3 panels, the unit can be built with $3w$ panels in modular pieces, together with the leftmost wall. So, Suzanne's rule states that the total number of panels required for w wards is

$$p_2(w) = 1 + 3w.$$

Suzanne also notices that there is a constant increase in the number of walls needed, with three new walls needed for each additional ward. This constant rate of increase adds to her confidence, since her rule is clearly a rule for a linear function.

Suzanne and Roberto are concerned because their respective rules look different. Wanting to present the "right" rule to their boss, they spend the morning arguing back and forth, each showing various specific cases where his or her rule gives the correct answer. Unable to agree, they check their rules by constructing tables of values, as shown below.

Roberto uses his rule $p_1(w)$ to produce the following table:

w	$2w+2 + (w{-}1)$	$p_1(w)$
2	$(4{+}2) + (2{-}1) = 6 + 1$	7
4	$(8{+}2) + (4{-}1) = 10 + 3$	13
6	$(12{+}2) + (6{-}1) = 14 + 5$	19
8	$(16{+}2) + (7) = 18 + 7$	25
10	$(20{+}2) + (9) = 22 + 9$	31
12	$(24{+}2) + (11) = 26 + 11$	37
14	$(28{+}2) + (13) = 30 + 13$	43

Suzanne chooses the same inputs. Her rule, although requiring somewhat less computation, still yields the same values as Roberto's!

w	$1 + 3w$	$p_2(w)$
2	$1 + 3(2) = 1 + 6$	7
4	$1 + 3(4) = 1 + 12$	13
6	$1 + 3(6) = 1 + 18$	19
8	$1 + 3(8) = 1 + 24$	25
10	$1 + 3(10) = 1 + 30$	31
12	$1 + 3(12) = 1 + 36$	37
14	$1 + 3(14) = 1 + 42$	43

By this time, the two workers are beginning to believe that their rules are not as different as they appear to be. However, the tables give only a limited number of data pairs, (input, output). Might results be different for other pairs? To check a larger range of pairs, the workers construct graphs for their respective rules.

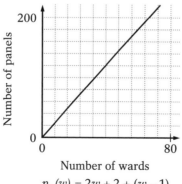

Number of wards

$p_1(w) = 2w + 2 + (w - 1)$

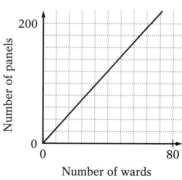

Number of wards

$p_2(w) = 1 + 3w$

Even though only Suzanne's rule has the familiar form of a linear function, both rules graph as straight lines. Moreover, they seem to graph as the same straight line!

The workers now suspect that all of the data pairs, (input, output), for their two rules are exactly the same. In mathematics, function rules that yield exactly the same set of data pairs are called **equivalent**. Merely by studying the above tables and graphs, which display only limited and incomplete sets of data pairs, Suzanne and Roberto cannot determine with certainty whether or not their respective function rules are equivalent.

As you have seen, a rule for a function is frequently written in a form like $p_1(w) = 2w + 2 + (w-1)$. The symbols $2w + 2 + (w-1)$ are used to calculate the function outputs for given inputs. These symbols form an **algebraic expression**, often called simply an *expression*. The function rule $p_2(w) = 1 + 3w$ gives another example of an expression, namely $1 + 3w$. An expression may be **evaluated** for specific values of the variables appearing in it. We say that two expressions are **equivalent to each other** if, for each assignment of numerical values to the variables in the expressions, the values of the two expressions are equal.

Exploration I

SITUATION 1.2

Soon after Suzanne and Roberto start developing their rules, a coworker stops by to offer some assistance. The coworker suggests what seems to be an easy solution, namely: "If you want to know the number of panels needed, just multiply the number of wards by 4. For example, if you have one ward, then the number of panels needed is one times 4, or 4. Your expression ought to be $4w$."

1. The expression $4w$ offered by the third worker is another way of talking about a function rule $p_3(w) = 4w$. Do you think this function rule is equivalent to either of the two rules found by Roberto and Suzanne? Create and record a table or tables that give evidence to support your decision.

    a. According to the tables, do you believe that the rule $p_3(w) = 4w$ is equivalent to Roberto's rule, $p_1(w) = 2w + 2 + (w-1)$?

b. Do the tables support the conclusion that the rule $p_4(w) = 4w$ is equivalent to Suzanne's rule, $p_2(w) = 1 + 3w$?

c. Justify your answers to parts a and b.

Create a graph or graphs that give evidence to support your decision whether the expression $4w$ is equivalent to either Roberto's expression or Suzanne's expression. Sketch your graphs on your paper. Use as many diagrams as you think necessary.

d. Do the graphs support the conclusion that the expression $p_4(w)$ is equivalent to Roberto's expression $2w + 2 + (w - 1)$?

e. According to the graphs, do you believe that the expression $4w$ is equivalent to Suzanne's expression $1 + 3w$?

f. Justify your answers to parts d and e.

SITUATION 1.3

A second coworker offers still another way to look at the problem: count four panels for each ward and then subtract the number of overlapping panels—two for each ward except the first and last. So this rule is $p_4(w) = 4w - (2(w - 2))$.

2. Is the second coworker's function rule equivalent to either of the two rules found by Roberto and Suzanne? Create and record a table or tables that give evidence to support your decision.

a. Do the tables support the belief that the rule $p_4(w) = 4w - (2(w - 2))$ is equivalent to Roberto's rule, $p_1(w) = 2w + 2 + (w - 1)$?

b. Do the tables support a claim that the rule $p_4(w) = 4w - (2(w - 2))$ is equivalent to Suzanne's rule, $p_2(w) = 1 + 3w$?

c. Justify your answers to parts a and b.

Create a graph or graphs giving evidence to support your decision whether the second coworker's rule is equivalent to either Roberto's or Suzanne's. Sketch your graphs on paper. Use as many diagrams as necessary.

d. Based on your observations of the graphs, do you believe that the rule $p_4(w) = 4w - (2(w - 2))$ is equivalent to either Roberto's rule or Suzanne's rule?

e. Justify your answer.

3. Which of the rules, p_1, p_2, p_3, or p_4, correctly determines the number of panels needed for a unit of wards at a pet hotel? Justify your answer.

4. For each of the rules, p_1, p_2, p_3, or p_4, that is not appropriate, explain what is wrong with the reasoning that was discussed in presenting the rule. For each inappropriate rule tell how the line of reasoning used to develop it could be adjusted to produce an appropriate rule.

Exploration II

SITUATION 1.4

The management of the pet-hotel chain decides to consider a different arrangement of wards. Instead of one row of wards, each unit will have two rows of two-meter by two-meter wards. This arrangement is illustrated by the partial floor plan below.

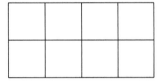

1. Analyze this situation using the following discussion.

    The number of two-meter panels of wall needed for a unit again depends on the number of wards included in the unit. The management decides that there always should be an even number of wards in this configuration.

    The members of the staff who are searching for a rule decide to gather some data and try to look for a pattern. Here are some of the sketches they make.

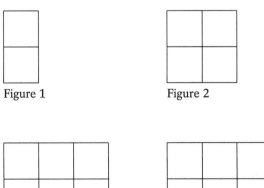

Figure 1   Figure 2

Figure 3   Figure 4

a. Roberto suggests looking for a pattern in the number of panels shown vertically, then looking for a pattern in the number of panels shown horizontally, and finally combining these results. Copy and complete the table below with the appropriate data.

Number of wards	Number of vertical panels	Number of horizontal panels
2	4	3
4	6	
6		
8		

Find a pattern in the data, and extend your table to larger cases. Find a rule describing the number of vertical panels as a function v of the number of wards w in a unit.

Find a rule describing the number of horizontal panels as a function h of the number of wards in a unit.

Combining his results, Roberto finds a rule for the total number of panels. Using the rules you have found for $v(w)$ and $h(w)$, write Roberto's rule giving the total number of panels as a function t_1 of the number of wards w in a unit.

b. Suzanne suggests trying to find a rule by studying data from smaller cases. Using Figures 1 through 4 and generalizing for larger cases, copy and complete the following table.

Number of wards	2	4	6	8	10	12	14	16
Total number of panels	7							

Using the pattern of data from the table, write a rule describing the total number of panels as a function t_2 of the number of wards.

c. Compare the function rules $t_1(w)$ and $t_2(w)$ for the total number of panels. Gather evidence that helps you answer the question, *Are these two function rules equivalent?* Answer this question and write a reasonable argument that justifies your answer.

2. Another staff person who analyzes the two-row problem develops an argument to support a different rule. The number of panels can be computed by taking the total possible number of walls, namely $4w$, and subtracting the number of overlapping walls. The staff

member calculates the number of overlapping horizontal panels to be w, because there is one horizontal wall between every pair of adjacent wards. Since all but two wards have an overlapping right-hand wall, this person reasons that the number of overlapping vertical panels is $w - 2$. The rule is

$$t_3(w) = 4w - w - (w - 2).$$

However, the staff person notices that this rule does not check for the simplest case—the case of just two wards as shown in Figure 1—since instead of the correct number 7, the rule predicts 6 panels as follows:

$$\begin{aligned} t_3(2) &= 4(2) - 2 - (2 - 2) \\ &= 8 - 2 - 0 \\ &= 6. \end{aligned}$$

Find the flaw and describe what is wrong with the reasoning or work. Then find a correct rule based on the method of subtracting the number of overlapping walls from $4w$. Explain your formula.

Exploration III

You saw in the pet-hotel situations that equivalent expressions produce the same tables and graphs. For each of the following pairs of expressions, explore whether the pair seems to be equivalent or not equivalent, and explain the evidence supporting your decision.

The first group of expressions involve linear functions.

1. $3x + 2$ and $2 + 3x$

2. $1.2a - 4.5$ and $4.5 - 1.2a$

3. $3(g + 2)$ and $3g + 2$

4. $3k + (44 - k)$ and $(3k + 44) - k$

5. $4(v + 8)$ and $(8 + v)4$

6. $3 - (s + 7)$ and $3 - s + 7$

7. $-\left(L - \frac{1}{2}\right)$ and $\frac{1}{2} - L$

The next group for you to consider includes quadratic functions.

8. $q^2 + 2$ and $2 + 2q$

9. $0.5r^2 - r$ and $(r^2 - 2r)0.5$

10. $3j^2 + 2j$ and $5j^2$

11. $(e + 2)(3 - e)$ and $(3 - e)(e + 2)$

The last group involves rational functions.

12. $\dfrac{x^2 + 2x}{x}$ and $\dfrac{3x^2 + 6x}{3x}$　　　**13.** $\dfrac{3x^2 + x}{10 + x}$ and $\dfrac{3x^2}{10}$

Making Connections

Look at your responses in Exploration III. What do you notice about particular pairs of expressions that may help you to predict whether they are equivalent before you construct any tables or graphs?

Exercises

The method of discovering a pattern by studying several small cases (like the approach used in Exploration II) is a strategy that is often successful when looking for geometric patterns. Use this strategy to solve Exercises 1 and 2.

1. Following are the first five figures in a sequence of equilateral triangles, all formed by lines of equally spaced dots.

Figure 1　　Figure 2　　Figure 3　　Figure 4　　　Figure 5

Your goal is to find a rule giving the number of dots in a triangle as a function of n, the number of dots on each side of the triangle. You may be able to think of many methods for counting the dots in such a triangle.

  a. Choose a method that appeals to you, and describe it.
  b. Express the results of your method in the form of a function rule. Let $T_1(n)$ be the total number of dots in the equilateral triangle having n dots on each side.
  Compare your method with methods that other students have found. Record one alternative function rule and call it $T_2(n)$. Does your function rule $T_1(n)$ seem to be equivalent to the other function rule $T_2(n)$? Explain why.

   c. Now try to develop a function rule using the method of constructing a table and looking for a pattern. Copy and complete the table below.

n		2	3	4	5	6	7
Total number of dots		3					

   d. What do you expect would be the total number of dots in a triangle with $n = 10$? $n = 100$?

   e. Write the rule for the function T_3 developed by this method. Is the rule $T_3(n)$ equivalent to your rule $T_1(n)$? Is it equivalent to the rule $T_2(n)$? Explain your reasoning in each decision.

2. Each of the figures in the following sequence is a large triangle formed from small, congruent, equilateral triangles (called unit triangles, because each side has a length of one unit). Notice that in Figure 3, for example, the sides of three unit triangles lie on each side of the large triangle.

The following exercises will help you to find a rule giving the total

Figure 1   Figure 2   Figure 3   Figure 4

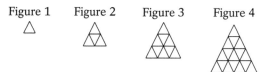

number of unit triangles in a large triangle as a function of the number of unit triangles lying on each of its sides.

   a. Describe a method to find the total number of unit triangles in a large triangle without counting all of them.

   b. Use a symbolic function rule to describe the total number of unit triangles as a function of n, the number of unit triangles on a side. Call your function rule $f_1(n)$.

   c. Now try to derive a function rule by constructing a table and looking for a pattern. Copy and complete the following table.

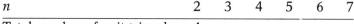

n		2	3	4	5	6	7
Total number of unit triangles		4					

   d. What do you expect would be the total number of unit triangles in a large triangle with $n = 10$? $n = 100$?

   e. Write the rule for your function obtained by using your table. Call the function $f_2(n)$.

   f. Is the function rule $f_2(n)$ equivalent to the rule $f_1(n)$? Explain your reasoning.

3. SITUATION Each team in a summer baseball league has one coach, two assistant coaches, and 25 players. Each of the following rules describes the number of people in these teams as a function of the number of teams t. For each, explain how you can look at the situation in order to generate the rule.

    a. $P(t) = t + 2t + 25t$

    b. $P(t) = 3t + 25t$

    c. $P(t) = (1 + 2 + 25) t$

4. There may be other ways for a pet hotel to determine the number of panels needed to build a unit with a given number of wards. Decide whether each of the following methods and its related rule yields a correct number of panels in the two-row situation discussed in Exploration II. In each case, explain why the method is or is not valid.

    a. Count the number of panels needed for one row of wards; then double this number. Since the unit has w wards, there are $\frac{w}{2}$ wards in one row; each row has $\frac{w}{2} + 1$ panels shown vertically in the diagrams. It also has two rows of walls shown horizontally, each with $\frac{w}{2}$ panels, and so altogether $2 \cdot \frac{w}{2}$ horizontal panels. So the total number of panels is a function w_1 of the number of wards w and can be expressed using the rule

    $$g_1(w) = 2[2(\frac{w}{2}) + \frac{w}{2} + 1].$$

    b. Starting with a pair of panels at the left-hand end, two new wards are formed by attaching 5 panels, as shown in the diagram below.

    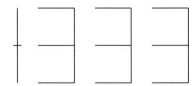

    So the number of panels is a function g_2 of the number of wards r and can be expressed using the rule

    $$g_2(w) = 2 + 5(\frac{w}{2}).$$

    c. There are $\frac{w}{2}$ columns. Each column has 2 wards made with 7 panels. So the total number of panels $g_3(w)$ needed for w wards can be given by the rule

    $$g_3(w) = \frac{w}{2}(2 \cdot 7).$$

5. In the figures of this exercise, each diagram consists of one or more small squares, each of whose sides is a segment one unit long. Answer the following to determine how many segments are in any diagram of this type.

   a. Imagine a sequence of large square diagrams, where Figures 1, 2, and 3 show the first three diagrams in the sequence. Notice that Figure 2 consists of 12 segments. If a large square has n segments on each outside wall, find a rule for the total number of unit segments in the large square, say $s(n)$, as a function of n.

   Figure 1   Figure 2   Figure 3

   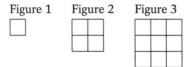

   Compare your rule with those of other students and determine whether the function rules are equivalent.

   b. Imagine a sequence of large rectangular diagrams, illustrated by Figures 4, 5, and 6. In each of these large rectangles, let L be the number of segments in the length of the rectangle and let W be the number of segments in its width. Then let $S(L, W)$ be the total number of unit segments in the rectangle. Notice in Figure 5 that the value of L is 5 and the value of W is 3. There are 38 segments altogether, so $S(5, 3) = 38$. Find a rule for the function $S(L, W)$. Explain how you derived your rule.

   Figure 4   Figure 5   Figure 6

   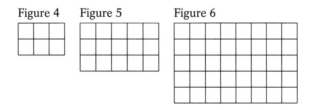

   Compare your rule with those found by other students and determine whether the function rules are equivalent.

   c. Can you use your rule $S(L, W)$ from part b to find the number of segments in the large squares in part a? Explain.

Making Connections

Consider the rules $T_1(n)$, $T_2(n)$, and $T_3(n)$ in Exercise 1. While answering the following questions, ignore the exceptional case of a "triangle" having only one dot, as shown in Figure 1.

1. Which of the rules best displays the relation as a linear function with a constant rate of change?

2. Explain how the *constant rate of change* comes into play in creating this sequence of dot triangles. Use a drawing or graph in your explanation.

2 Formal Symbolic Reasoning in Algebra

In many cases, we can get a fairly good idea of whether two expressions are equivalent by looking at their graphs or tables. If the expressions are not equivalent, it usually shows quickly in a list or in a graph of data pairs, (input, output).

The generation of tables and graphs may require considerable time, however, so it would be nice to be able to determine equivalence mentally without reliance on computing devices. A lot is known about mentally working with and producing equivalent symbolic rules. Knowledge of a small set of facts and procedures usually allows us to determine whether two symbolic rules are equivalent, and in many cases this mental work is quicker than generating and examining relevant tables or graphs.

Although you can analyze more complicated expressions with technology that uses these rules, it is helpful to be able to treat less complicated cases mentally. In fact, this mental work with equivalent symbolic rules has been one of the principal activities of introductory algebra courses. In many cases it still is, but the advent of symbolic manipulation programs has allowed us to focus on other concepts in algebra.

An additional benefit of using a set of facts and procedures to determine equivalence is that you can show equivalence for all inputs at once. Graphs and tables often give only a partial set of data pairs, (input, output), for a function, and input and output values are sometimes rounded off when they are reported in tables and graphs. Because of this, it is not always possible to tell whether two expressions are equivalent just by looking at their graphs or tables. In this section, we shall examine the equivalence of expressions through the use of a small set of properties

2.1   Properties of Operations

There is a surprisingly small and very familiar collection of properties that govern numerical computation. You have been using these properties since you first studied arithmetic. Because the variables with which we work in algebra represent numbers, the properties we need in our algebraic study are simply general rules describing the properties of operations on numbers. In your algebra experience prior to this chapter, you have already encountered most of these properties in your work with function rules.

Recall that in many situations involving a linear function, the output is the sum of a variable cost and a fixed cost. In Chapter 7, for example, hosting the Dapper Dan game costs $3 for every person attending the game in addition to $30,000 which is paid no matter what the attendance. When x people attend the game, you may wish to begin with the variable cost, $3x$ dollars, and add on the fixed cost. Then, for the total cost, you might write the expression

$$3x + 30\ 000.$$

A different way to treat the costs is to begin with the initial outlay of money, $30,000, and increase it by 3 dollars for every person attending the game. In this case, it seems more natural to select the expression

$$30\ 000 + 3x.$$

In fact, it does not matter whether we start with $3x$ and add 30 000 or whether we start with 30 000 and add $3x$. In general, the order in which we add two numbers does not affect their sum. This is one of two **commutative properties** for operations on real numbers. (Real numbers are the numbers on the number line.) There are commutative properties for both addition and multiplication.

Commutative Properties

Addition is a commutative operation:
For real numbers a *and* b, a + b = b + a.

For example:
4 + 5 = 9 and 5 + 4 = 9.
2.1 + 3.7 = 5.8 and 3.7 + 2.1 = 5.8.

Multiplication is a commutative operation:
For real numbers a *and* b, a × b = b × a.

For example:
23 × 4 = 92 and 4 × 23 = 92.
7.1 × 2.2 = 15.62 and 2.2 × 7.1 = 15.62.

The next property is the **associative property**, which says that grouping does not matter in a sum or a product when more than two numbers are involved. For example, consider a concessions manager whose vendors sell 500 bags of peanuts at one football game. The manager's costs are $75 to purchase the bags of peanuts, $50 to pay commissions to the vendors, and $275 to hire help and maintain licenses. The total costs can be computed in at least two different ways.

Method 1   To the sum of the the variable costs ($75 and $50), add the fixed costs ($275).

$$(75 + 50) + 275 = 125 + 275 = 400.$$

Method 2   To the cost of the peanuts ($75), add the sum of the other costs ($50 and $275) involved in selling them.

$$75 + (50 + 275) = 75 + 325 = 400.$$

The total cost is the same, no matter which of the two groupings is used. This equality illustrates one of two associative, or grouping, properties for operations on real numbers.

Associative Properties

Addition is an associative operation:

For real numbers a, b, *and* c, (a + b) + c = a + (b + c).

For example:
(12 + 4) + 21 = 16 + 21 = 37 and
12 + (4 + 21) = 12 + 25 = 37.

Multiplication is an associative operation:

For real numbers a, b, *and* c, (a × b) × c = a × (b × c).

For example:
(4× 2) × 5 = 8 × 5 = 40 and 4 × (2 × 5) = 4 × 10 = 40.

Another pair of particularly useful properties are the **inverse properties**, again for addition and multiplication. For addition, the inverse property assures us that, given *any* number a, the equation $a + x = 0$ has one root. The root is called the **additive inverse** of a. Furthermore, the additive inverse of a is unique—there is *exactly* one root. As examples, the additive inverse of 6 is –6, since 6 + (–6) = 0 and the additive inverse of –13.7 is 13.7, since (–13.7) + 13.7 = 0.

For multiplication, the inverse property is slightly different because of the exceptional nature of the number zero. According to this property, if *any non-zero* number a is given, the equation $a \times x = 1$ has one root. The **multiplicative inverse** of a non-zero number a is also unique. For example, the multiplicative inverse of 6 is $\frac{1}{6}$, since $6 \times \frac{1}{6} = 1$. The multiplicative inverse of –4 is –0.25 (or –1/4), since (–4)(–0.25) = 1.

Inverse Properties

Addition has an inverse property:

For every real number a, *there is a unique real number,* –a, *for which* a + (–a) = 0.

We call $-a$ the *additive inverse* of a or sometimes the *opposite* of a. For example:

The additive inverse, or opposite, of 2.1 is -2.1, and $2.1 + (-2.1) = 0$.
The opposite of -5 is $-(-5)$, or simply 5, and $-5 + 5 = 0$.

Multiplication has a (slightly restricted) inverse property:

For every non-zero real number a, *there is a unique real number,* $\frac{1}{a}$, *for which* $a \times \frac{1}{a} = 1$.

We call $\frac{1}{a}$ the *multiplicative inverse* of a or sometimes the *reciprocal* of a. For example:

The multiplicative inverse of 15 is $1/15$, and $15 \times (1/15) = 15/15 = 1$.
The reciprocal of $2/7$ is $\frac{1}{(2/7)}$, or simply $7/2$, and $2/7 \times 7/2 = 14/14 = 1$.

One of the essential properties of numbers pertains to expressions that involve both addition and multiplication. Such combinations can occur in a variety of settings. For example, suppose you are working at a job that pays a wage of \$6 per hour, and suppose that you work 23 hours one week and 37 hours the next week. How can you compute your paycheck for those two weeks? Should you first find the total number of hours by adding, or should you first find the separate weekly salaries by multiplying? We can examine these two choices:

First computing each week's salary and then adding them together, we get

$$(23 \times 6) + (37 \times 6) = 138 + 222 = 360.$$

First finding the total number of hours worked in the two weeks and then multiplying by the hourly wage, we get

$$(23 + 37) \times 6 = 60 \times 6 = 360.$$

Since the result is the same by either method,

$$(23 + 37) \times 6 = (23 \times 6) + (37 \times 6).$$

This example illustrates the vital **distributive property**, which interconnects the two operations of addition and multiplication.

Distributive Property

Multiplication is distributive with respect to addition:

> *For real numbers* a, b, *and* c, $(a + b) \times c = (a \times c) + (b \times c)$ *and*
> $a \times (b + c) = (a \times b) + (a \times c)$.

You may have noticed that all of the properties listed so far deal only with addition and multiplication, the two operations of primary importance. You may wonder whether there are corresponding properties for the other familiar operations, subtraction and division.

For example, is subtraction commutative? Definitely not, since order matters in subtraction. As a simple illustration, $8 - 5$ is not equal to $5 - 8$. Test several other pairs of numbers for yourself.

Instead of analyzing properties for the operations of subtraction and division, mathematicians prefer to relate these procedures to the more fundamental operations, addition and multiplication, respectively.

Subtraction

The operation of subtraction can be described in terms of addition and its inverse property:

> *For real numbers* a *and* b, $a - b = a + (-b)$.

For example:

> $27 - 9 = 18$ and $27 + (-9) = 18$.

You can see that this is a reasonable description. You can subtract b from a by finding the number x that combines with b by addition to give a as the sum. That is, you solve the equation $x + b = a$. The root of the equation is the difference $a - b$. You can check that $a + (-b)$ also fits in the equation: $[a + (-b)] + b = a + [(-b) + b] = a + 0 = a$.

Division

The operation of division is describable in terms of multiplication and its inverse property:

> *For real numbers* a *and* b, *where* b ≠ 0, $a \div b = a \times \frac{1}{b}$.

For example:

> $21 \div 3 = 7$ and $21 \times (1/3) = 7$.

As with subtraction, you can see that this is reasonable. You can divide a by b by finding the number x that combines with b by multiplication to give the product a. That is, you solve the equation $x \times b = a$. The root is the quotient $a \div b$. The product $a \times \frac{1}{b}$ also works. You can check by using the associative property for multiplication in the same manner as above.

In summary, we can treat every subtraction as an addition (adding the opposite) and every division as a multiplication (multiplying by the reciprocal). We can then use the properties of addition and multiplication for analyzing expressions involving any of the four basic operations.

Although the list of properties and rules shown here—Commutative, Associative, Inverse, Distributive, Subtraction, and Division—is a small one, it is a powerful one. For any given algebraic expression, these properties and basic arithmetic are often sufficient to generate many equivalent expressions!

Exercises

1. When computations involve several steps, there are often alternate ways to obtain answers. In each of the following situations, two different ways are given for computing the answer. Do both of the methods *always* produce the same answer even if the data in the situation are different? Justify your answer using the properties of operations.

   a. A hotel owner wants to know how much molding is needed to go around the walls of a 45-foot by 100-foot rectangular ballroom. He notices that two different methods lead to the same answer. He can first add the dimensions and then multiply by two, getting 290 feet. Or he can first double each dimension and then add the results, again getting 290 feet.

b. A student notices that she can compute the area of the right triangle shown here in two different ways.
She can multiply half of 8 by 6 ($4 \times 6 = 24$) or she can multiply half of 6 by 8 ($3 \times 8 = 24$).

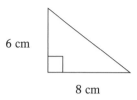

6 cm

8 cm

2. The properties developed in this section refer to addition and multiplication. Several of these properties do not have counterpart properties in subtraction and division.

For example, we have already investigated a possible commutative property for subtraction. We have noted the number $8 - 5$ is not equal to the number $5 - 8$. This one test is enough evidence for us to say that in general, the difference $a - b$ is not equivalent to the difference $b - a$, so subtraction is *not* a commutative operation!

a. Write a symbolic statement claiming that division has the commutative property. Try to find some pairs of real numbers for which the "property" does not work and several pairs of numbers where the "property" is satisfied. Explain why you believe your claim is or is not valid.

b. In a similar manner construct—and test—possible associative and distributive properties involving subtraction and division.

3. Certain common equivalences are direct applications of the basic properties. For each pair of expressions, complete the following.

First, predict whether or not the expressions are equivalent after testing a few different values for the variable. Second, if you believe the expressions are equivalent, establish the equivalence by using the properties developed in this section.

a. $-(3x)$ and $(-3)x$

b. $2x + 3x$ and $5x^2$

c. $4x + 7x$ and $11x$

d. $2x \times 3x$ and $6x$

e. $2x$ and x^2

2.2   Using Properties to Find Equivalent Expressions

The properties of Section 2.1 allow us to change a function rule into an equivalent expression. Recall that two rules are equivalent if they have exactly the same set of data pairs, (input, output). Sometimes the new expression is simpler than the original, and frequently the information displayed by the two forms is different. Each of the forms may have advantages, depending upon what information is desired.

In previous chapters, you have encountered symbolic rules that can be expressed in a variety of different forms. The properties of operations discussed earlier in this section can generally be used, as you will soon see, to produce equivalent expressions in interesting and useful forms.

The various properties are included in symbolic manipulation programs, and we can use these programs to produce equivalent forms. It is also desirable for you to be able to produce equivalent forms for some of the more frequently used types of expressions without the use of technology. Knowing the properties and a few pairs of equivalent forms will allow you to choose between mental or technological methods on the most common and frequent symbolic manipulations.

By the additive inverse property, we know that every real number has an opposite. The next two properties tell how the concept of opposite relates to either a sum or a product.

Opposite-of-a-Sum Property

The opposite of a sum of terms is the sum of their respective opposites:

For real numbers a *and* b, $-(a + b) = (-a) + (-b)$.

We may note that this is a type of "distributive" property—the operation of opposition is distributive with respect to addition.

For instance, take any two numbers, say -5 and 3. The opposite-of-a-sum property says that the opposite of the expression $(-5) + 3$ is equal to the sum of 5 (the opposite of -5) and -3 (the opposite of 3). That is, $-[(-5) + 3] = [-(-5)] + (-3)$. You can readily check this equality, since $-[(-5) + 3] = -(-2) = 2$ and $[-(-5)] + (-3) = 5 + (-3) = 2$. This property applies to any pair of real numbers. Try checking it for other pairs of numbers, being sure to test numbers of varied types such as positive, negative, zero, fractional, *etc.*

Opposite-of-a-Product Property

The opposite of a product of two factors is the product of either factor by the opposite of the other factor:

For real numbers a *and* b, $-(a \times b) = (-a) \times b = a \times (-b)$.

Look at an application of the property when we choose a negative number for a and a positive number for b. For example, suppose $a = -3$ and $b = 5$. Then $ab = -15$, $-a = 3$, and $-b = -5$. So

$$-(ab) = -(-15) = 15,$$
$$(-a)b = (3)(5) = 15, \text{ and}$$
$$a(-b) = (-3)(-5) = 15.$$

All three expressions have the same value. You should try other combinations of positive and negative numbers to convince yourself this will always happen.

Simplifying The distributive property is especially useful in transforming an algebraic expression into an equivalent form, which may be simpler than the given form. In particular, a form may be simpler because "like terms" have been combined. The following are a few examples of attempts at simplification.

Note: There is not universal agreement among mathematicians concerning what it means for one form to be "simpler" than another. Some have even suggested that "simplest form" is an outmoded concept in the age of automated computation.

Example 1. Find a simpler equivalent form for $3x + 5x$.

Answer: By the distributive property, for all real values of x, $3x + 5x = (3 + 5)x = 8x$. So a simpler equivalent form for $3x + 5x$ is $8x$.

Example 2. Find a simpler equivalent form for $3a - 4a$.

Answer: Since our properties for operations do not directly address subtraction, we recall that a difference can be expressed in terms of a sum and an opposite. Using the subtraction rule, we can rewrite $3a - 4a$ in the form $3a + [-(4a)]$. By the opposite-of-a-product property, $3a + [-(4a)]$ is equivalent to $3a + (-4)a$. The distributive property lets us rewrite $3a + (-4)a$ in the form $[3 + (-4)]a$. Again using the subtraction rule, $[3 + (-4)]a$ is equivalent to $(3 - 4)a$, which is simply $-1a$. Summarizing our results, the expressions $3a - 4a$ and $-1a$ are equivalent to each other.

This example illustrates a general principle. The steps in transforming $3a - 4a$ into $(3 - 4)a$ can be applied to numbers other than 3 and 4; that is, if r and s are any numbers, then $ra - sa$ is equivalent to $(r - s)a$. This general principle is another distributive property: namely, multiplication is distributive with respect to subtraction.

Example 3. Find a simpler equivalent form for $(2x + 3y) - (4x - 7y)$.

Answer: Transforming the given algebraic expression into a simple form requires many applications of the various properties you have been studying. Instead of writing our development in paragraph style, we may arrange the steps in a column format. The left column shows the step-by-step transformation of the expression; each expression is equivalent to the ones before and after it. The property used to rewrite an expression is indicated to the right of the new form. For example, the subtraction rule is used to rewrite $(2x + 3y) - (4x - 7y)$ in the equivalent form $(2x + 3y) + [-(4x - 7y)]$.

$(2x + 3y) - (4x - 7y)$	
$(2x + 3y) + (-[4x + (-7y)])$	subtraction
$(2x + 3y) + ([-(4x)] + [-(-7y)])$	opposite of a sum
$(2x + 3y) + ([-4]x + [-(-7)]y)$	opposite of a product
$2x + 3y + (-4)x + 7y$	associativity of addition
$2x + (-4)x + 3y + 7y$	commutativity of addition
$(2 + (-4))x + (3 + 7)y$	distributivity

Finally, using arithmetic and the important properties mentioned, we see that the original expression $(2x + 3y) - (4x - 7y)$ is equivalent to the simple expression $-2x + 10y$. Note that the two terms involving the variable x have been combined and the two terms involving y have also been combined; this is an example of the notion of combining **like terms.**

Combining like terms is not the only way to simplify an expression. In general, a simpler form is one that is easier to work with to find an answer to a question. The simpler form may vary, depending on the question. The following example does not have like terms to combine, but it still can be simplified.

Example 4.   Find a simpler equivalent form for $\frac{6x+4}{2}$.

Answer: For all real values of x, the following equivalences hold. By the division rule, $\frac{6x+4}{2} = (6x + 4) \times (1/2)$. By the distributive property, $(6x + 4) \times (1/2) = (6x) \times (1/2) + 4 \times (1/2)$. By the commutative property for multiplication, $(6x) \times (1/2) + 4 \times (1/2) = (1/2) \times (6x) + 2$. By the associative property for multiplication, $(1/2) \times (6x) + 2 = [(1/2) \times 6]x + 2$ or $3x + 2$. So $\frac{6x+4}{2}$ is equivalent to $3x + 2$.

In the following situation, properties involving opposites are used in order to combine like terms.

SITUATION 2.1

The cost of running a bed-and-breakfast inn depends upon the total number of overnight "guests" (a person who stays three nights is counted the same as three guests). An inn manager notes that the total monthly cost for operating a bed-and-breakfast inn is the sum of the pay for cleaning and linen services ($10 per guest per night), the cost of meals ($6.75 per guest per night), and the money paid for license and staff salaries (a total of $1745 per month in fixed costs). The total cost for a month is a function C of the number of guests n and can be expressed by the rule

$$C(n) = (10n + 6.75n) + 1745.$$
The manager charges each guest $55 per night.

We can use the properties of operations to state the rule for this function C in a simpler equivalent form. In particular, we can combine the terms $10n$ and $6.75n$ by the distributive property, obtaining:

$$C(n) = (10 + 6.75)n + 1745 \text{ or } C(n) = 16.75n + 1745.$$

The inn's revenue is also a function R of the number of guests n and can be expressed by the rule

$$R(n) = 55n.$$

The inn's profit is therefore a function P of the number of guests n and can be expressed by the rule

$$P(n) = 55n - (16.75n + 1745).$$

By the subtraction rule,

$P(n) = 55n + [-(16.75n + 1745)].$

By the property of the opposite of a sum,

$P(n) = 55n + [(-16.75n) + (-1745)].$

By the associative property of addition,

$P(n) = [55n + (-16.75n)] + (-1745).$

By the property of the opposite of a product,

$P(n) = [55n + (-16.75)n] + (-1745).$

By the distributive property,

$P(n) = [(55 + (-16.75)]n + (-1745)$ or $P(n) = 38.25n + (-1745).$

Finally, by the subtraction rule,

$P(n) = 38.25n - 1745.$

Here we have written in paragraph style the steps leading to the representation of the function rule $P(n)$ in the familiar form for a linear function. We might have used the column format illustrated in Example 3. In your own work you should try using both styles in various cases and decide which style you prefer.

The innkeeper could save some computational steps by finding the net profit provided by each extra guest, namely $55 - (10 + 6.75) = 38.25$ dollars, rather than treating separately the total expenses for maintenance and the total cost for food.

The accumulation of money in a bank account gives another example of how equivalent forms can make calculations easier.

SITUATION 2.2

To prepare for college costs, families often set aside money in college funds long before their children are of college age. Suppose that for each child in a family, a savings account is established on the date of the child's birth, and that the account pays interest at a rate of 8% per year. The amount of money in the account when the child is 18 years old (a typical age for students entering college) is a function of the number of dollars d deposited on the birthdate.

The amount of money in the account after one year is

$$d + (0.08 \times d).$$

This sum may be rewritten as

$$(1 \times d) + (0.08 \times d)$$

which, by the distributive property, is equivalent to the expression

$$(1 + 0.08\,) \times d \text{ or simply } 1.08d.$$

Clearly the expression $1.08d$ is less complicated than the original form, and it allows us to compute the account balance with a single multiplication. Instead of calculating $0.08 \times d$ and adding the product to d, we simply multiply 1.08 by d.

Using this simple principle and applying the distributive property again, we can compute the balance after two years.

$$\begin{aligned}
\text{Two-year balance} &= \text{One-year balance} + (0.08 \times \text{One-year balance}) \\
&= [1 \times (1.08\ d)] + [0.08 \times (1.08\ d)] \\
&= (1 + 0.08) \times (1.08\ d) \\
&= 1.08 \times (1.08\ d).
\end{aligned}$$

By the associative property of multiplication, this last expression for the balance after two years may be written as

$$(1.08 \times 1.08) \times d \text{ or simply } (1.08)^2 \times d.$$

So the distributive and associative properties allow us to verify that the algebraic expression $(d + 0.08 \times d) + 0.08 \times (d + 0.08 \times d)$ is equivalent to $(1.08)^2 \times d$. The latter expression not only looks less complicated but requires less time and effort to calculate.

By similar reasoning, we can use the properties of operations to demonstrate that the balance in the savings account at the end of 18 years is

$$(1.08)^{18} \times d \text{ or } 3.996 \times d.$$

So, to find the accumulated amount after 18 years, simply multiply the amount of the initial deposit by about 3.996. This particular equivalence can save many computational steps.

Exploration

SITUATION 2.3

The expense in dining out involves not only the cost of the meal, but also a tip and usually tax. The current convention is that a tip should be at least 15% of the cost of the meal.

It is not always convenient to have available a calculator in order to figure a tip, but some properties of operations make this calculation easy mental arithmetic. You just find 10% of the bill and add half of that result to it; the total is the tip.

For a bill of $23.00, for example, first find 10% of 23. This is 2.30, since $0.1 \times 23 = 2.30$. Half of 2.30 is 1.15, so the sum is $2.30 + 1.15 = 3.45$. Adding the tip of $3.45 to the initial bill of $23.00, the total is $26.45.

1. Show that this mental shortcut applies to any restaurant bill. For the sake of simplicity, assume that there is no tax on the bill.

   a. Try a few examples for yourself just to get a feel for the shortcut. Check to see that the shortcut works for each of your examples.
   b. Now represent the shortcut with algebraic symbols. For example, if the cost of the meal is B, then the tip is $0.15B$. Use the fact that $0.15 = 0.10 + 0.05 = 0.10 + \frac{1}{2}(0.10)$ in your argument.

SITUATION 2.4

When a large group of people requests that the cost of their meals be placed on a shared bill, restaurants often require waitstaff to add a 15% tip on the bill automatically. Suppose a waiter is calculating the total charge for a large group in a state in which an 8% sales tax applies. She regularly calculates the tax first and then computes the tip based on the total. One of the members of the group suggests that she should calculate the tip first. Which of these choices results in a lower total cost to the group?

2. First, calculate a few examples with different restaurant bills. For each of your examples, do the following.
   a. Find the total bill if the waiter calculates the tip and then computes the tax based on the restaurant bill including the tip.

    b. Determine the total bill if the waiter calculates the tax first and then computes the tip based on the restaurant bill including the tax.

    c. Compare the two methods. Which is less expensive for the group?

3. Consider the two methods.

Method I: Calculate the tip first, then pay tax on the restaurant bill including the tip.

Method II: Calculate the tax first, then pay the tip on the restaurant bill including the tax.

    a. Make a general conjecture about the two methods. Should the group insist on method I or method II?

    b. Test your conjecture using an argument based on equivalent expressions. Support your argument in three ways: using graphs, tables, and properties of operations applied to algebraic expressions.

Exercises

In each of Exercises 1 through 9, determine the simplest equivalent form you can find.

1. $3x + 4x + 5y$

2. $5a - 6b + 7a + 4b$

3. $5x^2 - 2x + x^2$

4. $(q^2 + 2) - (2 + 2q)$

5. $(r^2 - 2r)\,0.5$

6. $(3k + 44) - k$

7. $-(L - \frac{1}{2})$

8. $\dfrac{21x + 14}{7}$

9. $(3 + e)(e + 2)$

10. In several situations explored in this chapter, different rules describe the same set of data pairs, (input, output). In each part of this exercise, analyze the possible equivalence of the given rules which you first saw in Section 1 of this chapter. Specifically:

–determine which of the given expressions are equivalent to one another;

–for any pair of expressions that you identify as not equivalent, explain your reasoning; and

–for any pair of expressions that are equivalent, use the properties of operations to show the equivalence.

    a. Each of the following expressions has been suggested to repre sent the number of two-meter panels needed to build a single-row unit, w wards long, for a pet hotel.

        Expressions: $2w + 2 + (w - 1)$ $1 + 3w$

                      $4w$ $4w - [2(w{-}2)]$

    b. Each of the following expressions might have been suggested to represent the number of dots in a dot-triangle with n dots on a side.

        Expressions: $3(n - 1)$ $3n - 3 + 2$

                      $2n + (n - 1)$ $3(n - 2) + 3$

                      $n + (n - 1) + (n - 2)$

    c. The number of two-meter panels needed to build a second type of pet-hotel unit having w wards (where w is an even number) might be represented by any one of the following expressions.

        Expressions: $(w + 2) + \dfrac{w}{2}$ $\dfrac{5}{2}w + 2$

                $2\{[2(\dfrac{w}{2})] + (\dfrac{w}{2} + 1)\}$ $2((2(\dfrac{w}{2})) + (\dfrac{w}{2} + 1))$

3 Equivalent Linear Expressions

In Chapter 3 you learned that a function of the form $f(x) = mx + b$, where m and b are numbers, is a linear function. Sometimes a function rule is not given in the familiar linear expression $mx + b$, although it actually describes a linear function. For these functions three important questions arise.

    1. What information about a situation can we learn from the differ ent forms in which the linear function may be given?

    2. How can we transform a linear function rule into a more famil iar or more useful form?

    3. How can we recognize a linear function when it is given in a less familiar form?

1.   *What information about a situation can we learn from the different forms in which the linear function may be given?*   A linear function expressed in the form $mx + b$ offers easy access to such information as the rate of change of the function (that is, the slope of its graph) and the output when the input is the number 0.   Recall that these two items of information are the numbers m and b, respectively.

Other forms in which linear functions appear may provide easy access to other sorts of information.   For example, a rule you have used to describe attendance at a talent show as a function of ticket price

$$a(p) = 800 - 100p$$

can be written in the equivalent form

$$a(p) = 100\,(8 - p).$$

From the original expression $800 - 100p$, it is easy to see that the attendance decreases at a rate of 100 people per dollar increase in ticket price and that the maximum attendance is 800 people.   From the equivalent expression $100(8 - p)$, it is easy to see that a price of \$8 assures that no one will attend.

Consider two equivalent expressions of the hot dog sales function. The profit in dollars is a function of the number of hot dogs sold with rule

$$P(n) = 0.50n - 450.$$

An equivalent form for this rule is

$$P(n) = 0.50(n - 900).$$

Having two different but equivalent forms for a single function rule allows us to choose between the expressions for different information. For example, we may be interested in answering questions like the following:

–How many hot dogs need to be sold in order to break even?

–What is the profit if no hot dogs are sold?

–For each hot dog sold, by how much does the profit increase?

The break-even problem can be analyzed easily by using the expression $0.50(n - 900)$.   This form readily shows that the profit is 0 when $n = 900$.

The expression $0.50n - 450$ is better for recognizing that a \$450 loss occurs when no hot dogs are sold.

It appears that either rule makes clear the rate of increase in the profit; namely, that the profit increases by \$0.50 for each additional hot dog sold.

Although the questions we may want to answer may vary from situation to situation, it is always helpful to use the most convenient equivalent form.

2. *How can we transform a linear function rule into a more familiar or more useful form?* For the least complicated cases, it is easy to determine whether a function rule is linear by some direct application of the properties of operations. Here, we use the column format to transform the given expression into the familiar form of a linear function.

a. Attendance at a talent show as a function a of the ticket price p in dollars can be expressed by the rule $a(p) = 800 - 100p$.

$800 - 100p$
$800 + (-100p)$ subtraction
$800 + (-100)p$ opposite of a product
$-100p + 800$ commutativity of addition

So a is a linear function of p, whose graph has slope -100 and intercept 800 on the vertical axis.

b. A concession manager's profit $P(n) = 0.75n - (0.15n + 275)$ in dollars is a function of the number n of bags of peanuts sold.

$0.75n - (0.15n + 275)$
$0.75n + [-(0.15n + 275)]$ subtraction
$0.75n + [-0.15n + (-275)]$ opposite of a sum
$[0.75n + (-0.15n)] + (-275)$ associativity of addition
$(0.75n - 0.15n) - 275$ subtraction
$0.60n - 275$ distributivity

So P is a linear function of n; its graph has slope 0.60 and intercept -275 on the vertical axis.

The simplification given above involves step-by-step applications of several properties. Notice that in effect we have combined like terms. The two terms $0.75n$ and $0.15n$, which involve the variable n, are combined to yield 0.60n. If we think of 0.75n as a short expression for 0.75n + 0, then we can combine the constant terms. The number 275 is subtracted from the unwritten number 0 to yield –275. So P(n) is simply 0.60n – 275. You should practice the short method of combining like terms in your work, but always be ready to justify your results in detail whenever asked to do so.

3. *How can we recognize a linear function when it is given in a less familiar form?* Some function rules that are not in the form $f(x) = mx + b$ are easy to recognize as equivalent to linear rules. Among these are expressions like $d - cx$, $r + sx$, and $kx - n$, where c, d, k, n, r, and s are numbers. You can quickly verify that each of these is equivalent to a familiar linear form. For example, $kx - n$ is equivalent to $kx + (-n)$.

For more complicated cases, symbolic manipulation technology can be helpful. Most symbolic manipulation tools have commands that simplify algebraic expressions. Although the particular conventions used vary from tool to tool, you should be able to determine whether an algebraic expression is linear by examining its simplified form.

Many times in earlier chapters, you combined separate function rules (such as Revenue and Cost) to find a new function (such as Profit). Some frequently occurring combinations of function rules give linear function rules. Several examples appear in the following exploration.

 ## Exploration

In the following exercises, use symbolic manipulation technology as an aid in your exploration.

SITUATION 3.1

Large organizations sometimes run conferences on university campuses. Last year at one university, the arrangers for one such meeting purchased room and board contracts from the Housing Department and from Food Services, respectively. There was no extra charge for the meeting rooms themselves.

The Housing Department charged $28.50 per participant, in addition to a fee of $85 to pay for Police Services. The Housing bill is a function H of the number of participants n and can be expressed by the rule

$$H(n) = 28.50n + 85.$$

Food Services charged $18.75 per participant, in addition to a flat fee of $35 to set up a morning coffee service. The Food Services bill is a function F of the number of participants and can be expressed by the rule

$$F(n) = 18.75n + 35.$$

1. The total bill for a client is the sum of the charges paid to the Housing Department and Food Services. One expression for this sum as a function of the number of participants is given by the rule

$$T_1(n) = (28.50n + 85) + (18.75n + 35).$$

   a. Express this rule $T_1(n)$ in a simpler form.
   b. To which family of functions (linear, quadratic, exponential, or rational) does the Housing bill function H belong?

   c. To which family of functions (linear, quadratic, exponential, or rational) does the Food Services bill F belong?
   d. To which family of functions does T_1 belong?

SITUATION 3.2

This year, the university has established a separate department to coordinate conferences. For a $1200 coordination fee, the department arranges for registration, equipment, and coordination of room and board arrangements. The cost to the client is the sum of the Housing Department bill, the Food Services bill, and the coordination fee.

2. a. Write a rule to express the client cost (this year) as a function T_2 of the number of participants n.
   b. If possible, write your rule in a simpler form.
   c. To which family of functions does T_2 belong?

SITUATION 3.3

The Conference Department is considering another way of charging for its services. Instead of a flat fee, the alternate plan is to charge 5% of the total housing and food bill. The cost to the client is still a function (call it T_3) of the number of participants n. This new function can be expressed by the rule

$$T_3(n) = (47.25n + 120) + 0.05(47.25n + 120).$$

3. a. Express $T_3(n)$ in a simpler form.
   b. To which family of functions does T_3 belong?

4. For each of the following, make a conjecture about the type of function that would be obtained. (You will test your conjectures in Exercise 5.)

   a. Two linear functions are added.

   b. A constant is added to a linear function.

   c. A linear function is multiplied by a constant.

5. Test your conjectures from Exercise 4 by gathering and analyzing additional data.

   a. Start by selecting three linear functions of your own choice and call them $F(x)$, $G(x)$, and $H(x)$. Record your function rules.

   b. For each of the following, first write the function rule and then (if possible) record a simplified form.

      i. $F(x) + G(x)$

      ii. $F(x) + H(x)$

      iii. $G(x) + H(x)$

   c. Tell whether your examples in part b support or disprove the conjecture you made in Exercise 4a. Explain fully.

   d. For each of the following, first choose a constant. Next, write the function rule, and then (if possible) record a simplified form.

      i. $F(x)$ + (a constant of your choice)

      ii. $G(x)$ + (a constant of your choice)

      iii. $H(x)$ + (a constant of your choice)

   e. Tell whether your examples in part d support or disprove the conjecture you made in Exercise 4b. Explain fully.

   f. For each of the following, first write the function rule and then (if possible) record a simpler form.

      i. $5 \times F(x)$

      ii. $0.4 \times H(x)$

      iii. $100 \times G(x)$

   g. Do these examples support or disprove the conjecture you made in Exercise 4c? Explain.

6. Based on your work in Exercise 5, adjust (if necessary) the conjectures you made in Exercise 4 and complete the following sentences.

   a. The sum of two linear functions is....

   b. The sum of a linear function and a constant is ....

   c. When a linear function is multiplied by a constant....

7. What kind of function rule results when a linear function is added to a multiple of another linear function? Conduct an exploration to help you make a conjecture, and record your conjecture along with a few examples.

Making Connections

Look at your three conjectures in Exercise 6. Demonstrate one of them graphically by using one or more linear functions and a constant of your own choosing, if necessary. Explain how your graphs illustrate your conjecture.

Give a graphical argument explaining why your conjecture should be true no matter what linear functions and constant may be chosen.

Exercises

For each of the function rules in Exercises 1 through 8:
    a. identify whether or not the given function rule is linear;
    b. if the function rule is linear, tell how you know it is linear; and
    c. if the function rule is linear, use the properties of operations to transform it into the standard linear form, $f(x) = mx + b$.

1. $f(x) = 45 - 56x$

2. $g(y) = (3y + 8) + (0.02y + 6)$

3. $h(k) = 9 + (87 - 52k)$

4. $u(t) = 5t(3t + 8) + 17(5 - 4t)$

5. $j(w) = 5w - 89 + 6(78w + 45) - (3w - 8)$

6. $r(d) = 3d + 0.78d + 5(3600d - 0.0089)$

7. $w(v) = 2(v + 3) + v(v + 3) + 6(v + 3)$

8. $b(a) = 3(a + 7) + 6(a + 7) - 4(a + 7) + 5a$

The rules given in Exercises 9 and 10 are linear function rules that you have encountered in your previous work. For each of these function rules:

    a. use the properties of operations to transform the rule into the standard form $f(x) = mx + b$;

    b. determine which of the two forms, the original form or the standard form, is better for answering the given question; and

    c. explain the reasons for your choice.

9. A waiter's daily pay (in dollars) can be considered a function P of the total amount of money (in dollars) in meal checks c. One possible rule for such a function is

$$P(c) = 0.15(c + 120).$$

Question: What is the pay if there are no customers at all?

10. The altitude of a plane (in meters) can be considered a function A of the time t (in minutes) since the plane's descent began. Suppose one particular plane's altitude can be expressed by the rule

$$A(t) = 120(3 - t).$$

Question: What is this plane's altitude when it starts its descent?

The rules given in Exercises 11 and 12 are also linear function rules that you have encountered in your previous work. For each of these function rules:

    a. use the number properties to transform the rule into the standard form $f(x) = mx + b$;

    b. write one question that can more easily be answered with the original form of the rule—then explain why you think the original form is better; and

    c. write one question that can more easily be answered by the standard form of the function rule—then explain why you think the standard form is better.

11. Suppose that the cost (in dollars) of renting video equipment is a function c of the number of days d for which the equipment is rented. One such function might be expressed by the rule

$$c(d) = 80 + 8(d - 7).$$

12. The temperature in degrees Celsius is a function C of the temperature F in degrees Fahrenheit and can be expressed by the rule

$$C(F) = \frac{5}{9}(F - 32).$$

13. Several students decide to start a car-washing business. They split into an exterior team and an interior team. The exterior team does the washing and waxing, while the interior team vacuums and cleans the interior and the trunk. Since the jobs are quite separate, the teams have separate rules for determining their costs. The exterior team paid $22.35 for a hose and pail, while the interior team purchased a vacuum cleaner for $69.95. Rags, soap, and water cost the exterior team about $1.45 per car while the interior team pays about $0.35 per car for rags, window cleaner, and electricity for the vacuum.

   a. Write a rule for the exterior team's cost (in dollars) as a function C_E of the number of vehicles washed v.

   b. Write a rule for the interior team's cost (in dollars) as a function C_I of the number of vehicles washed v.

   The interior team decides that a fair charge for their services is $3.50 per car, while the exterior team charges $4.50 per car.

   c. Write a rule for the exterior team's revenue (in dollars) as a function R_E of the number of vehicles washed.

   d. Write a rule for the interior team's revenue (in dollars) as a function R_I of the number of vehicles washed.

   e. Write a rule for the exterior team's profit (in dollars) as a function P_E of the number of vehicles washed.

   f. Write a rule for the interior team's profit (in dollars) as a function P_I of the number of vehicles washed.

   g. Which of the rules that you have found in this exercise are linear function rules?

   h. Suppose the two teams decide to split their expenses and revenue. Without actually simplifying the rules, determine whether the total cost, the total revenue, and the total profit are linear functions. Explain each of your conclusions.

4   Quadratics in Equivalent Forms

In Section 3 you have learned that a linear function often appears in a form different from the familiar expression $mx + b$. In some cases those other forms are more useful than the standard form. You have also learned in earlier sections how to decide whether two expressions are equivalent and how to change an expression into an equivalent, more useful form. In this section you will apply your knowledge to quadratic and higher-degree polynomial functions.

SITUATION 4.1

A student government association, anxious to raise money, has a lucky opportunity. Two recent graduates of the school belong to a rap music group called *e PI i*, and they offer to give a special show at the school with all proceeds going to the SGA fund.

Doing some market research, an SGA committee finds that the relation between ticket price x dollars and the number of tickets sold $s(x)$ might be $s(x) = 1500 - 75x$. Their next task is to determine a ticket price that gives the most revenue. They recognize that ticket revenue is equal to ticket sales times ticket price and they write the revenue function

$$R(x) = (1500 - 75x)x.$$

One committee member suggests that they should study a graph of the function R to learn more about this relation between price and revenue. They generate the picture shown here.

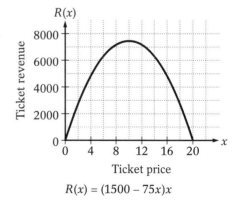

$$R(x) = (1500 - 75x)x$$

This picture gives answers to several questions very quickly. First, it shows that the maximum revenue can be expected to be about $7500 at a ticket price of $10. Second, no revenue at all is predicted at ticket prices of $0 or $20. Apparently they can't expect anyone to pay $20 for a ticket!

After looking at these results for a short time, one student remarks, "Of course, those results make perfect sense."

1. The revenue is zero if $x = 0$ or if $x = 20$, because substituting either one of those values of x in the revenue function gives a factor 0:

$$R(0) = (1500 - 75 \times 0) \times 0 \qquad R(20) = (1500 - 75 \times 20) \times 20$$
$$= 1500 \times 0 \qquad\qquad = (1500 - 1500) \times 20$$
$$= 0 \qquad\qquad\qquad = 0 \times 20$$
$$\qquad\qquad\qquad = 0$$

2. The revenue function is a quadratic function in a different form, because the distributive property guarantees that $(1500 - 75x)x$ is equivalent to $1500x - 75x^2$, which is equivalent to $-75x^2 + 1500x$.

   The graph of a quadratic function is a parabola, which is a symmetric curve with its maximum or minimum point midway between the zeroes of the function.

This situation illustrates several very important and useful facts about quadratic functions and their equivalent forms. First, the rule for a quadratic function can occur in a form other than the familiar expression $ax^2 + bx + c$. One common form is as a product of two linear expressions. This is called the **factored form**, because each expression is a factor of the product. Second, when the alternative form is a product of two linear factors, it is easy to determine the zeroes of the quadratic function: simply find the values that make the individual factors zero. Third, after locating the zeroes of a quadratic, it is easy to sketch the graph and to find the maximum or minimum point.

4.1 Factoring Quadratics

The ability to move back and forth between the standard form and the factored form of a quadratic function rule is so useful that many technological tools that perform symbol manipulation have commands for these transformations. For instance, commands with names like FACTOR or EXPAND may perform as follows:

Given the quadratic	FACTOR produces
$x^2 + 5x + 6$	$(x + 2)(x + 3)$
$x^2 - 7x - 30$	$(x - 10)(x + 3)$

Given the quadratic	EXPAND produces
$(2x - 1)(x + 5)$	$2x^2 + 9x - 5$
$(8 - 3x)(x + 4)$	$-3x^2 - 4x + 32$

The following exploration gives you some experience in using the factored forms of quadratic or higher degree polynomials.

Exploration I

For each of the functions in Exercises 1 through 9, use symbol manipulation technology to find an equivalent expression in factored form and report this as part a. Use this factored form to complete parts b through d below.

    a. Rewrite the function in factored form.

    b. Find and record all the zeroes of the function.

    c. Sketch the graph of the function.

    d. Record the maximum or minimum point of the graph of the function.

1. $f(x) = 3x^2 - 18x$ **2.** $g(x) = x^2 + x - 6$

3. $h(x) = x^2 - 6x + 9$ **4.** $g(x) = -x^2 + 4x + 12$

5. $f(x) = x^2 + x + 4$ **6.** $g(x) = 5x^2 + 15x$

7. $h(x) = 2x^2 + x - 15$

The next two examples are cubic and quartic functions. See how these cases are like, or different from, the quadratic cases.

8. $f(x) = x^3 - 2x^2 - 5x + 6$ **9.** $g(x) = x^4 + 6x^3 - 4x^2 - 24x$

The next experiment compares the factored forms, zeroes, and minimum points of three specific quadratics. Look closely at the three rules to see if you can find an explanation for any patterns that you observe.

10. $f(x) = 3x^2 + 3x - 18$

    $g(x) = x^2 + x - 6$

    $h(x) = 0.5x^2 + 0.5x - 3$

    a. Write the factored forms of each of these functions.

    b. Find the zeroes of each function.

    c. Sketch the graph of each function. You may want to sketch all three on the same diagram.

    d. Find the minimum point on each graph.

Making Connections

Study the examples you have just worked as you try to complete the following statements to summarize properties of quadratic and cubic polynomials.

1. A quadratic polynomial with rule in the factored form $f(x) = c(x + a)(x + b)$ has zeroes....

2. The quadratic function described in Statement 1 has a maximum value at $x =$ ___ when....

   It has a minimum value at $x =$ ___ when....

3. A function with rule in the form $g(x) = c(x + a)(x + b)(x + d)$ is a cubic polynomial with zeroes....

You have seen how symbol manipulation technology produces equivalent factored-form expressions for quadratic, cubic, and quartic polynomial functions. You also have seen how those factored forms can reveal properties of such a function and its graph. Now try the following exercises that reverse the procedure.

Exploration II

In each of the following you are asked to write a function rule. Record the rule as a product of linear factors, so that the resulting function has the given zeroes. You may start with the factors and then use the EXPAND command of your symbol manipulation technology to find the expanded form. To check your rule, use the SOLVE command of your symbol manipulation technology to find the zeroes.

1. Write a rule for a quadratic function that has output 0 when $x = 0$ and when $x = 7$. Record the rule in both factored and expanded form.

2. Write a rule for a quadratic function that has output 0 when $x = -4$ and when $x = 5$. Record the rule in both factored and expanded form.

3. Write a rule for a quadratic function whose zeroes are at $x = 3$ and $x = 7$. Record the rule in both factored and expanded form.

4. Write a rule for a quadratic function whose output is 0 only when $x = 12$. Record the rule in both factored and expanded form.

5. Write a rule for a cubic function that has output 0 when $x = 0$, when $x = 7$, and when $x = -2$. Record the rule in both factored and expanded form.

To remind you why the factored form of a function rule may be useful or interesting, consider the following application. The basic situation should look familiar, but the recommended solution procedure is new.

SITUATION 4.2

When a golfer hits a shot, the height of the ball varies as the ball approaches the green. For an approach shot of perhaps 100 meters, the height (in meters) as a function H of time t in flight (in seconds) might fit a rule like this:

$$H(t) = -5t^2 + 22.5t.$$

Write this rule as a product of two linear expressions. Use your symbol manipulation technology, if necessary. Then use the two factors to determine the time when the ball returns to the ground, the time until the ball reaches its maximum height, and the maximum height.

The next set of exercises gives you further practice in understanding the relation between equivalent forms of quadratic function rules, especially the expressions that are products of linear factors.

Exercises

In each of Exercises 1 through 6, you are given a rule for a quadratic function written as a product of linear factors. Use what you know about such expressions to find the zeroes of the function and its minimum (or maximum) output. Sketch the graph of each function.

1. $f(x) = (x + 4)(x - 3)$

2. $g(x) = (x - 5)(x - 2)$

3. $h(x) = x(x + 4)$

4. $j(x) = (3x - 5)(2x + 4)$

5. $k(x) = (4 - 2x)(8 - x)$

6. $m(x) = (4 - 2x)(x - 2)$

In each of Exercises 7 through 11, write a rule in factored form for a quadratic function that has the given zeroes. For example, in Exercise 7 write a rule, call it $f(x)$, where $f(6) = 0$ and $f(-2) = 0$.

7. $x = 6$ and $x = -2$
8. $x = -3$ and $x = 12$

9. $x = 0$ and $x = -21$
10. $x = 0.5$ and $x = 2.5$

11. $x = 1$ and $x = -1$

12. Explain the property of operations that justifies each step in the following argument.
   a. $(x + 3)(x + 5)$ is equivalent to $(x + 3)(x) + (x + 3)(5)$.
   b. $(x + 3)(x) + (x + 3)(5)$ is equivalent to $(x^2 + 3x) + (x \cdot 5 + 15)$.
   c. $(x^2 + 3x) + (x \cdot 5 + 15)$ is equivalent to $(x^2 + 3x) + (5x + 15)$.
   d. $(x^2 + 3x) + (5x + 15)$ is equivalent to $x^2 + (3x + 5x) + 15$.
   e. $x^2 + (3x + 5x) + 15$ is equivalent to $x^2 + 8x + 15$.

13. Use the reasoning illustrated in Exercise 12 to write a standard form rule for each of the following quadratic expressions.
   a. $(x + 2)(x + 5)$ b. $(x - 7)(x + 3)$ c. $(2x + 9)(x - 3)$

In each of Exercises 14 through 18, you are given a pair of expressions. Decide which of these pairs are equivalent. You might make your decisions either by comparing the values that the expressions in a pair give for several different inputs or by reasoning based on properties of operations.

14. $x^2 - 4x + 3$ and $(x + 3)(x - 1)$

15. $x^2 - 4x$ and $(x - 4)(x)$

16. $2x^2 + 5x - 12$ and $(2x - 3)(x + 4)$

17. $x^3 - 4x^2$ and $(x)(x)(x - 4)$

18. $x^2 - 4$ and $(x - 2)(x - 2)$

19. **SITUATION**   Suppose that the rap group *e PI i* from Situation 4.1 offered to put on a show for another school, asking that their expenses of \$4800 be paid from ticket revenue. Market research indicates the profit as a function of ticket price might be given by the rule $P(x) = (300 - 75x)(x - 16)$. Use this profit function rule to find:
   a. the break-even prices for the rap show;
   b. the ticket price that gives maximum profit for the show; and
   c. the maximum profit.

4.2   Simplifying Quadratic Expressions

In many situations the rule for an important function is found by combining two (or more) other rules. It is often possible to use properties of numbers and operations to simplify those composite rules.

SITUATION 4.3

The Evergreen Nursery Company sells Christmas trees every year at its two locations in Denver. The company plans to advertise a single price for any tree at either location. Since demand at the two locations is different, the company has to consider two different revenue functions. If x is the price (in dollars) for a tree, revenue might be expressed by the rules

$$R_1(x) = (480 - 12x)x \text{ at location 1 and}$$
$$R_2(x) = (420 - 15x)x \text{ at location 2.}$$

You can check that the optimal price at location 1 is $20 per tree, but at location 2 it is only $14 per tree. The big question is: *What single price per tree offers the best result for both lots combined?*

The natural way to consider this question is to search for the price that gives the greatest total revenue from the two locations. The total revenue function has rule

$$R(x) = R_1(x) + R_2(x) = (480 - 12x)x + (420 - 15x)x.$$

This expression is equivalent to the much simpler form

$$R(x) = 900x - 27x^2$$

which can be factored as follows:

$$R(x) = 9(100 - 3x)x.$$

The zeroes of this function are at $x = 0$ and $x = 33\frac{1}{3}$, so the maximum revenue occurs when $x = 16\frac{2}{3}$. The result is best when the price is about $16.67 per tree.

Just as symbol manipulation technology can factor or expand expressions as needed, many such tools also have commands to simplify complex algebraic expressions. As a user of this type of technology, you should know what kind of result to expect from a simplification, as well as how to do easy cases without going to the trouble of finding a calculator or computer!

Exploration III

The questions that follow are designed to reveal some of the more important principles governing simplification of expressions. In each case you are given a question about combining functions. Test the given sample expressions using the SIMPLIFY command of your symbol manipulation program. You may wish to consider other examples of your own design. After you have simplified the samples, use the results to write an answer to the question posed.

1. When two quadratic expressions are combined by addition or subtraction, what sort of expression results?
   Sample expressions:
   $$(3x^2 + 4x - 2) + (8x^2 - 6x + 12)$$
   $$(-3x^2 + 14x + 13) + (2x^2 - 6x - 13)$$
   $$(3x^2 + 4x - 2) + (-3x^2 - 6x)$$
   $$(3x^2 + 4x - 2) - (8x^2 - 6x + 12)$$
   $$(3x^2 - 2) + (8x^2 - 6x + 12)$$

2. When a quadratic expression and a linear expression are combined by addition or subtraction, what sort of expression results?
   Sample expressions:
   $$(3x^2 + 4x - 2) + (-6x + 12)$$
   $$(-3x^2 + 14x + 13) + (-14x - 4)$$
   $$(3x^2 + 4x - 2) + (-6x)$$
   $$(3x^2 + 4x - 2) - (-6x + 12)$$
   $$(3x - 2) - (8x^2 - 6x + 12)$$

3. When two quadratic expressions are combined by multiplication, what sort of expression results?

   Sample expressions:
   $$(3x^2 + 4x - 2) \times (8x^2)$$
   $$(-3x^2 + 14x + 13) \times (2x^2 - 6x)$$
   $$(3x^2 - 2) \times (-3x^2 - 6x)$$
   $$(3x^2 + 4x - 2) \times (8x^2 - 6x + 12)$$
   $$(3x^2 - 2) \times (8x^2 - 6x + 12)$$

   Try some other test cases of your own design.

4. When a quadratic expression and a linear expression are multiplied together, what sort of expression results?

Sample expressions:
$(3x^2 + 4x - 2) \times (8x)$
$(-3x^2 + 14x + 13) \times (2x - 6)$
$(3x^2 - 2) \times (-5x - 6)$
$(4x - 2) \times (8x^2 - 6x + 12)$
$(15x) \times (5x^2)$

Try some other test cases of your own design.

5. When a quadratic expression and a cubic expression are multiplied, what sort of expression is the product?

Sample expressions:
$(3x^2 + 4x - 2) \times (8x^3)$
$(-3x^2 + 14x + 13) \times (2x^3 - 6x^2)$

Try other test cases of your own design.

6. When a quadratic expression and a cubic expression are combined by addition or subtraction, what sort of expression results?

Sample expressions:
$(3x^2 + 4x - 2) - (8x^3)$
$(-3x^2 + 14x + 13) + (2x^3 - 6x^2)$

Try other test cases of your own design.

7. What general patterns can you find that help you predict the type of expression that results from combining any two polynomials by addition, subtraction, or multiplication?

4.3   Rules for Simplifying Expressions

Exploration III probably has given you some ideas about what to expect when two polynomial expressions are combined and the result is simplified using your technological tool. To check the results of symbol manipulation technology, it helps to know the basic procedures for simplification. Of course, knowing those procedures also allows you to do easy cases without technology.

Example 1. Consider the task of simplifying the following sum of two quadratics:

$$(3x^2 + 5x + 12) + (8x^2 + 7x + 9).$$

Since these expressions are combined by addition and the terms in each expression are also combined by addition, we can apply the commutative and associative properties of addition to rearrange the sum in the following equivalent form:

$$(3x^2 + 8x^2) + (5x + 7x) + (12 + 9).$$

The distributive property allows us to write this expression in yet another equivalent form,

$$(3 + 8)x^2 + (5 + 7)x + (12 + 9)$$

which simplifies to the familiar quadratic form as

$$11x^2 + 12x + 21.$$

Example 2. When a combination involves subtraction, more care must be used in rearranging and simplifying terms of expressions. The basic idea is to convert subtractions into equivalent additions. For example,

$$(3x^2 - 5x + 12) - (8x^2 + 7x - 9)$$

is equivalent to

$$(3x^2 - 5x + 12) + (-8x^2 - 7x + 9),$$

which is equivalent to

$$[(3x^2) + (-8x^2)] + [(-5x) + (-7x)] + (12 + 9),$$

which simplifies to

$$-5x^2 - 12x + 21.$$

Compare this approach with the principle of combining like terms. If we apply this method to the given expression, we combine terms by subtracting the coefficients and obtain

$$(3 - 8)x^2 + [(-5) - 7]x + [12 - (-9)],$$

which after numerical simplification becomes (just as before!)

$$-5x^2 - 12x + 21.$$

Cases that involve sums or differences of linear and quadratic expressions, two cubic expressions, quadratic and cubic expressions, and so on, are handled in a similar way. They require careful use of number system properties.

Example 3. When an expression is a product of two quadratic expressions (or some other pair of polynomials), use the distributive property. For instance,

$$(3x^2 + 5x + 12) \cdot (8x^2 + 9)$$

is equivalent to

$$(3x^2 + 5x + 12) \cdot (8x^2) + (3x^2 + 5x + 12) \cdot (9),$$

which is equivalent to

$$(3x^2)(8x^2) + (5x)(8x^2) + (12)(8x^2) + (3x^2)(9) + (5x)(9) + (12)(9),$$

which is equivalent to

$$24x^4 + 40x^3 + 96x^2 + 27x^2 + 45x + 108,$$

which is finally equivalent to

$$24x^4 + 40x^3 + 123x^2 + 45x + 108.$$

A great deal of practice is required to develop skill in quickly simplifying polynomial products. You might decide that if you ever need to solve such a problem you would take the time to find symbol manipulation technology. The homework exercises that follow give practice with some of the easier cases. After you have done them, you should know well the type of polynomial that results from a given sum, difference, or product. The justification of the principles that help you make such predictions involves properties of operations in a fashion similar to the three examples just given.

Exercises

The polynomial function $f(x) = 4x^2 + 3x - 5$ has degree 2 because the highest exponent on any term is 2. The number 4 is called the leading coefficient of the polynomial because it is the coefficient of the term with the highest exponent. The polynomial function $g(t) = -9t^3 + 7t^2 - 1$ has degree 3 and the number -9 is the leading coefficient of this polynomial.

In each of Exercises 1 through 10, use the properties of polynomials you have discovered in the explorations of this section to determine the degree and the leading coefficient for the polynomial that results from simplifying the indicated combination of polynomials.

1. $(x^2 + 3x + 2) + (4x^2 + 1)$ 2. $x(5 - x)$

3. $(x + 2)(x - 7)$ 4. $(2x - 3)(4x + 6)$

5. $x^2(3x + 4)$ 6. $(5x^3 - 8x) - (4x^3 - 12x^2)$

7. $(x^2 + 4)(x^3 - 8)$ 8. $(3x - 5) + (4x^3 + 3x^2 + x - 5)$

9. $(2x^2 + 14x - 25) - (2x^2 + 12x + 23)$

10. $(4x^3 + 3x^2 + 4x - 5) + (5x^2 + 4x - 32)$

11. Write in simplest form a polynomial that is equivalent to the sum $(4x^3 + 3x^2 + 4x - 5) + (5x^2 + 4x - 32)$. Be prepared to explain the reasoning that leads to your result.

12. Write a polynomial in standard form (using descending powers of x) that is equivalent to the difference $(6x^2 + 14x - 25) - (2x^2 + 12x + 23)$. Be prepared to explain your reasoning.

13. Write a polynomial in standard form that is equivalent to the product $(x + 3)(x + 4)$. Be prepared to explain your reasoning.

14. Express the product $(x + 3)(x - 4)$ as a polynomial in standard form. Be prepared to explain the reasoning that leads to your result.

5 Equivalent Rational Expressions

In earlier sections you have seen that a single linear or quadratic function may be described by two or more rules that, although they look different, are actually equivalent: they produce the same data pairs, (input, output), and the same graph. Rules for rational functions can also appear in different but equivalent forms.

The examples and explorations of this section are designed to develop your ability to recognize equivalent expressions for rational function rules. You also will develop your ability to decide which equivalent forms are useful for answering the various questions that are posed in problem situations.

SITUATION 5.1

Each year the Big Apple Travel Agency organizes a spring vacation trip to the Caribbean. The trip includes a charter flight round-trip between New York and San Juan as well as room and meals at a special youth resort in Puerto Rico.

To reserve space, the agency must promise to pay $90,000 for the charter airplane and $105,000 for the resort. To help set prices to be charged for each student, the director of the agency asks an aide to find a rule that gives cost per student as a function of the number of students taking the trip. The aide finds two rules that seem different and is unsure which to use.

In developing one rule, the aide reasons that for x students the cost per student is $90\,000/x$ for the air charter and is $105\,000/x$ for the resort. The combined cost $C(x)$ (in dollars) per student can be written

$$C(x) = \frac{90\,000}{x} + \frac{105\,000}{x}.$$

Thinking in another way, the aide reasons that the total cost of the trip for the agency is $195,000$, so the cost $D(x)$ (in dollars) per student can be given by

$$D(x) = \frac{195\,000}{x}.$$

The agency director, a former mathematics teacher, decides to use this opportunity to give the aide a mathematics lesson. The director suggests that the aide should produce tables of sample values, (input, output), and graphs of the two function rules to help decide which rule might be better. The aide's graph is shown at the right, and the table follows.

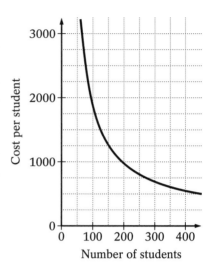

Number of students	$C(x)$	$D(x)$
50	$\dfrac{90\ 000}{50} + \dfrac{105\ 000}{50} = 1800 + 2100$	$\dfrac{195\ 000}{50} = 3900$
100	$\dfrac{90\ 000}{100} + \dfrac{105\ 000}{100} = 900 + 1050$	$\dfrac{195\ 000}{100} = 1950$
150	$\dfrac{90\ 000}{150} + \dfrac{105\ 000}{150} = 600 + 700$	$\dfrac{195\ 000}{150} = 1300$

This information convinces the aide that the two rules are equivalent, but the Agency director continues the lesson by remarking that familiar properties can be used to prove the equivalence. For any real number x:

$$\frac{90\ 000}{x} + \frac{105\ 000}{x} \text{ is equivalent to } 90\ 000\left(\frac{1}{x}\right) + 105\ 000\left(\frac{1}{x}\right),$$

$$\text{which is equivalent to } (90\ 000 + 105\ 000)\left(\frac{1}{x}\right),$$

$$\text{which is equivalent to } (195\ 000)\left(\frac{1}{x}\right),$$

$$\text{which is equivalent to } \frac{195\ 000}{x}.$$

To connect this formal proof to familiar mathematical ideas, the director observes that this combination of two fractional expressions into one is really just another type of distributivity—addition of fractions with common denominators. In general, if a and b are any real numbers, the expression $\dfrac{a}{x} + \dfrac{b}{x}$ is equivalent to $\dfrac{a+b}{x}$. The equivalence of these two expressions can be used to study certain relations among variables either by examining the behavior of $\frac{a}{x}$ and $\frac{b}{x}$ separately or by examining the behavior of the combined form, $\frac{a+b}{x}$.

Exercises

In each of Exercises 1 through 10, express the sum or difference as a single fraction that is equivalent to the given rational expression. Then check the equivalence by calculating and comparing the values of the two expressions—yours and the one given—for $x = -3$, for $x = 1$, and for $x = 10$.

1. $\dfrac{2}{x} + \dfrac{5}{x}$

2. $\dfrac{1.6}{x} + \dfrac{500}{x}$

3. $\dfrac{7}{x} - \dfrac{12}{x}$

4. $\dfrac{3x^2}{x} + \dfrac{4}{x}$

5. $\dfrac{11}{4x} - \dfrac{3}{4x}$

6. $\dfrac{3}{x} + \dfrac{x}{3}$

7. $\dfrac{2x}{4x+1} + \dfrac{12.4}{4x+1}$

8. $\dfrac{7}{2x+4} - \dfrac{3}{2(x+2)}$

9. $\dfrac{3}{2x} + \dfrac{7}{x}$

10. $\dfrac{3x}{x+1} + \dfrac{2}{x+1}$

In each of Exercises 11 through 16, write a sum or difference of two rational expressions that is equivalent to the given rational expression. **Note:** Each of these exercises has many different correct answers. You might try to write several answers in each case. For example,

$$\frac{2+x}{2x} = \frac{1}{2} + \frac{1}{x} = \frac{5+x}{2x} - \frac{3}{2x}.$$

11. $\dfrac{25}{x}$

12. $\dfrac{14}{x+2}$

13. $\dfrac{5}{x^2}$

14. $\dfrac{3x^2+2}{x}$

15. $\dfrac{17}{x-2}$

16. $\dfrac{1-x}{x^2}$

17. **SITUATION** Suppose that the director of the Big Apple Travel Agency is offered a better deal on the Caribbean vacation: only $75,000 for the charter flight round-trip and only $88,000 at the Puerto Rican resort.

    a. Write two equivalent function rules, each giving the cost per student for the trip as a function of the number of students making reservations.

    b. For each of the rules, explain what information about the vacation situation is given better by that rule than by the other one.

18. **SITUATION** On a special cross-country road rally, the goal is to travel as close as possible to a given speed. Suppose that the trip has four legs of 600, 750, 450, and 800 kilometers.

    a. For each leg of the trip, write a rule giving the time as a function of the speed x in kilometers per hour. Hint: If you travel 200 km at 40 km/hr, how long does the trip take?

    b. Write two equivalent rules giving the total combined time for the four legs of the trip as a function of the speed x in kilometers per hour.

19. Consider the three rational functions $f(x) = 2/x$, $g(x) = 3/x$, and $h(x) = 5/x$.

    a. Generate tables of values showing the outputs of these three functions for $x = 0$ to $x = 10$ in steps of 1. Record the tables on your paper.

    b. Plot the points from your tables in part a on a single coordinate grid, using a different color for each function. Use a scale of 0.5 on the vertical axis.

    c. What relations do you see among the vertical coordinates of the points on these three graphs?

Exploration I

Since rational functions occur in various forms, there are many different combinations of rational expressions that are equivalent. The next example illustrates one particularly important pair of equivalent forms.

SITUATION 5.2

As part of their economics course, Victoria and Eugene are managing a business that sells candy. They charge $0.50 for each box, but they must order the candy before the sale begins and cannot return any of the candy they order. If they order 500 boxes for $100, the total profit $TP(x)$ in dollars depends on the number of boxes sold x and can be expressed by the rule $TP(x) = 0.50x - 100$.

Thinking about prospects for profit, Victoria analyzes the function rule $TP(x)$. She quickly recognizes that it represents a linear function. She notes that the slope of its graph, 0.50, is positive and so the profit steadily increases as more boxes of candy are sold. The greatest possible profit occurs if all 500 boxes are sold. This profit is

$$TP(500) = 0.50(500) - 100 = 250 - 100 = 150.$$

Victoria also observes that the same function is expressed by the equivalent rule $0.50(x - 200)$. From this rule, she can see that the business breaks even when 200 boxes of candy are sold and can make money only when the number of boxes sold is greater than 200. For example, if 300 boxes are sold, the total profit is $50, since

$$TP(300) = 0.50(300) - 100 = 50.$$

After studying the total profit prospects, Victoria begins to wonder about an **average profit**. First, recalling her sample calculation $TP(300) = 50$, she reasons that 50 dollars profit for 300 boxes of candy means that the average profit is $50/300 \approx 0.17$ dollars per box. Wondering whether the same average applies in other cases, she tries to find a rule giving the **average profit per box** for any number of boxes. She reasons that

$$\text{Average profit} = \frac{\text{Total profit}}{\text{Number of boxes sold}}.$$

In symbolic form, Victoria's rule for the average profit is

$$VA(x) = \frac{0.50x - 100}{x}.$$

Eugene, who is also thinking about average profit, reasons in another way:

> Our income is $0.50 per box. As we sell x boxes we can distribute our investment cost of $100 over those x boxes at 100 / x dollars per box. The average profit per box is the difference between income per box and cost per box.

In symbolic form, Eugene's rule for the average profit is

$$EA(x) = 0.50 - \frac{100}{x}.$$

This rule looks similar to Victoria's, but not identical. Are the two rules equivalent?

1. You know that one way to begin checking for equivalence of two function rules is to produce a table of sample data pairs, (input, output). The following is an outline for a table. Complete the calculations for the given values of x and make a table of your own.

x	Victoria's rule	Eugene's rule
500	$\frac{0.50(500)-100}{500} = 0.30$	$0.50 - \frac{100}{500} = 0.30$
450		
400		
350		
300		
250		
200		
150		
100		
50		
0		

2. Use your graphing utility to study graphs of Victoria's and Eugene's rules for average profit per box. Sketch the results.

3. What do the tables and graphs suggest about the equivalence of the two rules for calculating average profit per box of candy sold?

While the tables and graphs might have convinced you about equivalence of the rules, they do not guarantee that the two rules give identical results for *any* input x. As in the case of $a/x + b/x$, reasoning based on properties of operations gives a more convincing argument.

1. By the division property, $\frac{0.50x-100}{x}$ is equivalent to $(0.50x - 100)(\frac{1}{x})$.

2. Applying distributive and associative properties, we conclude that $(0.50x - 100)(\frac{1}{x})$ is equivalent to $(0.50)(x)(\frac{1}{x}) - 100(\frac{1}{x})$.

3. Applying the division property and the inverse property that $x(\frac{1}{x}) = 1$ for any non-zero real number x, we conclude that $(0.50)(x)(\frac{1}{x}) - 100(\frac{1}{x})$ is equivalent to $0.50 - \frac{100}{x}$.

This chain of reasoning shows conclusively that the two proposed rules for calculating average profit per box of candy are equivalent.

The equivalence of these two function rules illustrates a general property relating rational expressions. For any real numbers a and b, $\frac{ax+b}{x}$ and $a + \frac{b}{x}$ are equivalent expressions.

The proof follows the same reasoning as the example:

$$\frac{ax + b}{x} \text{ is equivalent to } (ax + b)(\frac{1}{x}),$$

$$\text{which is equivalent to } ax(\frac{1}{x}) + b(\frac{1}{x}),$$

$$\text{which is equivalent to } a + \frac{b}{x}.$$

The next examples and questions should help you to see the kinds of problems for which each form is particularly helpful.

1. Consider the rational function $f(x) = 2 + \frac{6}{x}$, whose rule is equivalent to the expression $\frac{2x+6}{x}$. Use technological tools to produce a table of values for this function for $-10 < x < 10$ in steps of 1 and draw a graph over the same set of input values. Record both your table and diagram.

    Use the rules, along with your table and graph, to help you answer the following questions about the given function.

a. For which input values x is the function output undefined and why?

b. For which inputs x is the output equal to zero? How is that value easy to find using the rule form $f(x) = \frac{2x+6}{x}$ even without technological help?

c. What happens to the outputs of the function as x becomes very large? How is this pattern easy to see, even without technological help, by looking at the rule in the form $f(x) = 2 + \frac{6}{x}$?

2. Consider next the rational function g with equivalent rules

$$g(t) = 5 - \frac{3}{t} \text{ and } g(t) = \frac{5t - 3}{t}.$$

a. For which values t is this function undefined?

b. By studying the two equivalent rules for this function, find the values of t for which $g(t) = 0$. Explain how you found your answer and which rule is more helpful in this case.

c. What can you say about the outputs of the function when the inputs t are very large numbers? How is this pattern easy to see, even without technological help, by looking at the rule in the form $g(t) = 5 - \frac{3}{t}$?

d. Draw a graph of this function. Label the points where $g(t) = 0$.

The previous examples illustrate the principle that equivalent forms of rational expressions are often useful in answering the different questions that arise about such functions. In particular, for a rational function with rule in the form

$$f(x) = a + \frac{b}{x} \text{ or } f(x) = \frac{ax + b}{x},$$

the first form reveals that as x gets very large, $f(x)$ gets very close to a; the second form reveals that $f(x) = 0$ when $x = -b/a$. These patterns occur for logical reasons:

1. in the first expression $a + \frac{b}{x}$, the fraction $\frac{b}{x}$ is small if x is large; and

2. in the second expression $\frac{ax+b}{x}$ is 0 only when $ax + b = 0$. The only root of this simple linear equation is $x = -b/a$.

The exercises that follow give you some additional practice in applying these ideas.

Exercises

In Exercises 1 through 10 write each given rational function rule in a different but equivalent form. Then find the input values that give output 0 and find the trend in output values for very large inputs.

1. $f(x) = 7 + \dfrac{14}{x}$

2. $g(x) = 7 - \dfrac{14}{x}$

3. $h(t) = \dfrac{2t + 5}{t}$

4. $k(s) = \dfrac{2s - 5}{s}$

5. $m(r) = \dfrac{12 + 4r}{r}$

6. $j(r) = \dfrac{r - 5}{r + 4}$

7. $P(x) = \dfrac{5x - 2}{x}$

8. $d(t) = 50 + \dfrac{300}{t}$

9. $h(t) = \dfrac{300}{t} - 60$

10. $r(x) = 3x - \dfrac{27}{x}$

11. For each of the following functions make a rough sketch showing the overall pattern you expect in the graph of the function.

    a. $f(x) = 2 + \dfrac{1}{x}$ **b.** $g(t) = \dfrac{5t - 2}{t}$

12. Consider again Victoria and Eugene's candy business. The average profit per box is given by two equivalent rules

$$VA(x) = \frac{0.50x - 100}{x} \text{ and } EA(x) = 0.50 - \frac{100}{x}.$$

    a. How does the average profit per box change as the number of boxes sold increases? Which rule is more helpful in answering this question?

    b. What is the break-even point in sales for the business? Which rule is more helpful in answering this question?

 ## Exploration II

In this section you have been studying some important and useful cases of equivalent rational expressions. There are, of course, many others. The examples in this exploration show some other significant ideas and how you can use symbolic reasoning to analyze the situations.

1. Consider the function with rule $f(x) = 1.5x^2 / x$. Use your techno-logical tools to create a table, using inputs from $x = -10$ to $x = 10$ with steps of 1. Create a graph showing the pattern in data pairs, $(x, f(x))$, for this function.

Do the results of your table and graphing work surprise you? You might have expected the rule to produce a curved graph and to have a variable rate of change in output values. Instead, it displays a much simpler pattern—a general pattern that you usually expect only from a linear function. However, as your program may have warned you, there is a very important caution: the input value $x = 0$ gives an exception to this pattern!

The secret to this surprising result can be uncovered by reasoning based on important properties of operations. Using the division rule and the associative property of multiplication, we note that the following three expressions are all equivalent to one another:

$$\frac{1.5x^2}{x}, (1.5 \times x \times x)(\frac{1}{x}), \text{ and } (1.5 \times x)(x \times \frac{1}{x}).$$

By the inverse property for multiplication, if x is a non-zero number, then $x \times (1/x)$ is the number 1. The function rule $f(x)$ is not defined for the input value $x = 0$. So, with the exception of the input $x = 0$, $(1.5 \times x)[x \times (1/x)]$ may be expressed as $(1.5x)(1)$, or just simply $1.5x$. Thus the function rule $1.5x^2/x$ is almost identical to the simpler rule $1.5x$. We say that two functions are **equivalent almost everywhere** when the outputs for the two rules agree except for a limited number of in-puts. In our example here, the functions agree for every input except the number 0, which is allowed as an input for the simpler rule but not for the original rule.

The following are other examples where a moderately complex rational function rule is equivalent to a simpler rule almost everywhere—with some important limitations on inputs for which no output is defined. Use your table-generating, graphing, and symbol manipulation technol-ogy to study the examples and look for patterns.

2. Consider the function $g(x) = \dfrac{x^2 - 9}{x + 3}$.

    a. Produce and record a table and a graph for $g(x)$ with $-10 < x < 10$.
    b. What simpler function rule seems equivalent (almost everywhere) to the given rule for $g(x)$? Explain your reasoning for the choice.
    c. What number x is not an allowable input for $g(x)$?
    d. Use your computer symbol manipulation technology to simplify the given expression for $g(x)$. Record the result.
    e. Use your symbol manipulation technology to rewrite the original function rule so that the quadratic numerator is in factored form. The denominator should still be $x + 3$.

        How does this factored form help explain the table and graph you produced in part a?
    f. For which values of x are the outputs for the simpler rule you found in part d and the original rule different from one another?

3. Consider now the function $h(t) = \dfrac{5t + 10}{t + 2}$.

    a. Construct a table and draw a graph for $h(t)$ with $-10 < t < 10$.
    b. What simpler function rule seems equivalent to the given rational expression for $h(t)$ almost everywhere? Explain your reasoning for the choice.
    c. Use your symbol manipulation technology to simplify the given rule for $h(t)$. Record the result.
    d. For which values of t do the outputs for the simpler rule and the original rule not agree with each other?

4. Consider the function $k(x) = \dfrac{x^2 - 5x - 6}{x - 6}$.

    a. Construct a table and draw a graph for $k(x)$ with $-10 < x < 10$.
    b. What simpler function rule seems to be equivalent to the expression for $k(x)$ almost everywhere? Explain your reasoning for the choice.
    c. Use your symbol manipulation technology to simplify the expression for $k(x)$. Record the result.

d. Use your symbol manipulation technology to rewrite the original function rule so that the quadratic numerator is in factored form. The denominator should still be $x - 6$.
How does this factored form help explain the table and graph you produced in part a?

e. For which values of x are the outputs for the expression found in part c and for the function rule $k(x)$ different from one another?

The examples in this exploration only begin to illustrate one way that a rational expression might be *nearly* or *almost* equivalent to a much simpler function rule. This kind of "simplification" is often very helpful in studying a rational function, and the SIMPLIFY commands of most computer-algebra programs also help with the more complex cases you might encounter. Since this type of simplification often yields an expression that is not quite equivalent to the original form, it is important to note, and to exclude from consideration, input values that have undefined output values. As in the case with quadratic, cubic, or quartic polynomial functions, writing numerators and denominators in factored form reveals secrets to the behavior of many rational functions.

Exercises

In the exercises below you are given rational function rules with numerators and denominators in factored form. For each rule:
a. identify input values for which the output is undefined;
b. find all inputs that give output 0; and
c. find the simplest rule equivalent to the given rule—except possibly where the original rule is undefined.

1. $f(x) = \dfrac{(x - 2)(x + 3)}{x + 3}$

2. $g(x) = \dfrac{(x + 4.5)(x + 2)}{x + 4.5}$

3. $h(t) = \dfrac{t(t - 3.1)}{t}$

4. $j(s) = \dfrac{5s(s + 12)}{s}$

5. $k(s) = \dfrac{81s(s - 43)}{s - 43}$

6. $m(r) = \dfrac{(r - 3)(r + 4)r}{r + 4}$

7. $d(t) = \dfrac{(t + 4)(t - 5)(t + 1)}{(t + 1)(t - 5)}$

8. $f(x) = \dfrac{(x + 1)(x - 2)}{x - 1}$

9. $g(t) = \dfrac{5t}{t - 2} - \dfrac{10}{t - 2}$

10. $P(x) = 4 - \dfrac{3x + 9}{x + 3}$

11. $r(x) = \dfrac{x}{x}$

12. $k(t) = 1 - \dfrac{t^2}{t^2}$

13. $h(x) = 1 - \dfrac{x^2}{x(x+1)}$

6 Summary

In this chapter you have studied algebraic expressions, along with their graphical, numerical, and symbolic representations. You have also studied the equivalence of expressions.

Two expressions are equivalent to each other if for every input they yield the same output. In order to verify that a pair of expressions are equivalent, comprehensive tables and carefully drawn graphs often provide valuable insight, but usually the most convincing demonstration is by symbolic reasoning. On the other hand, you can conclude that expressions are not equivalent if you find just one input for which the respective outputs are different.

Transforming function rules into more familiar forms

Below are some important rules useful in establishing equivalence.

Relating to addition:

1. Commutative property: *For any real numbers* a *and* b, a + b = b + a.
2. Associative property: *For any real numbers* a, b, *and* c, (a + b) + c = a + (b + c).
3. Inverse property: *For every real number* a, *there is a real number, written* –a, *such that* a + (–a) = 0. *We sometimes call* –a *the opposite of* a.
4. Subtraction rule: *For any real numbers* a *and* b, a – b = a + (–b).
5. Opposite-of-a-sum property: *For all choices of real numbers* a *and* b, –(a + b) = (–a) + (–b).

Relating to multiplication:

1. Commutative property: *For any real numbers* a *and* b, a × b = b × a.
2. Associative property: *For any real numbers* a, b, *and* c, (a × b) × c = a × (b × c).

3. Inverse property: *For every real number* a *different from 0, there is a unique real number, written* $\frac{1}{a}$, *such that* $a \times \frac{1}{a} = 1$. *We sometimes call* $\frac{1}{a}$ *the reciprocal of* a.

4. Division rule: *For any real number* a *and any non-zero real number* b, $\frac{a}{b} = a \times \frac{1}{b}$.

5. Opposite-of-a-product property: *For all choices of real numbers* a *and* b, $-(a \times b) = (-a) \times b = a \times (-b)$.

Relating to both addition and multiplication:

1. Distributive property: *For any real numbers* a, b, *and* c, $(a + b) \times c = (a \times c) + (b \times c)$ *and* $c \times (a + b) = (c \times a) + (c \times b)$.

Recognizing expressions that are not in familiar form

You have learned to recognize a variety of expressions for linear, quadratic, or other polynomial functions without first expressing the function rules in standard form. In particular, certain combinations of linear functions yield linear functions. You have also learned about circumstances in which an expression is equivalent to a simpler expression almost everywhere and about some limitations associated with this idea.

Getting different information from different equivalent forms

You have considered how different but equivalent forms of a function rule help in answering different types of questions about the situations represented by those functions. For example, a rational function rule in the form $\frac{ax+b}{x}$ is useful for determining what input value yields the output zero, whereas the equivalent form $a + \frac{b}{x}$ is more suitable for examining the behavior of the function for large values of x.

Review Exercises

1. SITUATION   Four students, Mark, Lauren, Sal, and Taunita, sell programs at basketball games. Each buys 200 programs for a total cost of $40. The students must pay this cost before the doors open. Fans pay $0.50 for each program. Each of the students devised a rule for the amount of money collected, which depends on the number of programs p he or she sells.

   a. Mark's rule, expressed in dollars, is $M(p) = -40 + 0.5p$. Explain how he may have reasoned in choosing his rule.

    b. Lauren noticed that each program costs her $0.20. Her rule, also expressed in dollars, is $L(p) = (0.5 - 0.2)p - 0.2(200 - p)$. Explain the reasoning that Lauren used in choosing her rule.

    c. Are Mark's and Lauren's rules equivalent? Justify your answer, using an appropriate combination of tables, graphs, and symbolic manipulation.

    d. Explain what each of the two function rules indicates about the situation.

    e. Write each of the two rules in simple form.

    f. Sal argues that he earns 30 cents for every program sold but loses 20 cents for every program not sold. Explain his reasoning in choosing his rule, $S(p) = 30p - 20(200 - p)$.

    g. Are Mark's and Sal's rules equivalent? Justify your answer.

    h. Can you explain the reasoning involved when Taunita claims that she earns $T(p) = 0.5(p - 80)$ dollars?

    i. Compare the four rules by describing what advantages and what disadvantages each of them may have.

2. Consider the following function rules:

$$F(t) = -5t + 8t \qquad\qquad G(x) = (x + 3)(x - 1)$$

$$H(w) = \frac{w^2 - 9}{w - 3} \qquad\qquad R(y) = (y - 7) + (-2 - y)$$

$$K(b) = \frac{1}{b}(6b + b^2) \qquad\qquad S(r) = r^2 + (r^2 + 5)$$

$$T(n) = (4n - 8)(-0.25) \qquad\qquad Z(v) = \frac{2.14v^3 + 3.1v^2}{2v}$$

$$Q(e) = \frac{1}{2}(2e + 10)e \qquad\qquad P(m) = 7(m - 1) + 4m$$

    a. Which of these functions are linear?

    b. Which of these function rules are quadratic?

    c. Which, if any, of these rules is equivalent to a linear function almost everywhere? Equivalent to a quadratic function almost everywhere?

    d. With the help of the properties of operations, write each of these function rules in simplest form.

    e. How do your results in part d confirm or contradict your responses in parts a, b, and c?

3. Consider the following questions about equivalence of expressions.

    i.   Are $-3(a + 6)$ and $-3a + 6$ equivalent expressions?

    ii. Is $c + 1$ equivalent to $\frac{(c+1)(c-1)}{c-1}$?

    iii. Are $3b + 12$ and $(b + 4) \cdot 3$ equivalent?

  a. For which of these three questions is the use of tables most effective? Explain your choice.

  b. For which of these three questions is the use of graphs most effective? Explain your choice.

  c. For which of these three questions is the use of properties of operations most effective? Explain your choice.

  d. Summarize your answers to parts a, b, and c.

4. For each of the following functions, make—without computer assistance—a sketch of its graph, showing the overall pattern. Justify each answer.

  a. $f(x) = \dfrac{(x - 4)(x + 5)}{x + 5}$
             b. $q(x) = \dfrac{4x^2}{x}$

  c. $z(x) = 1210 - \dfrac{2000}{x}$

5. **SITUATION**   In 1990, construction began to expand Beaver Stadium, Penn State University's football stadium. Several fans suggested that it should be covered by a dome. One way to help the building contractors determine the suitability of adding a dome on the stadium is to use mathematics to describe the shape of the dome. The diagram shown here, representing a vertical cross-section of the stadium, shows two measurements that should be used as constraints. This drawing may suggest that a quadratic function might be an approximation to the relation expressing the height of the dome above the top of the walls in terms of the horizontal distance away from the south wall (point S in the diagram).

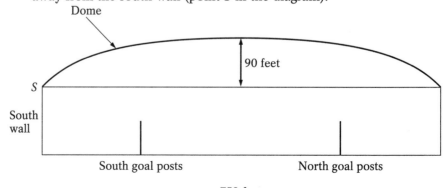

    a. Letting d feet be the horizontal distance from the south wall, think about the cross-section of the dome as part of the graph of a quadratic function with the origin at point S. Sketch this graph.

    b. What are the zeroes of this function?

    c. What is the maximum value of this function? At what distance d does the maximum occur?

    d. Write a function rule to represent this function in a form that gives information about its zeroes.

    e. Write the expanded form of your function rule. What information does this form give about the situation?

    f. The goal posts are approximately 182 and 542 feet from the south wall. Suppose you were standing beside the south goal posts looking directly above you at the dome. According to your function rule, how much higher than the top of the wall is the dome where you are? What if you were standing beside the north goal posts?

6. **SITUATION** The E-Z Clean company has advertised a new type of window, claiming "With the *N-Sol 8* window you can reduce your heating bill by 10%!" One customer reading the ad exclaimed, "If I use ten of these windows in my house, my bill will be nothing!"

    a. The customer believes that the heating bill for a house is a function of the number of special windows used in the house. Write a function rule that represents "10% off your heating bill for each window used". Do you think this is what the E-Z Clean company meant in the advertisement?

    b. According to your function rule, how much lower is the heating bill when a house has ten special windows than when it has no special windows? Would the customer's bill be "nothing"?

    c. Compare your function rule with the rules written by your classmates. Which of your rules are equivalent?

7. There is a "shortcut" for squaring numbers that end in 5, such as 45 or 95. The shortcut goes like this:

    (1) Drop the ending 5. (That is, subtract 5 and divide by 10). Example: 45 becomes 4.

    (2) Add 1 to the result of Step 1. $4 + 1 = 5$.

    (3) Multiply the results of Steps 1 and 2. $4 \times 5 = 20$.

(4) Multiply the result of Step 3 by 100. $20 \times 100 = 2000$.
(5) Add 25 to the last product. $2000 + 25 = 2025$.

You can easily verify that 45^2 does equal 2025!
a.   Use this shortcut to find the value of 95^2 and $10\ 005^2$.

Why does this shortcut work? We can verify its validity for all integers that end in 5 through the use of equivalent expressions. We start by representing any number ending in 5 as $10a + 5$, where a is an integer.
b.   What is the value of a when the number to be squared is 45? When the number is 95? When the number is 10 005?
c.   Represent the result of each of the five steps of the shortcut by writing its expression in terms of a.
d.   If the number ending in 5 is $10a + 5$, then the number squared is $(10a + 5)^2$. Show that the expression $(10a + 5)^2$ is equivalent to the expression written as Step 5 in part c.

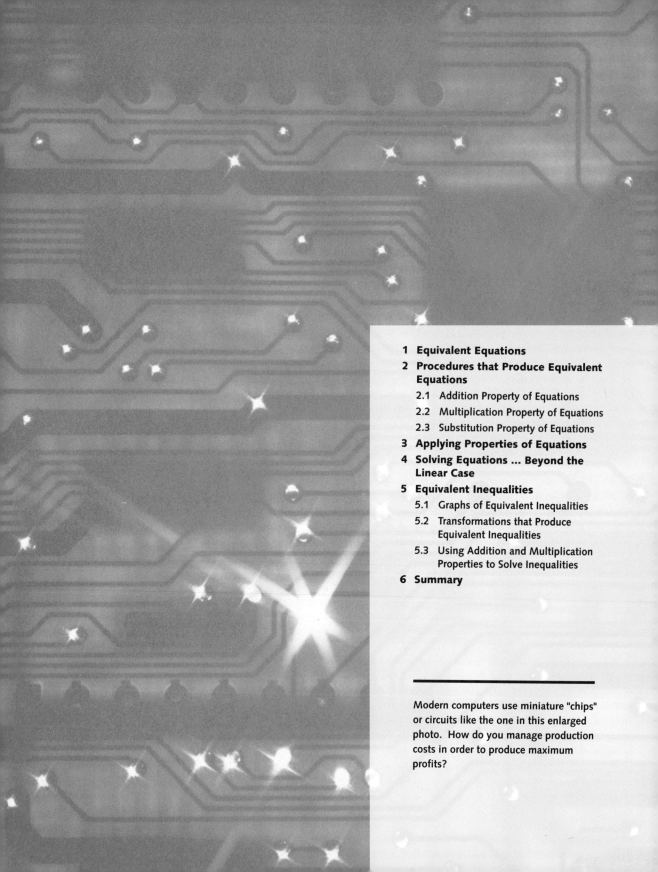

Modern computers use miniature "chips" or circuits like the one in this enlarged photo. How do you manage production costs in order to produce maximum profits?

9 Symbolic Reasoning: Equations and Inequalities

In Chapter 8 you learned that for many functions the data pairs, (input, output), can be calculated by several different but equivalent rules. For example, the quadratic expressions

$$x^2 - 21x + 98 \text{ and } (x - 7)(x - 14)$$

are **equivalent** because for every input x they produce the same output.

Some other pairs of algebraic expressions are **equivalent almost everywhere**; they give the same output for almost all input values. The expressions

$$\frac{x^2 - 21x + 98}{x - 14} \text{ and } x - 7,$$

for example, yield the same output for each input except $x = 14$.

For some pairs of algebraic expressions only a few inputs give the same outputs from both expressions. The diagram and table below suggest that the expressions $7x^2 + 6x$ and $x^2 + 36$ share only the data pairs $(-3, 45)$ and $(2, 40)$.

x	$f(x)$	$g(x)$
−5	145	61
−4	88	52
−3	45	45
−2	16	40
−1	1	37
0	0	36
1	13	37
2	40	40
3	81	45
4	136	52
5	205	61

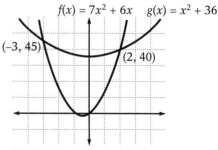

$f(x) = 7x^2 + 6x \quad g(x) = x^2 + 36$

The horizontal (input) scale unit is 1 and the vertical (output) scale unit is 10.

Finding the data pairs, (input, output), shared by the two algebraic expressions is the same as solving the equation $7x^2 + 6x = x^2 + 36$. You know how to use tables, graphs, and computer-algebra programs to find the roots for such equations.

In this chapter you will learn the principles behind procedures used by computer-algebra programs. Then you will be able to solve many of the simpler equations and inequalities without relying on this technology.

1 Equivalent Equations

Some equations are so simple you can solve them with only a little careful thought, but for many equations the roots are not at all obvious. In those cases, you can use your knowledge of algebra to reduce the given equation to an equivalent simpler equation.

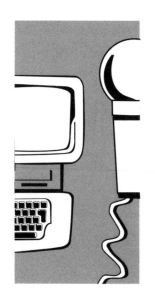

SITUATION 1.1

A computer parts manufacturer has created a device that allows computer users to enter words and numbers by speaking to the computer, rather than by typing at a keyboard. A market research study suggests that revenue for the product, in dollars, is expected to be a function r of the number sold n with rule $r(n) = -0.08n^2 + 600n$. Projected cost (in dollars) for the design and manufacture of the device is also a function c of the number sold with rule $c(n) = 200n + 150\,000$.

The relation between revenue and cost for production and sale of this new voice-input device is shown in the following diagram. Depending on the value of n, the revenue can be greater than, equal to, or less than the cost.

One of the critical tasks for this situation is determining the break-even points—the number sold for which revenue and cost are equal. This means solving the equation

$$r(n) = c(n) \text{ or } -0.08n^2 + 600n = 200n + 150\,000.$$

The diagram showing the graphs of $r(n)$ and $c(n)$ suggests break-even points near 500 and 4500. Using equation-solving technology such as a SOLVE command, you can find more accurate integer values $n = 408$ and $n = 4592$.

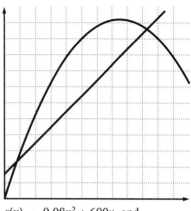

$r(n) = -0.08n^2 + 600n$ and
$c(n) = 200n + 150\ 000$

The horizontal (input) scale unit is 500 and the vertical (output) scale unit is 100,000.

You may remember from earlier work on similar problems that there is another way to think about the meaning of a break-even point. The revenue and cost functions can be combined to give a single function, called profit, with rule

$$p(n) = r(n) - c(n), \text{ or}$$
$$p(n) = -0.08n^2 + 400n - 150\ 000.$$

The graph for this function is shown here. Breaking even occurs when the profit is zero, which corresponds to the points where the graph crosses the horizontal axis. The diagram shows two such points, whose respective n-coordinates are near 500 and 4500. Your computer-algebra program can give you the integer values $n = 408$ and $n = 4592$. These are the same roots as those for the equation $r(n) = c(n)$.

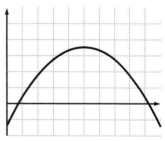

$p(n) = -0.08n^2 + 400n - 150\ 000$

The horizontal (input) scale unit is 500 and the vertical (output) scale unit is 100,000.

Because the two equations in this situation

$$-0.08n^2 + 600n = 200n + 150\ 000 \text{ and}$$
$$-0.08n^2 + 400n - 150\ 000 = 0$$

have identical solutions, they are called **equivalent equations.**

The equivalence can be portrayed graphically by plotting the graphs of all three functions on a single diagram (see the following). Notice that the values of n at the points where $r(n) = c(n)$ are the same as the two values of n where $p(n) = 0$.

Note: In some situations, an equation may have solutions that are not real numbers. These solutions will not be shown by a graph such as the one here. In this text, you will encounter only equations in which all solutions are real numbers.

You may decide that the second of the two break-even equations seems considerably more familiar and perhaps easier to solve than the first. The next exploration should help you discover some of the steps that enable you to replace an equation with a simpler equivalent equation.

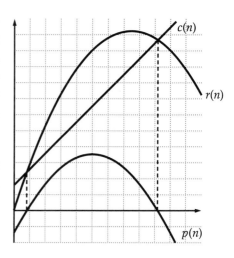

The horizontal (input) scale unit is 500 and the vertical (output) scale unit is 100,000.

 ## Exploration

The equations $x^2 = 2x + 3$ and $3 - x^2 = -2x$ are equivalent. You can verify this fact by using a computer-algebra SOLVE command to produce the roots $x = 3$ and $x = -1$ in each case or by using a graphing utility. To examine graphical evidence for the equivalence, it helps to produce two diagrams, one for each equation, as shown here.

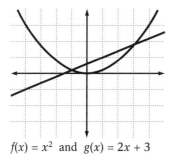

$f(x) = x^2$ and $g(x) = 2x + 3$

The horizontal (input) scale unit is 1 and the vertical (output) scale unit is 5.

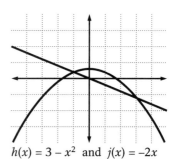

$h(x) = 3 - x^2$ and $j(x) = -2x$

The horizontal (input) scale unit is 1 and the vertical (output) scale unit is 5.

The two diagrams involve quite different function graphs. However, the input values at the points of intersection of the graphs are the same in both diagrams. This shows that the numbers x for which $x^2 = 2x + 3$ are the same as those x for which $3 - x^2 = -2x$. The two equations are equivalent.

Use your symbol manipulation and graphing programs to test equivalence of equations in the questions that follow.

1. Determine whether the following pairs of equations are equivalent. Justify each conclusion in the following ways. For each technique, explain why you believe the equations are or are not equivalent.

   i. Find the roots using pen-and-paper techniques or using a computer-algebra SOLVE command.

   ii. Sketch diagrams like those above.

   a. $(1/3)x + 5 = x^2 + 6x + 3$ and $(1/3)x + 5 = x^2 - 6x + 3$
   b. $(1/3)(x - 5)^2 - 3 = x + 9$ and $x^2 - 10x + 16 = 3x + 27$
   c. $2x + 1 = -3x + 6$ and $6x + 3 = -9x + 18$
   d. $x^2 - 22x + 112 = 14 - x$ and $x^2 - 2x - 2 = 4 - x$

2. Equivalent equations sometimes occur in different mathematical models that are related to a single situation, as illustrated by the following.

SITUATION 1.2

The freshman class plans to sponsor a Spring Break trip. The class officers believe that $250 is a reasonable price to charge. Three different travel agents offer the following cost proposals.

1. $200 per student going on the trip, plus a $600 service charge for making all reservations.
2. $225 per student going on the trip, plus a $100 service charge for making all reservations.
3. $230 per student going on the trip, with no additional charge for making reservations.

Use this information to answer the following.

   a. For each of the agency offers, write a rule giving the freshman class profit as a function of the number of students n going on the trip.

    b. Suppose the class officers set a goal of making \$400 profit for the class treasury. For each travel agency, write and solve an equation for which the root corresponds to the number of students required to produce exactly this profit under the agent's proposal.

    c. Are all three equations equivalent to each other? Explain your reasoning.

    d. Copy and complete the diagram shown here, which represents the three profit equations. Explain how the completed diagram shows the equivalence of the equations you discussed in part c.

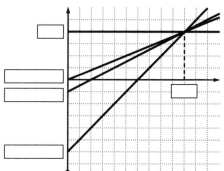

3. Any two equations that have exactly the same roots are equivalent to each other. When you are asked to solve an equation, it is often helpful to replace it with a simpler equation that is equivalent to the given equation. The trick is choosing that simpler and equivalent equation.

    a. Use technological tools or mathematical reasoning to solve the following equations. As you proceed through the list, keep a lookout for any pairs of equations that are equivalent. Try to see why those pairs are equivalent.

    i. $x^2 + x - 6 = 0$ **ii.** $3x = 6$

    iii. $\dfrac{7x}{3} + 5 = \dfrac{x + 24}{3}$ **iv.** $x^3 + 18 = 9x + 2x^2$

    v. $2x = 3$ **vi.** $(x - 3)(x + 2) = 0$

    vii. $\dfrac{-16x}{3} + 12 = 4$ **viii.** $168 - 28x^2 = 28x$

    ix. $(x - 2)(x + 3) = 0$ **x.** $(7x + 21)(4x - 8) = 0$

    xi. $2.25x + 5 = 0.25x + 8$ **xii.** $\dfrac{14x^2 + 14x}{3} + 2 = 30$

    xiii. $x^2 + x = 6$ **xiv.** $x = 9x - 12$

    xv. $168 + 28x = 28x^2$ **xvi.** $5x^2 + 5x = 30$

    xvii. $3x - 6 = 0$

b. Next list the sets of equivalent equations you found in part a. Note that some lists may contain several equations. The first entry in one list is given for you.
1. $3x = 6$ is equivalent to....

Making Connections

Examine each set of equivalent equations according to your lists in Exercise 3b. Describe similarities that might help explain why the equations in a list are equivalent.

Describe differences among equations in different lists that might help explain why these equations are not equivalent.

Exercises

For each pair of equations given in Exercises 1 through 8, use what you know about equivalent equations, algebraic expressions, and functions to determine whether the given equations are equivalent. In each case, explain your reasoning.

1. $3x + 11 = 23$ and $4x + 7 = 23$

2. $4.5x = 11.25$ and $11.25x = 4.5$

3. $3x - 11 = 23$ and $3x + 11 = 23$

4. $x^2 - 36 = 0$ and $x = 6$

5. $4x - 24 = 36$ and $\frac{x+4}{3} - \frac{x}{4} = \frac{76-3x}{12}$

6. $(x + 3)(x - 4) = 0$ and $2x = 8$

7. $x^2 - 5x = 14$ and $(x - 2)(x + 7) = 0$

8. $5x = 10$ and $4x + 5 = -3$

9. Following are two diagrams; the diagram at the left shows the solution of the equation $f(x) = g(x)$, while the right-hand diagram shows the solution of $h(x) = j(x)$. Explain how the diagrams do or do not

suggest that the two equations are equivalent. In both diagrams, the horizontal (input) unit scale is 1 and the vertical (output) unit scale is 5.

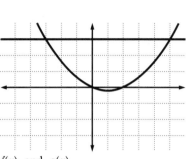

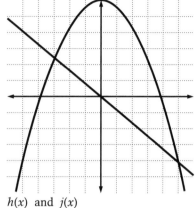

$f(x)$ and $g(x)$ $h(x)$ and $j(x)$

10. Examine the following table of inputs and outputs for the four functions $k(x)$, $m(x)$, $r(x)$, and $s(x)$. What can you conclude from the table about whether the equations $k(x) = m(x)$ and $r(x) = s(x)$ are equivalent to each other? Explain your conclusions.

x	$k(x)$	$m(x)$	$r(x)$	$s(x)$
−5	24	12	−24	−12
−4	15	10	−15	−10
−3	8	8	−8	−8
−2	3	6	−3	−6
−1	0	4	0	−4
0	−1	2	1	−2
1	0	4	0	0
2	3	6	−3	2
3	8	8	−8	4
4	15	10	−15	6
5	24	12	−24	8

2 Procedures that Produce Equivalent Equations

You have studied several methods for solving algebraic equations—by using numerical, graphical, and symbol-manipulation programs. Sometimes an equation may be solved rather quickly without these tools. One effective strategy for solving a given equation is to find an equivalent equation whose roots are easier to see.

For example, in the last section you may have seen that the linear equations

$$2x = 3, \frac{-16x}{3} + 12 = 4, \text{ and } 2.25x + 5 = 0.25x + 8$$

are all equivalent to each other. Each equation has the single root $x = 1.5$; however, you probably agree that

- –it is easier to see that $x = 1.5$ (or $x = \frac{3}{2}$) is the root for the equation $2x = 3$ than to see that it is the root for $\frac{-16x}{3} + 12 = 4$; and
- –it may be easier to recognize that 1.5 is the root for $\frac{-16x}{3} + 12 = 4$ than to see that it is the root for $2.25x + 5 = 0.25x + 8$.

The material presented in this section enables you to convert complex equations into simpler, equivalent equations whose roots may be seen more easily. Your experience with this technique for solving equations can provide you a wider variety of equation-solving methods.

2.1   Addition Property of Equations

We first focus on one of the simplest procedures for converting an equation into a more convenient equivalent equation.

SITUATION 2.1

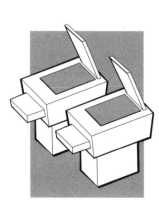

An office manager is deciding between two options to take care of the copying needs of the office.
- –Ace Copiers, the first company contacted, offers the option of leasing a copy machine for a fixed weekly fee of $50 and an additional charge of 2.1 cents ($0.021) per copy.

–For the same machine and comparable service, a second company, Lightning Printers, offers a fixed charge of $180 per week with an additional charge of 0.5 cents ($0.005) per copy.

The manager is interested in knowing which offer is more economical for the office.

Letting the variable n represent the number of copies the office requires during a week, the manager writes function rules for the two companies.

$$\text{Ace Copiers:} \qquad f(n) = 0.021n + 50$$
$$\text{Lightning Printers:} \quad g(n) = 0.005n + 180$$

Noting that the graphs of these linear functions are straight lines with different slopes (namely, 0.021 and 0.005), the manager realizes that the graphs intersect. One company offer is better for some values of n, but worse for others. It is necessary to identify the intersection point of the graphs. The office requires several thousand copies per week, so the manager chooses a suitable scale and generates the diagram shown here.

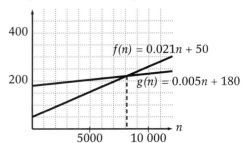

The diagram provides a graphical solution of the equation $f(n) = g(n)$. The value of n at the point where the two lines intersect is the root of the equation

$$0.021n + 50 = 0.005n + 180.$$

The critical value of n is approximately 8000. For further evidence consider the small table of values that follows.

n	$0.021n + 50$	$0.005n + 180$
6000	176	210
7000	197	215
8000	218	220
9000	239	225
10 000	260	230

The colored lines in the table confirm that the root of the equation is just over 8000. Using the SOLVE command on the office computer, the manager finds the exact root is 8125.

Inspecting the diagram and knowing the exact root, the manager concludes that Ace Copiers has the better deal when the number of copies per week required by the office is fewer than 8125, but Lightning Printers is better when n is more than 8125.

SITUATION 2.2

To entice customers during the summer season, Ace Copiers decides to eliminate its fixed charge of $50 per week. According to its advertisements, customers pay only for the copies they make! When Lightning Printers learns about the impending change, it immediately enters the price war by reducing its fixed charge also by $50.
The office manager, having analyzed the original fee schedule, wants to know how these reductions affect the relative advantages of the deals available from the two companies.

The manager adjusts the old function rules for the new charging policies of the two companies.

$$\text{Ace Copiers:} \quad h(n) = f(n) - 50$$
$$\text{Lightning Printers:} \quad j(n) = g(n) - 50$$

In order to know which new fee is better for the office, begin by finding the root of the relevant equation $h(n) = j(n)$, or $f(n) - 50 = g(n) - 50$.
How is the root of this equation related to the root of the earlier equation $f(n) = g(n)$? In other words, what happens to the solution when the manager subtracts 50 from (or adds −50 to) each side of an equation? The new equation

$$(0.021n + 50) + (-50) = (0.005n + 180) + (-50)$$

simplifies to

$$0.021n = 0.005n + 130.$$

A new table provides an estimate of the root of the new equation.

n	$0.021n + 50$	$0.005n + 180$
6000	126	160
7000	147	165
8000	168	170
9000	189	175
10 000	210	180

According to the new table, although the respective outputs for each input n are different from those in the previous table, the root of the equation again seems to be approximately 8000.

In order to compare graphically the new and the old pricing arrangements, consider the diagram shown here. This diagram shows that the intersection points depicting the solutions of the equations

$$0.021n + 50 = 0.005n + 180 \text{ and}$$
$$0.021n = 0.005n + 130$$

$f(n) = 0.021n + 50$
$g(n) = 0.005n + 180$
$h(n) = 0.021n$
$j(n) = 0.005n + 130$

appear to lie on the same vertical line.

Using a symbol manipulation program to solve $0.021n = 0.005n + 130$, the manager finds the root is $n = 8125$. This example suggests that adding the same number to both sides of an equation does not change the solution. In this case the second equation has fewer terms than the first and seems closer to an equation whose root might be more easily recognizable.

Do you think that *any* expression can be added to (or subtracted from) each side of *any* equation without changing its roots? In the following example, we analyze the effect of adding the same polynomial expression to both sides of an equation. Watch for the relation between the root of the original equation and the root of the new equation.

SITUATION 2.3

Instead of the first alternate pricing arrangement described in Situation 2.2, Lightning Printers is proposing to retain its fixed fee but eliminate its charge per copy. Ace Copiers decides that it cannot eliminate its charge-per-copy completely, but can at least match Lightning Printers by reducing its charge by 0.5 cents per copy. The office manager, still committed to choosing the better bargain, wonders how these reductions affect the number of copies for which the two offers are equal in cost.

In this second alternate situation the relevant equation is

$$(0.021n + 50) + (-0.005n) = (0.005n + 180) + (-0.005n)$$

or, more simply,

$$0.016n + 50 = 180.$$

The table below gives values for the expression $0.016n + 50$.

n	$0.016n + 50$
6000	146
7000	162
8000	178
9000	194
10 000	210

From the table we see that the output is about 180 when the input is about 8000. In other words, although the outputs are different from those in the earlier tables, the root of this new equation seems about the same, a number near 8000.

The effect on the graph of adding $-0.005n$ to each side of the original equation is shown here.

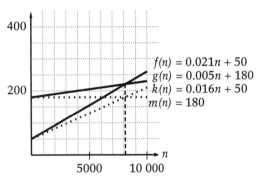

$f(n) = 0.021n + 50$
$g(n) = 0.005n + 180$
$k(n) = 0.016n + 50$
$m(n) = 180$

Once again, the solutions to the two equations appear to lie on the same vertical line. With a SOLVE command we verify that the same number $n = 8125$ is the root of the new equation $0.016n + 50 = 180$.

In this example, adding the same polynomial expression to both sides of an equation yields an equation that is equivalent to the original. Moreover, the resulting equation seems simpler than the first, and so we seem closer to a goal of working with an equation whose root is more recognizable.

Exploration I

A reasonable conjecture from the previous examples might be that adding the same number or the same polynomial expression to both sides of any equation gives an equivalent equation. However, you have tested only a few cases. Use your technological tools to explore the effect of adding other numbers or algebraic expressions to other types of equations. Test the given examples by doing the following.

    a. Find the roots of the given equation.

    b. Add the indicated number or expression to both sides of the equation. For tests 3 through 5, choose your own number or expression.

        i. Write the new equation.

        ii. Find the roots of the new equation.

    c. Compare your answers to parts a and b.ii. Are the two equations equivalent? Explain.

1. Equation: $x^2 + 5x - 3 = -3$
   Number to be added to both sides: 3

2. Equation: $x^2 + 5x - 14 = 5(x - 1)$.
   Expression to be added to both sides: $-5(x - 1)$

3. Equation: $7.3x + 11.9 = 84.9$
   Number or expression added to both sides: Choose your own.

4. Equation: $8x + 6 = 16 + 3x$
   Number or expression added to both sides: Choose your own.

5. Equation: $x^2 + 12 = 28$
   Number or expression added to both sides: Choose your own.

Now complete the following statement summarizing your observations in the preceding tests.

Addition Property of Equations: Adding the same number or polynomial expression to both sides of an equation....

    Be prepared to discuss with your classmates and your teacher why this is a reasonable property of equations!

2.2   Multiplication Property of Equations

In the previous subsection, you developed an addition property of equations. Usually, your goal in applying this procedure is to transform an equation into another that has the same solution but is easier to solve. This subsection describes another procedure useful for reaching the same goal.

Exploration II

The following experiments examine the effect of multiplying each side of an equation by the same number or polynomial expression.

Experiment 1. Investigate the effect of multiplication on the solution of

$$2x + 12 = 4x + 4.$$

You should explore both the graphical and numerical effects of this type of transformation.

Test 1. To get started, multiply each side of the equation by 2 and look for any change in the root.

    a. Copy and complete the following table of values using input values of your own choice. Complete seven rows in your table. Then record any results or patterns that you observe.

x	$2x + 12$	$4x + 4$	$2(2x + 12)$	$2(4x + 4)$

    b. Now investigate the graphical effect that multiplying by 2 has on the solution of the equation $2x + 12 = 4x + 4$. First create a diagram illustrating the solution of $2x + 12 = 4x + 4$. Then, on the same coordinate grid, superimpose the graphs that illustrate the solution of $2(2x + 12) = 2(4x + 4)$. Sketch the entire diagram on your paper, labeling each function and the key intersection points of the graphs.

    c. Using the table and the diagram you have just constructed, make a conjecture about the relation between the solutions of the equations

$$2x + 12 = 4x + 4 \text{ and } 2(2x + 12) = 2(4x + 4).$$

    d. Explain how the table and the graphs justify your answer.

Test 2. Are your conclusions true for multipliers that are fractions? In this test, you will multiply both sides of the equation by a positive number that is less than 1. Investigate the effect both on tables and on graphs.

    a. Copy and complete the following table of values using input values of your own choice. Then explain what you observe about the effect of multiplying by $\frac{1}{2}$.

x	$(1/2)(2x + 12)$	$(1/2)(4x + 4)$

b. Investigate how multiplication by $\frac{1}{2}$ influences the graph of the solution of the equation $2x + 12 = 4x + 4$. First sketch a diagram illustrating the solution of the given equation. Then, on the same diagram, superimpose the graphs that illustrate the solution of $\frac{1}{2}(2x + 12) = \frac{1}{2}(4x + 4)$.

c. Using the diagram you have just constructed, make a conjecture about the relation between the solutions of the equations

$$2x + 12 = 4x + 4 \text{ and } \frac{1}{2}(2x + 12) = \frac{1}{2}(4x + 4).$$

d. Explain how the diagram justifies your answer.

Test 3. What is the effect of multiplying each side of the equation $2x + 12 = 4x + 4$ by a negative number? Examine the effects on both graphs and tables by choosing two negative numbers of your own. For each number, record:

i. the negative number by which each side is multiplied.

ii. the new equation after multiplying.

iii. the roots of the new equation.

iv. any conclusions you can make.

On the basis of your experiments, complete the following conjecture about the effects of multiplying both sides of an equation by the same positive or negative number.

Multiplication Property of Equations: Multiplying both sides of an equation by the same *non-zero* number....

Be prepared to discuss your reasoning with your classmates.

Experiment 2. The previous experiment tested the effect of multiplying both sides of a linear equation by a positive or negative number. Is the effect the same for a quadratic equation?

Test 4. Investigate the effect of multiplication on the solution of

$$x^2 + 2x = 5 - 2x.$$

Begin by finding the roots of this equation. Then select at least four different numbers to use as multipliers, being sure to choose at least one negative and at least one positive multiplier. For each number, record the following:

      i. the number by which each side is multiplied.
     ii. the new equation after multiplying.
   iii. the roots of the new equation.
   iv. any conclusions you can make.

Using one of your test cases as an illustration, show the effect of multiplication by a negative number on graphs that demonstrate the solution.

Experiment 3. The previous examples tested the effect of multiplying both sides of an equation by a specific number. What if the multiplier is a polynomial expression involving a variable? Do you believe the new equation is equivalent to the original in all cases? Complete the following tests. Conduct additional tests using equations and multipliers that you think are interesting.

Test 5.
   a. Find the roots of the equation $5x + 7 = 3x + 4$.
   b. Multiply each side of the equation by x and record the new equation.
   c. Find the roots of the new equation.

Test 6.
   a. Find the roots of the equation $2(x - 3) = 0$.
   b. Multiply each side of the equation by $x + 2$ and record the new equation.
   c. Find the roots of the new equation.

Test 7. Choose an equation and multiplier of your own. Find the roots of the equation before and after multiplying.

Examine your answers to your tests. What conclusions can you make?

Making Connections

You may have noticed a variety of interesting results in the graphs and tables of equations transformed by multiplication. Among these results, one of the most important is the effect of the multiplication transformation on the solution of an equation.

1. When both sides of an equation are multiplied by the same non-zero number, how are the solutions of the original equation and the transformed equation related?

2. The multiplication property of equations that you conjectured excluded multiplication by zero. How does multiplication by zero affect the solution to an equation?

3. Suggest a reasonable division property of equations.

4. If both sides of an equation are multiplied by a polynomial expression involving a variable, how is the solution of the transformed equation related to that of the original equation?

Exercises

1. Each of the following is a pair of equivalent equations, the second of which results from the first by a single application of the addition or the multiplication property of equations. For each pair, identify the number or algebraic expression along with the operation of addition or multiplication.
   a. $3x - 7 = 8$ and $3x = 15$
   b. $4 - w = 6 - 2w$ and $4 + w = 6$
   c. $36 + 0.1z = 0.45z + 40$ and $3600 + 10z = 45z + 4000$
   d. $5t - 3(5 + t) = 9t + 18$ and $-3(5 + t) = 4t + 18$

2. When using the addition and multiplication properties of equations, it is important to be able to select properties that produce equivalent equations that are "simpler" than the equations with which you start. For each of the following equations:
   i. tell which property of equations you might use first, and
   ii. tell why your choice produces a simpler equivalent equation.

   a. $4.5h - 6 = 15$ b. $2.5g = 8$
   c. $3.5x + 7 = 7x$ d. $0.5y + 5 = 3 - 2y$
   e. $0.25d = 13$ f. $2x = x - 6$
   g. $3(a + 2) = 6a - 9$ h. $0.5(4r - 3) = 4(3r - 12)$

2.3   Substitution Property of Equations

In Chapter 8, you studied many cases in which the algebraic rule for a function can be replaced by another equivalent expression without changing the data pairs, (input, output), for the function. It seems reasonable to wonder whether a similar kind of substitution can be made in an equation (or inequality), without changing the solution.

Exploration III

In each of the following pairs of equations, an expression on one side of the first equation has been replaced by a different expression to produce the second equation. For each pair:

    a.   identify the substitution (tell which expression has been replaced by which other expression);

    b.   determine whether the equations are equivalent by finding their respective solutions; and

    c.   tell why it seems reasonable that the substitution does, or does not, lead to a second equation that is equivalent to the first equation.

Test 1. $2(f + 1.5) = 36$ and $2f + 3 = 36$

Test 2. $8 + 3m + 21m - 4 = 35 + 56m$ and $8 + 20m = 35 + 56m$

Test 3. $3(2x + 5) - 2x = 8x - 6$ and $6x + 5 - 2x = 8x - 6$

Test 4. $9(6 - 3m) = 8m - 8$ and $54 - 27m = 8m - 8$

Test 5. $(d + 4)(d - 4) = 0$ and $d^2 - 16 = 0$

Test 6. $49 + 2h = 4h + h^2 + 25$ and $49 + 2h = 4h + (h + 5)^2$

Test 7. $x^2 + 7x + 12 = 0$ and $(x + 3)(x + 4) = 0$

Test 8. $10 - 6.2h = -3(h - 5)$ and $10 - 6.2h = 15 - 3h$

Use your results to write a *Substitution Property of Equations*. Discuss your reasoning with your classmates.

3 Applying Properties of Equations

The addition, multiplication, and substitution properties of equations can help you to solve many problems without using computing technology. A crucial part of the process is deciding when and how to use these properties. This section gives some suggestions about when to apply the addition and multiplication properties and when to replace an expression by another.

Remember, in the examples that follow, the goal at each step is to choose a procedure yielding an equation that is less complicated, and closer to one with recognizable roots, than its predecessor.

Example 1. Use the properties of equations to solve $3f + 5 = f - 11$.

There are several reasonable first steps. Two of these possibilities are:

a. Addition property of equations—add the number –5.

 This seems like a legitimate first step since adding –5 to both sides of the equation eliminates the term 5 on the left side. Moreover, the number –5 can be combined with the term –11 on the right side, so that the resulting equation is less complicated than the original.

b. Addition property of equations—add the expression $-f$.

 It also seems reasonable to start by adding $-f$. Doing this eliminates the term f from the right side of the equation and permits a combining of terms on the left side of the equation.

For this example, we will use path a.

Sample solution path.

1. By the addition property (adding –5), we get
$$(3f + 5) + (-5) = (f - 11) + (-5).$$

2. The left expression $(3f + 5) + (-5)$ is equivalent to $3f + [5 + (-5)]$ and to $3f + 0$, while the right expression $(f - 11) + (-5)$ is equivalent to $f + [(-11) + (-5)]$ and to $f - 16$. So the substitution property gives us the equivalent equation $3f = f - 16$.

3. It now seems reasonable to eliminate the term f from the right side. We can do this by adding $-f$ to both sides. Applying the addition property, we get $(3f) + (-f) = (f - 16) + (-f)$.

4. Two applications of the substitution property (one on each side of the equation) yield the much simpler equivalent equation $2f = -16$.

5. Here the root is visible. That is, $f = -8$.

Example 2. Use the properties of equations to solve $\frac{-16x}{3} + 12 = 4$.

There are several reasonable first steps. Two of the possibilities are:

a. Addition property of equations—add the number -12.

   This seems an appropriate first step, because adding -12 to both sides of the equation eliminates the term 12 from the left side. On the right side the number 4 and the term -12 can be added; so the resulting equation is less complicated than the original.

b. Multiplication property of equations—multiply by the number 3.

   This is also a good choice, since multiplying by 3 and then simplifying eliminates the fraction on the left side of the equation.

For this example, we will demonstrate two different solution paths, one for each of the proposed first steps.

Sample solution path a.

1. By the addition property (adding the number -12), we write $(\frac{-16x}{3} + 12) + (-12) = (4) + (-12)$. On the right side the sum is -8. The sum on the left is equivalent to $(\frac{-16x}{3}) + [12 + (-12)]$ by an associative property and then to $(\frac{-16x}{3}) + 0$ by an inverse property. Substitution yields the equivalent equation $\frac{-16x}{3} = -8$

2. An equation without fractions is simpler. We next apply the multiplication property (multiplying by the number 3) to obtain $(\frac{-16x}{3}) \times (3) = (-8) \times (3)$. The right product is -24. The left product is equivalent to each of the following:

   $[(-16x) \times (1/3)] \times (3)$ by the division property,
   $(-16x) \times [(1/3) \times (3)]$ by an associative property, and
   $(-16x) \times (1)$ by an inverse property.

   Finally, the substitution property yields the equivalent equation $-16x = -24$.

3. Now we may divide by -16 in order to remove the coefficient of the variable x. Dividing each side of an equation by -16 is the same as multiplying by its reciprocal $\frac{-1}{16}$. So, by the multiplication property, we write $(-16x) \times (\frac{-1}{16}) = (-24) \times (\frac{-1}{16})$. By the substitution property (applied on each side of the equation), $x = \frac{3}{2}$

Because we have used only solution-preserving properties, each of the equations we have produced is equivalent to the others. That is, $x = \frac{3}{2}$ is the only root of each of the following four equations:

$$\frac{-16x}{3} + 12 = 4$$
$$\frac{-16x}{3} = -8$$
$$-16x = -24, \text{ and}$$
$$x = \frac{3}{2}.$$

Notice that generally any application of either the addition or the multiplication property is followed by use of the substitution property, in order to replace one or more expressions by less complicated expressions. Whenever you replace an expression by an equivalent one, you should be ready to justify your choice on the basis of your work in Chapter 8. Rather than continuing to write such details, we ask you to convince yourself that the substitutions are acceptable.

Sample solution path b.

In order to illustrate a different style of presentation, we will arrange the steps in column format below. (You may recall this style from our discussion of equivalent expressions in Chapter 8.) Any two successive equations printed on the left are equivalent to each other. The equivalence is based on one or more applications of the property listed at the right, on the line beside the second equation.

$\frac{-16x}{3} + 12 = 4$	
$(\frac{-16x}{3} + 12) \times 3 = (4) \times 3$	Multiplication
$-16x + 36 = 12$	Substitution
$(-16x + 36) + (-36) = (12) + (-36)$	Addition
$-16x = -24$	Substitution
$(-16x) \times (\frac{-1}{16}) = (-24) \times (\frac{-1}{16})$	Multiplication
$x = \frac{3}{2}$	Substitution

Of course, any solution path leads to the same root, namely, $\frac{3}{2}$.

Notice that repeated use of the substitution property enables us to simplify the various algebraic expressions that arise. As you gain experience, these simplifications should gradually become automatic for you—so automatic that occasionally you may not even bother to mention that you are applying the substitution property after an addition or a multiplication.

Example 3. Use the properties of equations to solve $\frac{m}{3} = \frac{3m-4}{2} - 12$.

A reasonable first step is to multiply by 6, since 6 is the least common denominator of the fractions in the equation; in this way we eliminate all the fractional expressions. Multiplying each side of the given equation by 6 (and of course simplifying), we get

$$2m = 9m - 84.$$

Next, subtracting $9m$ from both sides (and simplifying), we get

$$-7m = -84.$$

Finally, dividing by -7, we obtain the root of the equation,

$$m = 12.$$

As you can see from these examples, solving an equation often is the application of some reasonable sequence of properties of equations. Each choice of a property is made in order to obtain an equivalent equation whose solution may be more visible than that of the original. Often this sequence takes the form of:

1. selecting and applying the addition or the multiplication property;

2. substituting one or more appropriate equivalent expressions; and

3. assessing the result to see whether the roots are visible.

The diagram shown here outlines this process.

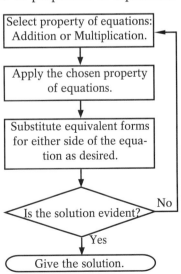

Exploration

The three examples given so far have suggested several strategies for solving equations without use of a SOLVE command. However, to become skillful in that process you should practice on other examples illustrating various possible situations. This exploration gives you an opportunity to study a variety of examples.

Before you begin to study the next example and the problems that follow, there are several points to note.

–We recommend (except in quite simple cases) use of a symbol manipulation program to calculate the results of addition, multiplication, or substitution as you proceed toward a solution.

This allows you to concentrate on the strategy rather than the mechanics of equation solving.

–Since different symbol manipulation programs may give different simplified forms, the results in our examples may not be identical to those you find; but they should be equivalent.

–Because we want you to discover solution strategies you could perform without using computing technology, we shall not use the one-step commands such as SOLVE available in most symbol manipulation programs.

1. The example below, already completed for you, demonstrates how you can record and report the steps of your solution procedure. Remember, in this exploration *do not use a one-step command such as SOLVE.*

    Solve $4x - 6 = x$.

Operation	Output
Enter $4x - 6 = x$	$4x - 6 = x$
Add $-x$	$4x-6+(-x) = x+(-x)$
Simplify	$3x - 6 = 0$
Add 6	$3x - 6 + 6 = 0 + 6$
Simplify	$3x = 6$
Multiply by $\frac{1}{3}$	$3x \left(\frac{1}{3}\right) = 6\left(\frac{1}{3}\right)$
Simplify	$x = 2$

    The following graphs show how you can verify that any two equations in a solution path are equivalent. In this case the diagram compares the equations $3x = 6$ and $3x - 6 = 0$. You do not need to produce graphs like this for all the equations in your solution paths for the problems that follow, but it might be interesting and helpful to check several pairs of equations in each.

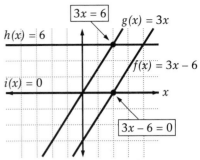

Now solve the following six equations using the properties of equations and the strategies demonstrated in earlier examples.

The horizontal (input) scale unit is 1 and the vertical (output) scale unit is 2.

a.   Solve $x - 7.006 = 0.5x - 6.888$.

b.   Solve $3.7b + 7 = 7.2b$.

c.   Solve $4.25y + 213 = 24\,000 - 0.25y$.

d.   Solve $\frac{2a}{4} + 3 = 0.25(3a + 2)$.

e.   Solve $(7x)/3 + 5 = (x + 24)/3$.

f.   Solve $\frac{9x}{4} - 11 = \frac{3x - 96}{12} - 6$.

2.   For most equations there are several possible solution paths. For part a and part b, do the following.

   i.   Solve along each of two different paths; and

   ii.   Compare the two solution paths. Which one, if either, is shorter? Which is easier?

a.   Solve $5x + 17 = 29 - 3x$.

b.   Solve $\frac{3r}{4} + 6 = 2r - 1$.

3.   Select two equations from either path for the solution of $5x + 17 = 29 - 3x$ in Exercise 2a. Use a graphing utility to make a diagram showing how the two equations are equivalent. Sketch the diagram and label all the significant features.

4.   As you gain more experience in solving equations by using the properties of equations, you should be developing strategies for finding efficient solution paths. Get together with a pair or group of students and compare the solution paths each of you followed for each equation in Exercise 1. Discuss your strategies with regard to how easy they are to remember, how easy they are to apply, and how much they contribute toward finding solutions in an efficient manner. After your discussion, write down a few guidelines you may want to remember in solving equations of this sort.

Exercises

1. In solving an equation, it is crucial to decide which property of equations to use at which time. For each of the following equations, identify one or more reasonable first choices for solving the equation. Explain why each choice is a reasonable first step. You do not need to solve the equation.

   a. $6.5a = 52$

   b. $43 - 5x = 15x + 3$

   c. $5.5a + 9 = 130$

   d. $13d + 187 = -73$

   e. $4(3f - 4) + 10 = -40$

2. The solutions of the following equations are represented in the diagrams shown here. Copy the diagrams on your paper. The horizontal (input) scale units and vertical (output) scale units are 1 in both diagrams.

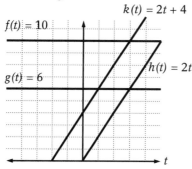

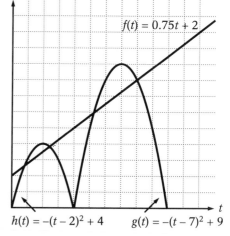

   For each equation, label with the designated letter the points on the appropriate diagram that correspond to the roots of the equation.

   a. Label A: $0.75t + 2 = -(t - 7)^2 + 9$

   b. Label B: $-(t - 7)^2 + 9 = -(t - 2)^2 + 4$

   c. Label C: $-(t - 2)^2 + 4 = 0.75t + 2$

   d. Label D: $2t + 4 = 10$

   e. Label E: $2t = 6$

   f. Label F: $2t + 4 = 6$

3. Recall the results from Examples 1 through 3 at the beginning of this section. You saw that: $3f + 5 = f - 11$ has one root, -8; $\frac{-16x}{3} + 12 = 4$ has one root, $\frac{3}{2}$; and $\frac{m}{3} = \frac{3m-4}{2} - 12$ has one root, 12. You may

have recognized that each of these equations is formed by setting one linear expression equal to another. Since the intersection of two distinct lines has only one point (or perhaps none at all), it makes sense that there is only one root to each of these equations.

Explain why it is reasonable that there are two roots for the equation $168 - 28x^2 = 28x$ and only one root for $2.25x + 5 = 0.25x + 8$.

Making Connections

Throughout this section, you have worked on choosing appropriate steps in solving linear equations. Summarize the most promising strategies for selecting steps in a solution path for a linear equation. To help identify some of the important cases, statements of a few of the more common strategies are started for you.

1. If an equation contains one or more fractions, try to start by....

2. If an equation contains one or more expressions in parentheses....

3. If the variable appears on both sides of an equation....

4. State a strategy of your own.

4 Solving Equations ... Beyond the Linear Case

In Section 3, you examined the process of solving linear equations by applying three basic properties of equations. For equations that involve quadratic or higher degree functions, the same properties are helpful, but often we need another very important principle. The following examples illustrate the strategy of combining this new principle with the familiar properties in order to solve polynomial equations.

The first example shows how to solve a simple case involving a product of linear expressions. Then each succeeding example introduces a new part of the general procedure, to deal with more complex cases.

Example 1. Find the roots of the equation $(x - 5)(x + 6) = 0$.

First, notice that one side of this equation is the number 0 while the other side is a polynomial expressed as a product of factors. To solve the equation, we use a very important characteristic of the number zero.

Zero-Product Property

> The only way that a product of two (or more) numbers can be zero is if at least one of the numbers is zero.

So, $(x - 5)(x + 6) = 0$ only when $x - 5 = 0$ or when $x + 6 = 0$. The roots of these linear equations, respectively, are $x = 5$ and $x = -6$. As you easily can see, when the expression is evaluated at either 5 or –6, one of the factors becomes 0:

$$[(5) - 5][(5) + 6] = 0 \times 11 = 0, \text{ and } [(-6) - 5][(-6) + 6] = -11 \times 0 = 0.$$

This solution method is fairly easy to apply. Simply set each factor equal to 0 and solve the resulting equations. The diagram on the right illustrates this method.

> Solve the equation.
>
> Set factors equal to 0.

Of course, not all equations are presented in factored form. Sometimes, an equation in which one side is 0 and the other side is a polynomial, but not given as a product of factors, can be put into this form. Example 2 shows how to solve some such equations.

Example 2. Find the roots of the equation $2r^2 + 10.4r - 30.66 = 0$.

First, notice that one side of this equation is 0 while the other side is a polynomial expression not in factored form. Applying the substitution property, we *factor* the polynomial expression to obtain

$$2(r - 2.1)(r + 7.3) = 0.$$

Applying the zero-product property, we note that the product

$$2(r - 2.1)(r + 7.3)$$

can be equal to zero only if either

$$r - 2.1 = 0 \text{ or } r + 7.3 = 0.$$

Solving these two linear equations, we obtain the roots of the original equation $r = 2.1$ and $r = -7.3$.

Notice that although the polynomial on the left side of the original equation has three factors, 2, $r - 2.1$ and $r + 7.3$, only two of them can possibly be equal to 0.

This example has extended the procedure used in the first example, to include cases that require factoring a polynomial expression.

> Solve the equation.
>
> Set factors equal to 0.
>
> Factor the expression.

The method illustrated in Example 2 requires that one side of the given equation be the number 0. However, a polynomial equation that is not given in this form can, by using some familiar properties, be transformed into an equivalent equation of the desired type.

Example 3. Solve the equation $3t^2 + 2.8t - 5.66 = 2t^2 - 4.1t - 16$.

Unlike the equations discussed in Examples 1 and 2 above,

$$3t^2 + 2.8t - 5.66 = 2t^2 - 4.1t - 16$$

is not in the form of a polynomial expression equal to zero. Our first step is to transform the equation into this style. Using the addition property we subtract $2t^2 - 4.1t - 16$ to get

$$(3t^2 + 2.8t - 5.66) - (2t^2 - 4.1t - 16) =$$
$$(2t^2 - 4.1t - 16) - (2t^2 - 4.1t - 16).$$

Applying the substitution property gives

$$t^2 + 6.9t + 10.34 = 0.$$

Again using the substitution property, we factor the polynomial that is on the left side of the equation to get

$$(t + 4.7)(t + 2.2) = 0.$$

According to the zero-product property, the product $(t + 4.7)(t + 2.2)$ has the value 0 only if one of the factors $t + 4.7$ or $t + 2.2$ is zero; so the solution of our relatively complicated equation reduces to the problem of solving a pair of linear equations

$$t + 4.7 = 0 \text{ and } t + 2.2 = 0.$$

The roots of these two linear equations are also the roots of the original equation, namely, $t = -4.7$ and $t = -2.2$.

This example has extended the procedure to include finding an equivalent equation having 0 as one of its sides.

It is interesting to examine a graphical interpretation of the steps that we have been developing. Below is the solution of an equation with the steps illustrated both symbolically and graphically.

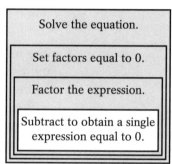

Solve the equation.

Set factors equal to 0.

Factor the expression.

Subtract to obtain a single expression equal to 0.

Example 4. Solve the equation $8a^2 - 20a - 40 = 3a^2 + 15a + 50$.

We begin symbolically by subtracting $3a^2 + 15a + 50$ from each side of the equation:

$$(8a^2 - 20a - 40) - (3a^2 + 15a + 50) =$$
$$(3a^2 + 15a + 50) - (3a^2 + 15a + 50).$$

This choice of a first step gives us (after we simplify each side of the equation) an equation with 0 on its right side:

$$5a^2 - 35a - 90 = 0.$$

We can see in the following graph (Figure 1) that the equations $8a^2 - 20a - 40 = 3a^2 + 15a + 50$ and $5a^2 - 35a - 90 = 0$ have the same roots.

As the second step in our solution path, we multiply symbolically by $\frac{1}{5}$ to get $(5a^2 - 35a - 90) \times \frac{1}{5} = (0) \times \frac{1}{5}$ or

$$a^2 - 7a - 18 = 0.$$

We see from Figure 2 that the roots of the two equations $5a^2 - 35a - 90 = 0$ and $a^2 - 7a - 18 = 0$ are the same.

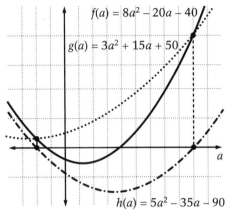
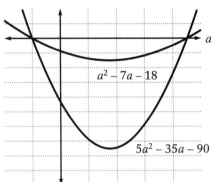

The horizontal (input) scale unit is 1 and the vertical (output) scale unit is 100.

Figure 1

The horizontal (input) scale unit is 2 and the vertical (output) scale unit is 20.

Figure 2

In order to finish solving the original equation, we factor $a^2 - 7a - 18$ then apply the zero-product property. The factored form is

$$(a - 9)(a + 2) = 0.$$

We can see from this last equation (as well as from the graphs) that the roots of these equivalent equations are the numbers 9 and –2.

Exploration

The problems of this exploration give you a chance to practice the new general strategy for solving equations. Solve each of the following equations in two ways.

a. First, follow the procedure outlined in the preceding examples. Use your computer-algebra program to help with the mechanics of adding, subtracting, multiplying, factoring, or simplifying expressions. *Do not use a one-step SOLVE command.*

b. Second, use a one-step SOLVE command.

Record the commands or procedures that you use and the roots you obtain for each equation.

1. $(x + 8)(x - 2) = 0$ 2. $3p^2 - 31p - 22 = 0$

3. $1.4t + 7 = 0.5t^2 - 0.8t$ 4. $200k^2 - 3000k = 2525k$

5. $0 = (d - 4)(d + 7)$ 6. $q^2 + 3q - 4 = 2q^2 - 5q + 12$

7. $(w - 49)(w - 23)(w + 112) = 0$

8. $20g^2 + 6.4g - 16 = 4.8 + 2.4g - 20g^2$

Exercises

1. For each of the following equations:
   –Find the solution; and
   –Show (by substitution) that each root makes the equation true.

   For example, the equation
   $$(3x - 6)(x + 1) = 0$$
   has roots $x = 2$ and $x = -1$. Substituting $x = 2$,
   $$(3x - 6)(x + 1) = [3(2) - 6][(2) + 1]$$
   $$= (6 - 6)(3)$$
   $$= 0(3)$$
   $$= 0.$$

   A similar substitution would be necessary to show $x = -1$ also makes the equation true.
   a. $(x - 4)(2x + 6) = 0$ b. $(3a + 12)(8 - 4a) = 0$
   c. $9 + 6r = 15$ d. $4f + 5 = 2f - 9$
   e. $4(g - 7) = 3(g + 2)$ f. $x^2 (5x - 15) (3x + 6.9) = 0$
   g. $(p^2 - 9)(6p + 14) = 0$

2. One solution path for the equation $z^2 + 9 = 8z - 3$ follows, along with the related graphs.
   a. For each of the steps in the procedure, explain why you might choose that step (e.g., you might want to subtract a suitable expression from each side of the equation in order to get 0 on one side).
   b. Explain the relation between the graph and steps i, iii, and iv in the solution path.

i. $z^2 + 9 = 8z - 3$
ii. $(z^2 + 9) - (8z - 3) =$
     $(8z - 3) - (8z - 3)$
iii. $z^2 - 8z + 12 = 0$
iv. $(z - 6)(z - 2) = 0$
v.   Either $z - 6 = 0$ or
     $z - 2 = 0$.
vi. The roots are $z = 6$
     and $z = 2$.

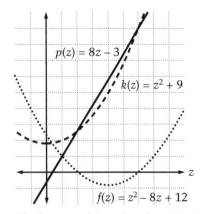

The horizontal (input) scale unit is 1 and the vertical (output) scale unit is 5.

5   Equivalent Inequalities

In many practical problems the questions of interest are modeled by algebraic inequalities rather than by equations. We may be interested in knowing when a profit is positive, when one cost is less than another, or when one temperature or height is at least as high as another.

As you already know, inequalities can be solved by study of graphs or tables of values for the related functions. It is also possible to replace a given inequality by a different but equivalent inequality for which the solution is easier to see.

Two (or more) inequalities are equivalent to one another if their solutions are the same. In this section, you should learn to recognize equivalent inequalities when they are represented in graphs, in tables, and in symbolic rules. You also should learn to solve inequalities by using basic rules for generating equivalent inequalities.

5.1   Graphs of Equivalent Inequalities

When analyzing business situations, we have often searched for break-even points—the points at which company profit is $0. Actually, it is usually more important to find the input values that produce a positive profit.

SITUATION 5.1

The New World Automobile Company has just announced release of its newest model, the Pioneer. Annual cost to produce this car, in dollars, is a function c of the number of cars manufactured n and can be expressed by the rule

$$c(n) = 73\ 687\ 500 + 1750n.$$

The company's annual revenue from this car, in dollars, is a function r of the number of cars manufactured and sold. Assuming each car manufactured is sold, this function can be expressed by the rule

$$r(n) = 30\ 800n - 2n^2.$$

What level of annual production and sales of Pioneers assures that the model is profitable for New World? The company makes money if the sales revenue exceeds the production cost, so we are interested in the values of n for which the following inequality is true:

$$30\ 800n - 2n^2 > 73\ 687\ 500 + 1750n.$$

From the graphs shown here, we can see that for the Pioneer revenue exceeds cost between the points at which the two graphs intersect. The company seems to make money for any number of cars between about 3000 and 11,000.

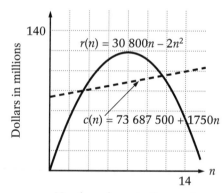

Number of cars in thousands

The solution of the inequality is defined by the intersection points of the revenue and cost graphs. We can find those intersection points by solving the equation

$$30\ 800n - 2n^2 = 73\ 687\ 500 + 1750n.$$

A symbol manipulation program gives $n = 3275$ and $n = 11\ 250$ as the roots of this equation. Looking at the relation between the two functions in the graph, we conclude that the New World Automobile Company can make money on this new model of car if it manufactures and sells any number between 3275 and 11,250 Pioneers.

An alternate way to think of the revenue being greater than the cost is to think of the profit as being positive. In a fashion similar to our work with equivalent equations, we may think of the following three inequalities as being equivalent to one another.

Revenue > Cost

Revenue – Cost > 0

Profit > 0

Using the rules for revenue and cost, we see that profit is a function of the number of cars manufactured, n, and can be expressed by the rule

$p(n) = r(n) - c(n)$
$= (30\ 800n - 2n^2) - (73\ 687\ 500 + 1750n)$
$= -2n^2 + 29050n - 73\ 687\ 500.$

As you might expect, the graph shows that $p(n) > 0$ when n is between about 3000 and 11,000.

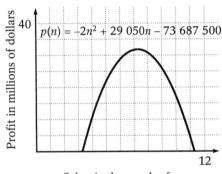

In your study of equivalent equations, you saw that one way to solve an equation such as $p(n) = 0$ is to factor $p(n)$ and to apply the zero-product property. To solve $p(n) > 0$, we can factor $p(n)$ and seek values of n for which the polynomial $p(n)$ is positive. The following steps illustrate this procedure.

Since the factored form of the expression $-2n^2 + 29\ 050n - 73\ 687\ 500$ is $-2(n - 3275)(n - 11\ 250)$, the inequality $p(n) > 0$ is equivalent to $-2(n - 3275)(n - 11\ 250) > 0$.

Once again, the values of n that yield a positive profit are between 3275 and 11 250. In other words, $p(n) > 0$ and $r(n) > c(n)$ have exactly the same solutions. Because of this, we can say

$-2n^2 + 29050n - 73\ 687\ 500 > 0$ and
$30\ 800n - 2n^2 > 73\ 687\ 500 + 1750n$

are equivalent inequalities.

You can see how the solution to the inequality can be found from the factored form of the profit function by testing various values of n in the profit rule. Copy the following table, and use your technological tool, as needed, to complete the entries.

n	-2	$(n - 3275)$	$(n - 11\,250)$	$-2(n - 3275)(n - 11\,250) > 0?$
0	-2	$-3\,275$	$-11\,250$	no
1 000	-2	$-2\,275$	$-10\,250$	no
3 000	-2	-275	$-8\,250$	no
4 000	-2	725	$-7\,250$	yes
6 000	-2			
8 000	-2			
10 000	-2			
11 000	-2			
12 000	-2			
15 000	-2			

Do you see a pattern that allows you to predict which inputs n result in positive profit, and which result in negative profit? Consider one special case:

> If $n = 4000$, then $(n - 3275)$ is positive, $(n - 11\,250)$ is negative, and (of course) -2 is negative. Then $p(4000)$ is the product of two negative numbers and one positive number, and so $p(4000)$ is positive.

Check your table to see the pattern of signs in the factors that give negative profit.

Exploration I

1. The diagram at the right shows the graphs of the two functions $f(u) = 6u + 3$ and $g(u) = 27 - 3u^2$; it can be used to study two inequalities

    $$6u + 3 < 27 - 3u^2 \text{ and}$$
    $$6u + 3 > 27 - 3u^2.$$

    As shown by the diagram, $6u + 3 < 27 - 3u^2$ if u is between -4 and 2, but $6u + 3 > 27 - 3u^2$ when u is greater than 2 or is less than -4.

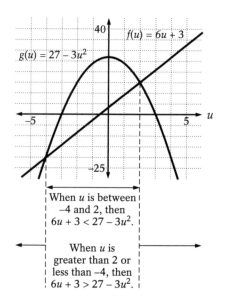

Use the same ideas to answer the following questions. In each of parts a and b below, a diagram showing the graphs of two functions can be used to study two inequalities. For each diagram:

   i. write the two inequalities suggested by the diagram;

   ii. describe the solutions to those inequalities; and

   iii. mark those solutions on the diagram.

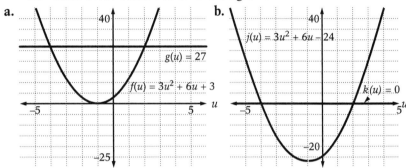

a. $g(u) = 27$, $f(u) = 3u^2 + 6u + 3$

b. $j(u) = 3u^2 + 6u - 24$, $k(u) = 0$

   c. Compare the solutions for your inequalities in parts a and b to the original sample pair $6u + 3 < 27 - 3u^2$ and $6u + 3 > 27 - 3u^2$. Note similarities and differences, and try to explain each pattern you observe.

2. Use graphs to determine which of the following pairs of inequalities could be equivalent. For each pair:

   i. sketch appropriate (rough) graphs and label your diagrams to show the solutions to the inequalities; and

   ii. indicate whether the inequalities seem to be equivalent or are not equivalent.

   a. $2d - 8 < 10$ and $d - 4 < 5$ **b.** $4 - 3k < -4$ and $-4 + 3k > 4$

   c. $2 - 5t > -13$ and $-5t > -15$ **d.** $x^3 - 2x > 0$ and $x(x^2 - 2) < 0$

   e. $a^2 < 11 - 10a$ and $11 - 10a > a^2$

Making Connections

Look back over the inequalities and graphs that you have been studying. Do you see any patterns that help you identify equivalent inequalities from the form of their symbolic expressions? Do you see any operations that might transform an inequality into a simpler equivalent form? Record these patterns and operations, along with any other observations you wish to make.

5.2 Transformations that Produce Equivalent Inequalities

Just as there are properties of equations that yield equivalent equations, there are also properties of inequalities that yield equivalent inequalities. In this section, we shall explore some properties relating to inequalities.

Understanding some of the principles underlying the transformation of one inequality into an equivalent inequality can enhance your understanding of the symbolic manipulation processes which technological tools use. It also allows you to solve rather simple inequalities without turning to such tools.

SITUATION 5.2

Two car rental companies advertise their rates, which consist of a weekly rental fee in addition to a charge based on the number of miles driven. We Are Cars charges $35 per week and 22 cents per mile. Federated Cars charges $85 per week and 6 cents per mile.

Which rental company has the better price?

Considering the weekly fee as a fixed cost and the mileage rate as a variable cost, we find the following linear rules for the charges (in dollars) as functions of the number of miles driven by a renter, n.

$$\begin{aligned} \text{We Are Cars:} \qquad & w(n) = 35 + 0.22n \\ \text{Federated Cars:} \qquad & f(n) = 85 + 0.06n \end{aligned}$$

The question, "Which rental company has the better price?" is not easy to answer immediately. We must first answer the question, "For which mileages driven is the fee charged by Federated Cars higher than the fee charged by We Are Cars?" This new question can be answered symbolically by solving the inequality $f(n) > w(n)$, or

$$85 + 0.06n > 35 + 0.22n.$$

The graph of the linear function $w(n)$ is a straight line with slope 0.22; the graph of the linear function $f(n)$ is a line with slope 0.06. As the following diagram shows, these two lines intersect. At the point of intersection the value of n appears to be slightly more than 300. So, We Are Cars offers a better deal to the renter who plans to drive no more than about 300 miles per week, while Federated Cars has a better offer for higher weekly mileages.

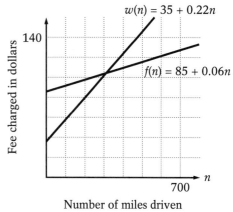

SITUATION 5.3

The two car rental companies are competing for the same customers and a price war begins. We Are Cars starts the war by eliminating its weekly charge of \$35. Although Federated Cars cannot afford to eliminate its entire weekly charge, it decides to match the \$35 decrease announced by We Are Cars.

The new charge by We Are Cars consists of only the mileage fee, which is given by the function rule $v(n) = 0.22n$. The weekly fee for Federated Cars is now \$50, and the rental charge is given by $g(n) = 50 + 0.06n$. If, under these new conditions, we are still interested in knowing when Federated Cars has a higher charge than We Are Cars, we should solve the inequality $g(n) > v(n)$, or

$$50 + 0.06n > 0.22n.$$

Information about this new inequality and the original inequality is given in the following table and diagram. Look at these representations to determine whether the two inequalities seem to be equivalent to each other.

n	$85 + 0.06n$		$35 + 0.22n$	$50 + 0.06n$		$0.22n$
50	88	>	46	53	>	11
100	91	>	57	56	>	22
150	94	>	68	59	>	33
200	97	>	79	62	>	44
250	100	>	90	65	>	55
300	103	>	101	68	>	66
350	106	<	112	71	<	77
400	109	<	123	74	<	88
450	112	<	134	77	<	99
500	115	<	145	80	<	110
550	118	<	156	83	<	121
660	121	<	167	86	<	132

In the table, values are compared for the functions $f(n) = 85 + 0.06n$ and $w(n) = 35 + 0.22n$. If $n < 300$, *then* $f(n)$ is greater than $w(n)$, but if $n > 350$, then $f(n) < w(n)$. The inequality $85 + 0.06n > 35 + 0.22n$ is satisfied for each given entry n that is less than or equal to 300. Similar results seem to hold for the inequality $50 + 0.06n > 0.22n$ as we compare values of the functions $g(n) = 50 + 0.06n$ and $v(n) = 0.22n$. The diagram also suggests that the solution of the inequality $85 + 0.06n > 35 + 0.22n$ is the same as the solution of $50 + 0.06n > 0.22n$.

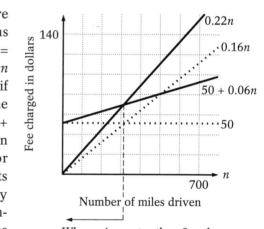

When n is greater than 0 and less than about 300, then $50 + 0.06n > 0.22n$. Similarly, for what appear to be the same values of n, $50 > 0.16n$.

We also can compare symbolically the two inequalities, $85 + 0.06n > 35 + 0.22n$ and $50 + 0.06n > 0.22n$. In fact, we can transform the first inequality into the second—by subtracting 35 from both the left and the right sides. It makes sense that if Federated Cars charges more than We Are Cars and if each company reduces its fee by the same amount ($35), then Federated Cars still is charging more than We Are Cars.

$$85 + 0.06n > 35 + 0.22n$$

implies that

$$(85 + 0.06n) - 35 > (35 + 0.22n) - 35,$$

which simplifies to

$$50 + 0.06n > 0.22n.$$

Adding the same number to (or subtracting the same number from) each side of the original inequality produces what appears to be an equivalent inequality.

What if we add the same algebraic expression to both sides of an inequality?

SITUATION 5.4

As the price war continues, Federated Cars decides on an additional reduction: offering free mileage but keeping the same fixed fee. This means decreasing its charge by 6 cents per mile. We Are Cars can match this offer by deducting 6 cents per mile from its mileage charge.

The effect of this strategy is to subtract $0.06n$ from each of the function rules $g(n)$ and $v(n)$. The new function rules follow.

We Are Cars: $\quad u(n) = v(n) - 0.06n$
$$= (0.22n) - 0.06n, \text{ or } u(n) = 0.16n$$

Federated Cars: $\quad h(n) = g(n) - 0.06n$
$$= (50 + 0.06n) - 0.06n, \text{ or } h(n) = 50$$

A prospective car renter now may be concerned with a new inequality, $h(n) > u(n)$. Since this new inequality

$$50 > 0.16n$$

is obtained by subtracting the same expression $0.06n$ from each side of the previous inequality

$$50 + 0.06n > 0.22n$$

the two inequalities are equivalent.

The diagram shown here gives graphical evidence of this equivalence, since the solutions appear to be the same. The solutions to both inequalities suggest that Federated Cars charges more than We Are Cars whenever the renter drives no more than about 300 miles. It seems that We Are Cars is a better bargain only for shorter trips!

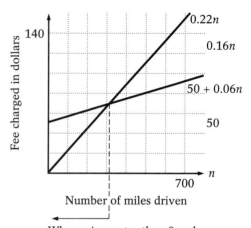

When n is greater than 0 and less than about 300, then $50 + 0.06n > 0.22n$. Similarly, for what appear to be the same values of n, $50 > 0.16n$.

The results in this situation suggest a property of inequalities that is analogous to the addition property of equations.

Addition Property of Inequalities

If the same number or polynomial expression is added to both sides of an inequality, then the resulting inequality has the same solution as the original.

The following exploration problems give further illustrations of this property and its use in solving inequalities.

 ## Exploration II

SITUATION 5.5

Three rental companies advertise their rates for renting recreational vehicles as consisting of a weekly rental fee in addition to a charge based on miles driven. World RV charges $35 per week and 31 cents per mile. ABC RV Rental charges $60 per week in addition to 16 cents per mile. Capital Cars has the largest fixed weekly charge of $100 and the smallest mileage charge of 6 cents per mile.

If n is the number of miles driven during a week of rental, then the function rules for the three companies are as follows.

World RV: $w(n) = 35 + 0.31n$
ABC RV Rental: $a(n) = 60 + 0.16n$
Capital Cars: $c(n) = 100 + 0.06n$

The graphs of these three functions are shown in the diagram at the right.

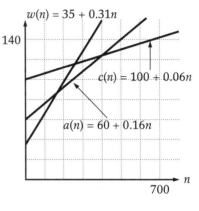

1. For each of the following questions about the rental charges:
    i.   write an inequality that can be used to answer the question; and
    ii. answer the question in a complete sentence that tells about the rental situation.
    a. Under which circumstances is renting an RV from World RV less expensive than renting one from ABC RV Rental?
    b. Under which circumstances is renting an RV from ABC RV Rental at least as expensive as renting an RV from Capital Cars?
    c. Under which circumstances is renting an RV from World RV more expensive than renting an RV from Capital Cars?

2. Which rental company is the best bargain? If none is clearly best, then explain the circumstances under which each one of the companies is best. Give your answers to the nearest mile.

3. During a special sales promotion campaign one week, World RV managers decide to eliminate the fixed charge completely for any RVs rented that week.
    a. Write a rule for the function w_2 that describes World RV's new rental rate.
    b. When World RV runs this special sales campaign, which rental company is the best bargain? If none is clearly best, then explain the circumstances under which each one of the companies is best. Give your answers to the nearest mile.

4. Capital Cars wants to eliminate ABC RV Rental (but not World RV) from the competition.

   a. In order to do this, what is the highest fixed cost Capital Cars can charge, while keeping the mileage rate at 6 cents per mile? Write a rule for the function c_2 describing the new charge structure of Capital Cars.

   b. Using your new function rule, write an inequality that describes the fact that Capital Cars has eliminated ABC RV Rental (but not World RV) from the competition.

   c. Explain how you found the new function rule. If you used computing technology for any steps in your process, explain why you decided to use it for those steps.

Exercises

1. Each of the following is a pair of equivalent inequalities. For each pair:

   i. indicate which inequality has a solution that is easier to see; and

   ii. explain your choice.

   a. $7x + 5 < 19$ and $7x < 14$

   b. $-8 < \frac{1}{4}a + 2$ and $-32 < a + 8$

   c. $6.5y > 4.5y + 60$ and $2y > 60$

2. For each of the following inequalities:

   i. using the addition property of inequalities and then appropriately simplifying expressions, write an equivalent inequality whose solution is more apparent than the original inequality; and

   ii. write the solution of the given inequality.

   a. $3x + 23 < 65$

   b. $210 + c > 2(c - 10)$

   c. $48 - 12a > 96 - 11a$

   d. $20(12 + 0.05t) + 120 < 360$

3. In Exploration I, you studied three equivalent inequalities. Copy and complete the following list of those inequalities. In the right-hand column, indicate what can be added to both sides of the first

inequality to generate the second, and indicate what can be added to each side of the second inequality to obtain the third.

Inequality	Add
$6u + 3 < 27 - 3u^2$	
$3u^2 + 6u + 3$ ____ 27	
$3u^2 + 6u - 24$ ____ 0	

The third inequality in this list can be solved symbolically by looking at its factored form. Use the fact that

$$3u^2 + 6u - 24 = 3(u + 4)(u - 2)$$

to determine the solution to this last inequality. Explain your reasoning.

Exploration III

As is the case with equations, solving inequalities without technological tools depends on being able to replace the given problem with a sequence of simpler, equivalent inequalities. In the previous exploration, you studied the effect of adding or subtracting a number or expression on both sides of an inequality.

In this exploration, you will examine the effects of multiplication on the solution of an inequality. You will find that the multiplication property of inequalities differs from the multiplication property of equations in a very striking way.

Notice first that if both sides of an inequality expressed with either "<" or ">" are multiplied by 0, we obtain $0 < 0$ or $0 > 0$, neither of which is true! Thus, we shall examine only multiplication by non-zero real numbers, splitting our investigation into two parts: multiplication by a positive number and multiplication by a negative number.

Example. What are the effects of multiplying both sides of $2x < 12 - 4x$ by the same non-zero number?

By looking at the following table and diagram, notice that $2x < 12 - 4x$ seems to be true whenever x is less than 2.

Table A

x	$2x$		$12 - 4x$
6	12	>	-12
5	10	>	-8
4	8	>	-4
3	6	>	0
2	4	=	4
1	2	<	8
0	0	<	12
-1	-2	<	16
-2	-4	<	20
-3	-6	<	24
-4	-8	<	28

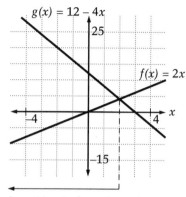

$2x < 12 - 4x$ for values of x less than 2.

Now investigate the effect on the solution of this inequality if both sides are multiplied by the same number.

Case 1. Multiplication by a positive number

For the inequality $2x < 12 - 4x$ discussed above, consider the effect of multiplying each side by $\frac{1}{2}$.

1. Using Table A as a model, copy and complete the following table. Fill each blank in the third column with "<", "=", or ">", whichever symbol fits.

Table B

x	$(2x)(1/2)$	$(12 - 4x)(1/2)$
6	—	
5	—	
4	—	
3	—	
2	—	
1	—	
0	—	
-1	—	
-2	—	
-3	—	
-4	—	

2. Compare Table B with Table A. What seems to happen when both sides of the inequality $2x < 12 - 4x$ are multiplied by $\frac{1}{2}$?

3. Explain, in terms of tables, why your conclusion seems likely to generalize to multiplication by any positive number. It may be helpful to begin by looking at a single row.

4. Generate graphs of the two functions that show the solution of the inequality $\frac{1}{2}(2x) < \frac{1}{2}(12 - 4x)$. Sketch the graphs on your paper, then use the graphs to describe the solution.

5. From your graphs, what appears to happen when both sides of the inequality $2x < 12 - 4x$ are multiplied by $\frac{1}{2}$?

6. Describe what happens when an inequality is multiplied on both sides by the same positive number. Explain your reasoning.

Case 2. Multiplication by a negative number

To analyze the effect of multiplying both sides of an inequality by a negative number, we can start by investigating the negative multiplier –1.

1. Copy and complete the following table (in the same style as Table A) and then look for the effect of multiplying by –1.

Table C

x	$(-1)(2x)$	$(-1)(12 - 4x)$
6		—
5		—
4		—
3		—
2		—
1		—
0		—
–1		—
–2		—
–3		—
–4		—

2. Compare Table C with Table A. What seems to happen when both sides of the inequality are multiplied by –1?

3. Use the results in Tables A and C to complete the following inequality so that it is equivalent to $2x < 12 - 4x$.

   $(-1)(2x)$ ____ $(-1)(12 - 4x)$

4. Sketch a diagram showing the solution of the inequality $2x < 12 - 4x$. On the same diagram, sketch the functions involved in the equivalent inequality you wrote in Exercise 3.

5. What do you notice about the relation between the graphical representations of the two inequalities you sketched above?

6. Do you believe that this pattern always applies when both sides of an inequality are multiplied by –1? Explain why or why not.

7. Based on conclusions you make from your graphs and tables, complete the following statement about any two functions f and g:

   $f(x) < g(x)$ is equivalent to $(-1)f(x)$ ____ $(-1)g(x)$.

8. What happens when both sides of an inequality are multiplied by a negative number other than –1? For example, what happens when both sides of the inequality $2x < 12 - 4x$ are multiplied by the number –3? One way to look at multiplication by –3 is to think of it as multiplication by 3 followed by multiplication by –1. Using this line of reasoning, complete the following statements.
   a. Multiplying by 3:
      $2x < 12 - 4x$ is equivalent to $6x$ ____ $36 - 12x$.
   b. Multiplying by –1:
      $6x$ ____ $36 - 12x$ is equivalent to $-6x$ ____ $-36 + 12x$.
   c. Combining your answers to parts a and b: $2x < 12 - 4x$ is equivalent to $-6x$ ____ $-36 + 12x$.

9. Based on your answers to Exercise 8, complete the following statement.

   If f and g are two real-valued functions and c is a *negative* real number, then $f(x) < g(x)$ is equivalent to $c \times f(x)$ ____ $c \times g(x)$.

Making Connections

Complete each of the following statements with either the symbol $<$ or the symbol $>$. For each finished statement, give an example for that part of the **multiplication property of inequalities**.

1. If $f(x)$ and $g(x)$ are functions and c is a positive number, then:
    a. $f(x) < g(x)$ is equivalent to $c \times f(x)$ ____ $c \times g(x)$.
        Example: $4x < x + 5$ is equivalent to $8x < 2x + 10$.
    b. $f(x) > g(x)$ is equivalent to $c \times f(x)$ ____ $c \times g(x)$.

2. If $f(x)$ and $g(x)$ are functions and d is a negative number, then:
    a. $f(x) < g(x)$ is equivalent to $d \times f(x)$ ____ $d \times g(x)$.
    b. $f(x) > g(x)$ is equivalent to $d \times f(x)$ ____ $d \times g(x)$.

Exercises

Some of the following pairs of inequalities are equivalent. For each pair of equivalent inequalities, explain the property that might be used to transform the first inequality into the second. For every other pair, explain why they are not equivalent inequalities.

1. $2d - 8 < 10$ and $d - 4 < 5$

2. $4 - 3k < -4$ and $-4 + 3k > 4$

3. $2 - 5t > -13$ and $-5t > -15$

4. $a^2 + 10a > 11$ and $2a^2 + 20a < 22$

5. $x^3 - 2x > 0$ and $x(x^2 - 2) < 0$

5.3 Using Addition and Multiplication Properties to Solve Inequalities

Solving inequalities symbolically is like solving equations symbolically, in that you must select and apply various properties.

Example. Recall the rental-car situation, Situation 5.2. Finding when Federated Cars charges more than We Are Cars involves solving the inequality

$$85 + 0.06n > 35 + 0.22n.$$

Sample solution path a.

1. A reasonable first step is to apply the addition property, giving
$$(85 + 0.06n) - (0.06n) > (35 + 0.22n) - (0.06n).$$

2. By the substitution property, $85 > 35 + 0.16n$.

3. Next, we again use the addition property and get
$$(85) + (-35) > (35 + 0.16n) + (-35).$$

4. By the substitution property, $50 > 0.16n$.

5. We now use the multiplication property to get
$$(50) \times \frac{1}{0.16} > (0.16n) \times \frac{1}{0.16}.$$

6. A third application of the substitution property gives $312.5 > n$.

This analysis shows that if mileage n is less than 312.5, then Federated Cars charges more than We Are Cars. Compare the precision of this result with that of our answer to the same question in Subsection 5.2, obtained from tables and graphs.

Sample solution path b.

1. A reasonable alternate first step is to subtract 85 (rather than $0.06n$) from both sides of the given inequality; this gives $(85 + 0.06n) - (85) > (35 + 0.22n) - (85)$, which is simply $0.06n > 0.22n - 50$.

2. As another application of the addition property, we subtract $0.22n$ from both sides of the inequality to obtain $(0.06n) - (0.22n) > (0.22n - 50) - (0.22n)$ or, in simplified form, $-0.16n > -50$.

3. By the multiplication property—and note that the negative multiplier $\frac{-1}{0.16}$ requires reversal of the direction of the inequality—we obtain $(-0.16n) \times \frac{-1}{0.16} < (-50) \times \frac{-1}{0.16}$.

4. Finally, substituting equivalent expressions on each side of this inequality gives us the solution, namely $n < 312.5$.

The result is the same, no matter what legitimate sequence of steps we apply. The exercises that follow give you some practice in planning operations to transform an inequality into a simpler equivalent form from which the solution is easy to see.

Exercises

1. Use the properties of operations as well as the addition and multiplication properties of inequalities to solve each of the following inequalities.
    a. $2.5a + 4 > -24.1$
    b. $10 > 14 - 8z$
    c. $9 < -3(2c + 3)$

2. Find a different solution path to solve the inequality in Exercise 1c.

3. Choose two equivalent inequalities from your solution path for Exercise 1b. Illustrate in two different ways (by constructing a table and by drawing a diagram) that these inequalities are equivalent to one another.

4. The following situations are drawn from your previous work with linear functions. For each question, write an inequality (or inequalities) that can model the question. Then use symbolic reasoning to solve each inequality.
    a. Renting movies from video clubs usually requires a membership fee along with a daily rental charge. Club A offers free club membership and has a daily charge of $3.50 per cassette. Club B has a yearly membership charge of $15 and a daily charge of $2.50 per cassette. Assuming that a person rents all of his or her cassettes from the same club, which club is the better choice?
    b. Health and fitness clubs usually charge an annual fee plus a charge for each visit. People joining these clubs visit them with different frequencies. Analyze the fee schedules for the clubs described below. Formulate a general scheme for deciding which club to join, based on the number of visits per year planned by a potential member.
        –The Healthy-and-Fit Club charges $250 per year membership fee and $2 per visit.
        –The Zodiac Health Club charges $195 in annual membership fees and $2.50 per visit.
        –The Olympia Health Club charges $208 in annual fees and $3.50 per visit.

    c. Temperature in degrees Celsius can be considered a function of the temperature F in degrees Fahrenheit. This function can be expressed by the rule $C(F) = (5/9)(F - 32)$. For what temperatures is the Celsius reading higher than the Fahrenheit reading?

6 Summary

This chapter concentrated on reasoning about equations and inequalities. If you know that the roots of two equations are the same, you know that the equations are equivalent. Similarly, if the solutions of two inequalities are the same, the inequalities are equivalent.

Graphs can be used to illustrate equivalence. For example, from the diagram below, we may conclude that the equations

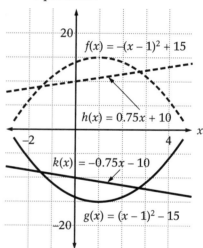

$$0.75x + 10 = -(x - 1)^2 + 15 \text{ and}$$
$$-0.75x - 10 = (x - 1)^2 - 15$$

are equivalent.

The same diagram illustrates the fact that

$$0.75x + 10 < -(x - 1)^2 + 15 \text{ and}$$
$$-0.75x - 10 > (x - 1)^2 - 15$$

are equivalent inequalities. Note carefully the direction of the inequality symbols in these two inequalities.

    Equations and inequalities have properties which allow us to change an equation into an equivalent equation or to produce an equivalent inequality from a given inequality. We have studied several of the more common of such properties in this chapter.

Addition Properties of Equations and of Inequalities

If $k(x)$ and $j(x)$ are two algebraic expressions and if c is a real number or an algebraic expression that has a value at every real number, then:

$k(x) = j(x)$ is equivalent to $k(x) + c = j(x) + c$;

$k(x) < j(x)$ is equivalent to $k(x) + c < j(x) + c$; and

$k(x) > j(x)$ is equivalent to $k(x) + c > j(x) + c$.

Multiplication Properties of Equations and of Inequalities

If $k(x)$ and $j(x)$ are two algebraic expressions and if d is a *positive* real number, then:

$k(x) = j(x)$ is equivalent to $d \times k(x) = d \times j(x)$;

$k(x) < j(x)$ is equivalent to $d \times k(x) < d \times j(x)$; and

$k(x) > j(x)$ is equivalent to $d \times k(x) > d \times j(x)$.

If $k(x)$ and $j(x)$ are two algebraic expressions and if b is a *negative* real number, then:

$k(x) = j(x)$ is equivalent to $b \times k(x) = b \times j(x)$;

$k(x) < j(x)$ is equivalent to $b \times k(x) > b \times j(x)$; and

$k(x) > j(x)$ is equivalent to $b \times k(x) < b \times j(x)$.

We often solve an equation by using the above properties of equations to generate an equivalent equation whose solution may be seen more easily. Similarly, we can solve inequalities by using the above properties of inequalities. Reasonable selection, sequencing, and application of these properties enable us to solve a variety of equations and inequalities.

Review Exercises

Exercises 1 through 6 ask you to review the relations among equations and inequalities. All six exercises are concerned with the following situation.

SITUATION

Each year many high school graduating classes take a class trip. At Lake-Lehman High School, the graduation class goes to Washington, D.C. The number and quality of the sights and activities the seniors see or do during the trip depends on how much money the class has in its treasury. To assure a senior trip that will be really special, each class begins raising money during its freshman year.

Suppose that the current freshman class is planning its sophomore year fundraiser. Among the many activities from which this class may choose are a talent show and a school play. For the talent show, the revenue (in dollars) can be estimated using a function T of the number of tickets sold n, expressed by the rule

$$T(n) = -0.04n^2 + 37n.$$

For a play, the revenue (in dollars) can be estimated using a function P of the number of tickets sold n, expressed by the rule

$$P(n) = -0.02n^2 + 18n.$$

1. Notice that the variable n, which represents the number of tickets sold for a fundraising activity, should be an integer and should not be negative.
    a. What are the zeroes of the quadratic function T?
    b. What values of n are reasonable inputs for the function T?
    c. What values of n are reasonable inputs for the function P?
    d. In the following discussion, the functions T and P will be compared. What numbers n are acceptable when we are investigating both $T(n)$ and $P(n)$ together?

2. Consider the following question:

    For what attendance levels do the talent show and the school play yield the same revenue?
    a. Write an equation that represents this question. In order to refer to this later, let us call it Equation I.
    b. With the help of your response to Exercise 1d, sketch a diagram (similar to the first diagram following Situation 1.1 in this chapter) that can be used to illustrate the solution of Equation I.
    c. Again with the help of Exercise 1d construct a table with headings n, $T(n)$, and $P(n)$ that can be used to find the solution of Equation I.

    d. What are the roots of Equation I?

    e. Which (if any) of these roots is a worthwhile answer to the question that Equation I models?

3. A new equation is obtained by multiplying each side of Equation I by 100.

    Let the new equation be called Equation II.

    a. Write Equation II.

    b. How does the diagram for Equation I, as described in Exercise 2b, compare to a similar diagram for Equation II?

    c. How does the table for Equation I, as constructed in Exercise 2c, compare to a similar table for Equation II?

    d. Compare the roots of Equation I and the roots of Equation II.

4. Describe how the class fundraiser situation is related to each of the following:

    a. the left side of Equation I;

    b. the right side of Equation I;

    c. the left side of Equation II; and

    d. the right side of Equation II.

5. Analyze which fundraising activity might be better from the viewpoint of revenue received.

    a. For what numbers of tickets sold would the talent show produce a greater revenue than the play?

    b. For what attendances would the play produce a larger revenue than the talent show?

    c. How are your answers to parts a and b related? Why does this relation occur?

    d. Based on the two revenue functions, which fundraiser do you think the class should conduct? Defend your choice.

6. Suppose there are fixed costs of $2050 for the talent show and fixed costs of $350 for the play. Consider the following question:

    For what attendance levels do the talent show and the school play yield the same profit?

    a. Write an equation that represents this question concerning profit. Let us call this Equation III.

    b. What are the roots of Equation III?

    c. Which (if any) of these roots is an acceptable answer to the question that Equation III models?

    d. How does the solution to Equation III compare to the solution of Equation I and to the solution of Equation II? Explain why these relations are reasonable.

In constructing a solution path for each of the equations or inequalities in Exercise 7 below, there are several different first steps that are reasonable. For example, to solve $3c - 6 < 9(c - 1)$, you might start by applying:

–the addition property of inequalities, adding the number 6 to both sides of the inequality;

–the addition property of inequalities, adding the expression $-(3c - 6)$ to both sides of the inequality;

–the multiplication property of inequalities, multiplying each side of the inequality by the number $\frac{1}{3}$; or

–the distributive property, substituting for $9(c - 1)$ an equivalent expression.

7. For each of the following equations and inequalities, think of several, different, reasonable first steps for solution paths that depend on properties of real numbers, of equations, or of inequalities. Then do the following.

    i. Categorize the equations and inequalities by first steps. Often an equation or inequality can fit into more than one category.

    ii. Examine the equations and inequalities in each of your categories. Find patterns that help you identify other equations and inequalities that might fit into the same category. Describe the patterns you find.

    iii. Write some hints for yourself to help you identify reasonable first steps.

    a. $\frac{x+2}{3} - 2 = 5$ b. $\frac{1}{3}d = \frac{1}{3}(d + 6)$

    c. $3y + 2 > 2y + 3$ d. $4a - 8.2 < 6a$

    e. $6b < 2b + 4$ f. $6x - 8.2 < 4x$

    g. $3(t^2 + 4t + 3) = 3t^2 + 15t + 12$

    h. $4(x + 3) + 6 = 3(x + 3) - 2$

    i. $6(2f + 4) + 3f = f - 8$ j. $z^2 + 12z = z^2 + 12$

k. $-1 + \frac{2k+10}{4} = \frac{-9k+48}{3}$ **l.** $p^2 + p + 5 = p^2$

m. $\frac{g+2}{3} - 2 = \frac{1}{5}$ **n.** $2y + 4 > 6y$

o. $2z = 2(z + 6)$ **p.** $\frac{1}{3}(m - 2) = \frac{1}{2}(m - 2)$

q. $0 = 4(b + 3)$ **r.** $2m + 4 > 4m + 2$

s. $3(y + 3) + 2 = 4(y + 3) - 6$ **t.** $2(x^2 - 3x + 2) = 2x^2$

u. $4 + \frac{2n+4}{2} = \frac{6n+9}{3}$ **v.** $3(r - 4) = 2(r - 4)$

w. $0 = 0.2(g + 3)$ **x.** $-(2x - 3) + 2x = x + 1$

Exercises 8 and 9 ask you to compare different solution paths.

8. Consider the equation $42(x - 4) + 210 = 252$. Below are two different solution paths for this equation.

Path 1	Path 2
$42(x - 4) + 210 = 252$	$42(x - 4) + 210 = 252$
$(x - 4) + 5 = 6$	$42x - 168 + 210 = 252$
$x + 1 = 6$	$42x + 42 = 252$
$x = 5$	$42x = 210$
	$x = 5$

a. Explain why each of the steps performed above gives an equivalent equation.

b. Is one of the solution paths better than the other? Explain your decision.

9. For each of the equations or inequalities below:

   i. Solve the equation or inequality using the properties of equations or inequalities.

   ii. Solve the equation or inequality again, but this time *use a different solution path*—one starting with a different first step.

   iii. Find still another solution path—one with a *different number of steps* than the first.

   iv. Finally, compare your three solution paths. What are the advantages and disadvantages of each solution path?

a. $\frac{3x}{5} - 4 = \frac{3x+30}{15} - 12$

b. $\frac{3x}{8} + 3 > x - 2$

c. $0.75(52b + 28) = 2.25(44b + 20)$

d. $8123 - 4.2a < 7673 - 3.7a$

Index

Situations in text (by subject)

Situations in Exercises